SCIENCE BETWEEN EUROPE AND ASIA

BOSTON STUDIES IN THE PHILOSOPHY OF SCIENCE

Editors

ROBERT S. COHEN, *Boston University*
JÜRGEN RENN, *Max Planck Institute for the History of Science*
KOSTAS GAVROGLU, *University of Athens*

Editorial Advisory Board

THOMAS F. GLICK, *Boston University*
ADOLF GRÜNBAUM, *University of Pittsburgh*
SYLVAN S. SCHWEBER, *Brandeis University*
JOHN J. STACHEL, *Boston University*
MARX W. WARTOFSKY†, *(Editor 1960–1997)*

VOLUME 275

For further volumes:
http://www.springer.com/series/5710

SCIENCE BETWEEN EUROPE AND ASIA

Historical Studies on the Transmission, Adoption and Adaptation of Knowledge

Edited by

FEZA GÜNERGUN
Istanbul University, Istanbul, Turkey

and

DHRUV RAINA
Jawaharlal Nehru University, New Delhi, India

Editors
Feza Günergun
Department of the History of Science
Faculty of Letters
Istanbul University
34134 Istanbul
Turkey
gunerfez@istanbul.edu.tr

Dhruv Raina
School of Social Sciences
Jawaharlal Nehru University
110067 New Delhi
India
d_raina@yahoo.com

ISSN 0068-0346
ISBN 978-90-481-9967-9   e-ISBN 978-90-481-9968-6
DOI 10.1007/978-90-481-9968-6
Springer Dordrecht Heidelberg London New York

© Springer Science+Business Media B.V. 2011
No part of this work may be reproduced, stored in a retrieval system, or transmitted in any form or by any means, electronic, mechanical, photocopying, microfilming, recording or otherwise, without written permission from the Publisher, with the exception of any material supplied specifically for the purpose of being entered and executed on a computer system, for exclusive use by the purchaser of the work.

Printed on acid-free paper

Springer is part of Springer Science+Business Media (www.springer.com)

Contents

Introduction 1
Feza Günergun and Dhruv Raina

Part I On Technologies

Reflections on the Transmission and Transformation of Technologies:
Agriculture, Printing and Gunpowder between East and West 13
Christopher Cullen

The Ottoman Empire and the Technological Dialogue Between Europe
and Asia: The Case of Military Technology and Know-How in the
Gunpowder Age 27
Gábor Ágoston

General Observations on the Ottoman Military Industry, 1774–1839:
Problems of Organization and Standardization 41
Kahraman Şakul

Cultural Attitudes and Horse Technologies: A View on Chariots
and Stirrups from the Eastern End of the Eurasian Continent 57
Nanny Kim

Part II On Maps, Astronomical Instruments, Clocks and Calendars

Patchwork – The Norm of Mapmaking Practices for Western Asia
in Catholic and Protestant Europe As Well As in Istanbul Between
1550 and 1750? 77
Sonja Brentjes

The Ottoman Ambassador's Curiosity Coffer: Eclipse Prediction
with De La Hire's "Machine" Crafted by Bion of Paris 103
Feza Günergun

**The Clockmaker Family Meyer and Their Watch Keeping the
alla turca Time**... 125
Atilla Bir, Şinasi Acar, and Mustafa Kaçar

The Adoption and Adaptation of Mechanical Clocks in Japan......... 137
Takehiko Hashimoto

**Adoption and Resistance: Zhang Yongjing and Ancient Chinese
Calendrical Methods**... 151
Pingyi Chu

Part III   On Localizing, Appropriating and Translating New Knowledge

**Travelling Both Ways: The Adaptation of Disciplines, Scientific Textbooks
and Institutions**... 165
Dhruv Raina

**Between Translation and Adaptation: Turkish Editions
of Ganot's *Traité***.. 177
Meltem Akbaş

**Eclecticism and Appropriation of the New Scientific Methods
by the Greek-Speaking Scholars in the Ottoman Empire**.............. 193
Manolis Patiniotis

Part IV   On Medicine and Medical Practices

Conveying Chinese Medicine to Seventeenth-Century Europe.......... 209
Harold J. Cook

**Adoption and Adaption: A Study of Medical Ideas and Techniques
in Colonial India**... 233
Deepak Kumar

**How Electricity Energizes the Body: Electrotherapeutics and its Analogy
of Life in the Japanese Medical Context**......................... 245
Akiko Ito

What is 'Islamic' in Islamic Medicine? An Overview............... 259
Hormoz Ebrahimnejad

Index... 271

Contributors

Şinasi Acar Faculty of Arts, Anadolu University, Eskişehir, Turkey

Gábor Ágoston Department of History, Georgetown University, Washington, DC, USA, agoston@georgetown.edu

Meltem Akbaş Department of the History of Science, Faculty of Letters, Istanbul University, Istanbul, Turkey, akbas.meltem@gmail.com

Atilla Bir Faculty of Electrics and Electronics, Istanbul Technical University, Istanbul, Turkey, atilabir@gmail.com

Sonja Brentjes Department of Philosophy and Logic, University of Seville, Seville, Spain, brentjes@us.es

Pingyi Chu Institute of History and Philology, Academia Sinica, Taipei, Taiwan, kaihsin.2007@gmail.com

Harold J. Cook John F. Nickoll Professor, History Department, Brown University, Providence, RI, USA, halcook7@mac.com

Christopher Cullen Needham Research Institute, Cambridge, UK; Darwin College, Cambridge, UK, c.cullen@nri.org.uk

Hormoz Ebrahimnejad Faculty of Humanities, University of Southampton, Southampton, UK, h.ebrahimnejad@soton.ac.uk

Feza Günergun Department of the History of Science, Faculty of Letters, Istanbul University, Istanbul, Turkey, gunerfez@istanbul.edu.tr

Takehiko Hashimoto Department of History and Philosophy of Science, University of Tokyo, Tokyo, Japan, takehiko.hashimoto@gmail.com

Akiko Ito Department of World Languages and International Studies, Morgan State University, Baltimore, MD, USA, itoxx015@gmail.com

Mustafa Kaçar Department of the History of Science, Faculty of Letters, Istanbul University, Istanbul, Turkey, mkacar@istanbul.edu.tr

Nanny Kim Institute of Chinese Studies, Center of East Asian Studies, Heidelberg University, Heidelberg, Germany, nkim@sino.uni-heidelberg.de

Deepak Kumar ZHCES, Jawaharlal Nehru University, New Delhi, India, deepak_jnu@yahoo.co.in

Manolis Patiniotis Faculty of Philosophy and History of Science, Athens University, Athens, Greece, mpatin@phs.uoa.gr

Dhruv Raina School of Social Sciences, Jawaharlal Nehru University, New Delhi, India, d_raina@yahoo.com

Kahraman Şakul Department of History, İstanbul Şehir University, İstanbul, Turkey, kahramansakul@sehir.edu.tr

About the Contributors

Şinasi Acar received his M.Sc. degree in Istanbul Technical University (ITU), Faculty on Electric and Electronics in 1962. Thereafter he held positions in the industry and acted as the Director General of the Eston Company and the Zeytinoğlu Holding in Eskişehir between 1992–99. He worked as board member and director of the Nesim Foundation between 1984 and 2003. He is currently teaching the art of calligraphy in Eskişehir Anadolu University, Faculty of Fine Arts. He is the author of books and articles on the art of calligraphy, local handicrafts, history of art and technology in Turkey.

Gábor Ágoston is associate professor in the Department of History at Georgetown University, Washington, DC. His research focuses on early modern Ottoman military, economic and social history, and the comparative study of the Ottoman, Habsburg and Russian empires. He is the author of *Guns for the Sultan: Military Power and the Weapons Industry in the Ottoman Empire* (2005; Turkish, German, and Arabic editions, 2006, 2010), which challenges the sweeping generalizations of Eurocentric and Orientalist scholarship regarding Ottoman and Islamic societies. His latest book is the *Encyclopedia of the Ottoman Empire* (2009), which he co-authored with Bruce Masters.

Meltem Akbaş is a postdoctoral research assistant at the Istanbul University, Department of the History of Science. She completed her M.A thesis in 2002 and Ph.D. thesis in 2008 at the same department. Her dissertations focused on the reception of Einstein's relativity theory in Turkey and the institutionalization of physics in the Ottoman Empire in 19th century, respectively. Her research interests are the history of physics in Turkey and the Ottoman Empire in nineteenth and early twentieth century, particularly the role of textbooks and of physics education in appropriating new knowledge.

Atilla Bir received his B.Sc. and M.Sc. degrees in T.H. Karlsruhe – Electronics, Karlsruhe, Germany in 1966. He completed his Ph.D. at the Electrical Faculty in Istanbul Technical University (ITU) in 1975. He served as professor of 'system and control engineering' in ITU between 1989 and 2007. His research areas are Linear Control Systems Theory and Design, Multivariable

Systems, Robust Control, Optimal Control, Non-linear Control, Stochastic Control, Industrial Control and the History of Science and Technology. His retirement in 2007 enabled him to work extensively on the history of Ottoman astronomical instruments such as sundials and quadrants.

Sonja Brentjes is currently a researcher in a project of excellence at the University of Seville, Department of Philosophy and Logic, Spain and a guest professor of the project of excellence "Transfer of Knowledge Between Orient and Occident" at the Ludwig Maximilians University, Historicum, History of Science, Munich, Germany. Her publications focus on the history of mathematics, cartography and institutions in Islamic societies, the Mediterranean and early modern Europe and western Asia.

Pingyi Chu is a researcher at the Institute of History and Philology, Academia Sinica, Taiwan. His research focuses on the Jesuit science, history of medicine and historiography of sciences in China during the seventeenth and eighteenth centuries. His recent publications are "Scientific Texts in Contest, 1600–1800," and "Archiving Knowledge: A Life History of the Chongzhen lishu (Calendrical Treatises of the Chongzhen Reign)." Both reveal his interest in applying the insights from book history to study how scientific texts were produced, copied, and read in order to understand what we now label as "science" was perceived in late imperial China.

Harold J. (Hal) Cook has published extensively on medicine in early modern England and the Netherlands, with a particular interest in connecting it to developments to economic history as well as the history of ideas. In recent years this has led him to explore global history as well, which affected his award-winning *Matters of Exchange: Medicine, Science, and Commerce in the Dutch Golden Age* (Yale UP, 2007). He is currently exploring questions about processes of intellectual exchange and translation. Cook has also held many positions of administrative responsibility, most recently as Director of the Wellcome Trust Centre for the History of Medicine at UCL. He is currently the John F. Nickoll Professor of History at Brown University.

Christopher Cullen is originally trained as an engineer, and holds an MA from Oxford in Engineering Science. He has a Ph.D. in Classical Chinese from the School of Oriental and African Studies, University of London. He is Director of the Needham Research Institute (NRI), Honorary Professor of the History of East Asian Science, Technology and Medicine in the University of Cambridge, and is a fellow of Darwin College. His publications include *Astronomy and Mathematics in Ancient China: the Zhou bi suan jing*, Cambridge 1996, and *The Dragon's Ascent*, Hong Kong 2001. His latest book is *The Suan shu shu* 筭數書 *'Writings on Reckoning'*, NRI, Cambridge 2004.

Hormoz Ebrahimnejad is lecturer of history at the Faculty of Humanities, University of Southampton. Specialised in the history of eighteenth and

About the Contributors

nineteenth centuries of Iran with an emphasis on the state structure. Ebrahimnejad's current research interests include the relationship between medicine and the state, the development of medical profession, and the history of hospitals in nineteenth-century Iran. He has widely published in the history of medicine including *Medicine, Public Health and the Qâjâr State* (Brill, 2004).

Feza Günergun (*née* Baytop) is originally trained as a chemical engineer and received her Ph.D. with a thesis on the chemical drugs used in Ottoman medicine in the 14th–17th centuries. Her current research focuses on the introduction of modern sciences to Turkey in the 18th–20th centuries particularly through the translation of European books, educational institutions, science journals and the Europe-trained students. She also published on the history of medicine, chemistry and weights & measures in Turkey. She is currently the Head of the Department of the History of Science, Istanbul University, the editor of the Turkish journal on the history of science *Studies in Ottoman Science*, and the founding member of the Turkish Society for History of Science.

Takehiko Hashimoto received his Ph.D. in the history of science and technology at Johns Hopkins University in 1991. Since 1991, he has been a faculty member at the University of Tokyo, and is now a Professor at the Department of History and Philosophy of Science of the University of Tokyo. His publications include *Chikoku no Tanjō (The Birth of Tardiness)* (co-edited with Shigehisa Kuriyama) (2001), *Egakareta Gijutsu Kagaku no Katachi (Engineering Drawing, Scientific Form: An Iconographic History of Science and Technology)* (2008), and *Historical Essays on Japanese Technology* (2009). He is now preparing a book on the prewar history of aeronautical engineering.

Akiko Ito received her Ph.D. in Japanese Studies from The Graduate University for Advanced Studies, Kanagawa, Japan. She conducted her postdoctoral research at the International Research Center for Japanese Studies, Kyoto, Japan and the Program in the History of Medicine, University of Minnesota. She is currently lecturer in the Department of World Languages and International Relations, Morgan State University, Baltimore, Maryland. Her work is focused on the comparative history of electrical technology, particularly on the application of electrical technology in medicine and biology.

Mustafa Kaçar is trained as a historian at the Department of History, Faculty of Letters, Istanbul University where he received his M.A. with a thesis on the Ottoman telegraph network. Then, he joined the Department of the History of Science, Istanbul University, where he completed his Ph.D. thesis on "The Development in the Attitude of the Ottoman State Towards Science and Education and the Establishment of the Engineering Schools (*Mühendishane*s)". He is currently professor at the Department of the History of Science, Istanbul University. His research focuses the history of science in Turkey (especially 17th–19th c.), the introduction of modern sciences to

Turkey; the history of engineering education and Ottoman scientific instruments.

Nanny Kim is a China historian. She is currently a research fellow at Heidelberg University in the research group "Monies, Markets and Finance in China and East Asia, 1600-1900" (director Hans Ulrich Vogel). Her interest lays in tracing transformations in the technologies of pre-modern road and water transport in the context of cultural and economic systems as well as environmental change. For her M.A. she worked on a late Qing illustrated magazine (Heidelberg University, 1994), for her Ph.D. she explored the images of Korea in Chinese writings during the late nineteenth and early twentieth centuries (SOAS, London, 2000).

Deepak Kumar teaches history of science and education at Jawaharlal Nehru University, New Delhi. He is known for his work *Science and the Raj* (2nd ed.), OUP, Delhi, 2007. His edited volumes include, *Science and Empire*, Delhi, Anamika Pub., 1992, *Technology and the Raj*, Sage Pub., New Delhi, 1995, and *Disease and Medicine in India*, Tulika Pub., New Delhi, 2002.

Manolis Patiniotis is assistant professor in history of science at the Faculty of Philosophy and History of Science, Athens University. He delivers courses on Scientific Revolution and the sciences during the Enlightenment. His research interests include the impact of Newtonian natural philosophy on various intellectual environments in the eighteenth century, the appropriation of the seventeenth and eighteenth-century natural philosophy by the scholars of the European periphery, and the application of information technology in historical inquiry. He is a founding member of the international community for the History of Science and Technology in the European Periphery (STEP).

Dhruv Raina is professor of history of science and education at Jawaharlal Nehru University, New Delhi. He studied physics at Indian Institute of Technology, Mumbai and received his Ph.D. in the philosophy of science from Göteborg University. His research has focused upon the politics and cultures of scientific knowledge in South Asia. He has co-edited *Situating the History of Science: Dialogues with Joseph Needham* (1999) and *Social History of Sciences in Colonial India* (2007). *Images and Contexts: the Historiography of Science and Modernity* (2003) was a collection of his papers contextualizing science and its modernity in India. He co-authored with S. Irfan Habib, the *Domesticating Modern Science* (2004) addressing the encounter between modern science and the "traditional sciences" in colonial India. He has published papers on related subjects in various journals.

Kahraman Şakul received his Ph.D. (An Ottoman Global Moment: War of Second Coalition in the Levant, 2009) from the Department of History, Georgetown University. He is currently an adjunct faculty of the Department of History, Istanbul Şehir University. In 2010, he acted as the interim director of

Georgetown University McGhee Center for Eastern Mediterranean Studies in Alanya, Turkey. His research interests include military and technological aspects of Ottoman reforms as well as the political culture of the Ottoman court. His recent article treats the Ottoman attempts to control the Adriatic frontier in the Napoleonic Wars. He contributed to the *Encyclopedia of the Ottoman Empire* (Eds. G. Ágoston and B. Master, 2009).

Introduction

Feza Günergun and Dhruv Raina

The transmission of scientific, technical and medical knowledge between Europe and Asia has been the subject of several historical studies during the past decades. The complexity of the transmission process which involves cross-cultural interactions, the encounter of essentially different traditions and practices, the contexts of diffusion and reception, the center-periphery relationship and many other intriguing issues, has ensured that the theme has remained a highly fertile field of research for many scholars and continues to capture the attention of inquisitive minds. While scholarly attention focused on the diverse means by which knowledge diffused outside Europe and to the process of its introduction, attempts to understand the adoption of the transmitted knowledge and its adaptation to its new habitat were also undertaken. Was the newly received technical knowledge used in its original form or implemented within the framework of traditional technology or practice, or did it undergo some transformations? The essays in the present volume aim to explore this multifaceted question for a wide range of themes related to map making, mechanical clocks, astronomical instruments, translations, agricultural and military technology as well as medical techniques and related scientific methods.

Historical studies on the 'travelling' of scientific and technological ideas from Europe to South and East Asia were gradually supplemented by research investigating the transmission of the knowledge produced in Western Europe to the European Periphery. As revealed by researches analyzing the 'transmission' of ideas, this concept proved inadequate in contextualizing the dissemination of the new sciences in the recipient cultures and societies. Instead, the notion of 'appropriation' was suggested as a more coherent and fruitful analytic

F. Günergun (✉)
Department of the History of Science, Faculty of Letters, Istanbul University, Istanbul, Turkey
e-mail: gunerfez@istanbul.edu.tr

D. Raina (✉)
School of Social Sciences, Jawaharlal Nehru University, New Delhi, India
e-mail: d_raina@yahoo.com

instrument. In addition, filtering processes were also considered salient for understanding the changes and the metamorphoses of these ideas.

The present volume endeavors to analyze the adoption and adaptation of scientific ideas in both Eastern and Western parts of Asia: the studies focus on how scientific, technical and medical knowledge and practices were re-shaped in China, India, Japan and in the Ottoman Empire of which Turkey and Greece were parts in the past. Lying on the boundary between Asia and Europe, the Ottoman Empire has been well situated for scientific and technological exchanges among the two continents. The Ottomans, in the early centuries (15th–17th centuries) of their reign had relied heavily on the legacy of medieval Islamic science they had acquired mostly through the mobility of scholars between the Ottoman cities and the scholarly centers such as Baghdad, Cairo, Damascus, Samarkand and Tabriz, some of which were later incorporated into the Ottoman territories. As Gábor Ágoston puts it, between the 15th and 17th centuries, this multi-ethnic and multi-lingual empire was an ideal place for technological dialogue though this was at the time predominantly of a military nature. Direct military conflicts with enemies in the Balkans, the translation of European treatises on firearms and the art of war, the trade of arms and most of all of military experts, technicians and craftsmen flocking to Ottoman territories from various parts of Europe were influential both in the transmission and adoption of technical know-how and facilitated the exchange of military techniques between Europe and the Ottomans. Another feature of note in this acculturation process was the role of the Ottomans in the diffusion of gunpowder technology in the Middle East and Asia. Eighteenth century reforms of Ottoman artillery, as Kahraman Şakul analyzes in his essay, included the production of lighter and handier guns and gun carriages to facilitate their movement, and gun calibers were standardized in the early nineteenth century. The necessary technical knowledge was again acquired both by recruited experts and the material purchased from Europe. Thus by the end of the eighteenth century, the Ottoman gunnery would bear strong French and Habsburgian imprints.

Eclecticism is argued to be among the major features of Ottoman transmission and adoption of European science. To ensure their military superiority in the campaigns against the Muslim countries of the Middle East and those of Christian Europe, priority was given to the importation and adoption of firearms technology, military arts and engineering. The Ottoman reforms of the eighteenth century also focused on the military: the schools and the industrial plants created during the movement for reform were planned as institutions for transmitting and imparting engineering sciences for the military. The decisions taken in the mid-nineteenth century to create modern civilian schools apart from modern military institutions and the traditional *medreses*, the colleges teaching Islamic theology and jurisprudence, should be considered as a turning point. These civilian schools aiming to train the cadres to be employed in transportation, communication and health services were also instrumental in the transmission of scientific and technical knowledge from Europe and their diffusion throughout the empire.

The civilian medical school was established following a debate on the possibility of teaching medical sciences in the Ottoman language. This led to the translation of many scientific textbooks, among others, Adolphe Ganot's famous physics book. Meltem Akbaş analyzes the two Turkish translations of *Traité* by Antranik Gırcikyan, and examines how a 'translated' book was culturally transformed, fulfilled educational requirements, and became an indispensable instrument in the process of coining scientific terminology. Interestingly enough, the prefaces to the two editions of Gırcikyan's reveal the different contexts of translation: The Turkish prefaces render not only the translator visible, but foreground the source and target text, audience, author and translator that together constitute the dynamic context within which the text is reproduced. The preface becomes then an important source for studying the transmission of science across cultures. Akbaş maintains that the translation also contributed to the recognition of physics as a newly developing discipline in the Ottoman Empire.

The centre-periphery narrative, according to Manolis Patiniotis is premised on the idea of an original home from where innovations flow to the periphery and the scientist at the periphery is portrayed as one who undertakes the cloning process – cloning without philosophical reflection. Patiniotis revises the notion of centre and periphery in terms of the domination of certain ways of problematizing scientific concerns over others, arguing in a specific context that with the break up of the Ottoman Empire a new Greek identity was constituted along the fracture lines of the empire where Greek communities lived. Religion and Greek education became axes for the construction of this new identity. Translation of works from the French enlightenment commenced in the nineteenth century, and this process was accompanied by a transformation of the teacher-priest into a secular scholar in consonance with the needs of new Greek society. In historiographic terms it is worth noting that the efforts of Greek intellectuals in the spread of modern science is linked with the narrative just not of the enlightened transformation of Greek society but with its national struggle against the Ottomans seen as impeding any steps towards scientific enlightenment. Patiniotis takes up the discussion of Diderot's ideas amongst these Greek intellectuals who harked back to the logical and metaphysical discourse of traditional philosophy even while discussing the new experimental method which they did in order to maintain the unity and universal character of philosophy. While they recognised the goals of the moderns as obtaining empirical knowledge they saw themselves as seeking true scientific knowledge. This complex engagement rejects the idea that the Greeks opposed modern scientific ideas and hence the marginality of Greek science.

The relevance of state founded educational and industrial institutions in the transmission and adoption of European science in the Ottoman Empire and their role in shaping Ottoman thought has influenced historical research into highlighting their mission and activities. The relatively large number of documents concerning their creation and functioning was another factor that canalized historians to move into the institutional history of science. Individual initiatives taken in the transfer of European science, on the other hand, were regarded as minor themes and aroused less interest among historians, letting most of them to fall into oblivion. The present volume includes among others, a study by Feza Günergun elaborating upon Mehmed Said Effendi's – the Ottoman ambassador in Paris – curiosity for European scientific instruments and its consequences. The ambassador's keen interest in scientific matters led him to purchase a 'machine' in Paris designed by P. de la Hire to predict the dates of the solar and lunar eclipses in the Gregorian calendar. The mathematician Sıdkı Efendi, who translated its manual into Turkish, explained how to use the instrument to find the Hijri dates of the eclipses, i.e. he adapted the device to the Hijri Calendar used by the Ottomans.

The clockmaker of the Ottoman Palace, Johann Meyer's design and manufacture of a watch which automatically keeps the *Turkish time* is a concrete example of technical improvement designed to meet local needs. According to the Ottoman time system the duration of the day extended from sunset to sunset and so mechanical clocks had to be adjusted every day at sunset. The essay by Atilla Bir, Şinasi Acar and Mustafa Kaçar relates how Meyer, at the end of the nineteenth century, developed a clock mechanism – with a continuous heart cam of 18 steps – to automatically adjust the clock. A similar development is witnessed in Japan during the Tokugawa or Edo period (1603–1868). Japanese craftsmen, who had learned clock making from the Jesuits, adapted the Japanese clocks to the seasonally variable time system by using weights and springs.

Takehiko Hashimoto focuses on the case of China and Japan. While clocks were introduced into China and Japan by the Jesuits the subsequent trajectory of innovation was contingent upon the different policies adopted towards the missionaries in the two countries. In the Japanese case, clocks were meant to indicate seasonal time. In the case of both Japan and Turkey initial instruction in clock making was imparted by craftsmen trained in Europe – there was thus a migration of skilled craftsmen with whom the conditions for the reproduction of technology travelled. The Jesuit mission also had craftsmen trained to repair their own watches and this was possibly an additional route through which the technology travelled and the skill subsequently reproduced. Whereas the clock reformed the system of keeping time in the West, in Japan it had to be adapted to a seasonally variable time system. The work then focuses on a Japanese clockmaker Hisashige Tanaka. The automata made by Tanaka unlike in the case of China were for public display and more importantly for practical purposes and not for the houses of the elite and their cabinets of curiosities. This also meant that the purchasing capacity of the rich in Japan did not match those of the rich in China. With the end of the Edo period, and the adoption of modern time in 1873, Japanese mechanical clock making more or less came to an end. As Japan and Turkey adopted universal time and the Gregorian calendar in 1873 and 1925 respectively, the adaptation of clocks to seasons proved redundant.

In the case of map making, the adoption and adaptation processes did not lead to the acquisition of new technical knowledge. Sonja Brentjes compares the maps made in Istanbul between the mid-16th and mid-18th centuries, pointing out that the processes involved did not lead to a stable standard of map making

Introduction

in Istanbul: differences in techniques, coordinates, terminology, units of physical or political geography etc., reflect that eclecticism was the prime work ethic that producers of manuscript maps subscribed too. Those who produced and reproduced the maps of Ottoman geographical manuscripts translated from European atlases, were craftsmen with particular skills in painting and writing, but devoid of mathematical methods for creating a map. The adoption and adaptation of the most prestigious atlases of the sixteenth- and seventeenth-century Europe, namely the *Atlas Minor* and *Atlas Maior* into the culture of the Ottoman capital did not lead to a spread of mathematical concepts and technical skills among those who participated most actively in these processes. The maps rather testified to the domination of these processes by Ottoman standards of producing illuminated literary and historical manuscripts.

Christopher Cullen's paper asks some cross-cultural questions about the role of technology in society and revisits the Needham question by interrogating the standard conception of technology and technological progress. Taking up three case studies of Chinese technological inventions, from the origins of wet-rice cultivation, he then proceeds to discuss the almost accidental invention of gunpowder in China and its marginal impact on Chinese society as opposed to the cascading changes that the introduction of linked weapons' technologies unleashed in Europe. Similarly, turning back to the history of printing, Cullen argues that printing in China enabled a reinforcement of ideological control, while in Europe, printing produced the very opposite effects. This exploration of the comparative context of technology in society wherein the social and political outcomes of technological change or innovation could be so radically different begs a re-framing of the Needham question, in ways where our notions of technology and technological progress are not dependent upon the European historical trajectory. Naturally, when confronted with the magnificence of Chinese technology, Needham could not help but ask as to why these developments did not lead up to the path that Europe pioneered, however problematic that question may seem today. Cullen recommends a change in our conception of technology from one of technology seen as 'lumps of matter or significant quantities of energy having their configurations and positions changed' to one where we focus upon 'the methods by which knowledge is validated, authorised, transmitted and subverted'.

These cross-cultural comparisons of the context of technology are quite central to the essays in the volume and most certainly, as just discussed, what was technologically salient in the West was not in China and vice versa. In the same vein Nanny Kim's paper looks at the domestication of horse, the chariot and stirrup from the Eastern end of the civilizational spectrum since little work has been done on the same. One of the advantages of this comparative perspective is that it throws into relief received understandings and possibly presses the need to revise the questions we need to ask. Kim argues that the two wheeled chariot rather than serving as an instrument in warfare had a more important role in the representation of the Chinese elite when state structures were expanding. The stirrup, of which so much is made by historians of technology, on the other hand appeared in this context as a minor innovation within the evolving culture of horse-riding. Kim seeks an alternate perspective on technologies travelling between civilizations and the peripheral cultures that transmitted them. In the case of the horse and chariot, the transmission of technology had little to do with the ethnic domination of a group in possession of the technology. Furthermore, the horse and chariot like in the case of the ox and cart moved with migrant populations and were then adapted to a variety of cultural settings and environments. Consequently, it could be argued that local innovations were often stimulated by the imagery of power and prestige that motored state building processes as well.

Sceptical about the Needham metaphor, Pingyi Chu sees the introduction of new knowledge as reordering power relations and hence causing uneasiness within the recipient culture. Examining the case of the initial response in China to Western European astronomy, Pingyi Chu suggests that this knowledge impacted upon religious and social life, but was gradually legitimated as being part of a long lasting Chinese tradition. Zhang Yongjing was a Confucian intellectual who rejected Western astronomy and sought to develop Chinese astronomy internally – more specifically he attempted to push the boundaries of traditional Chinese calendrical studies. Traditional historiography has seen these responses as conservative. Some of the issues discussed by Zhang such as the apparent size of the sun had been discussed by Ricci and Dias as part of a discussion on vision and Zhang wished to express his reservations about the new instruments and what they revealed – expressing thereby a mistrust of 'vision' itself as understood in different visual cultures. The belief in an ancient Chinese astronomical tradition was itself a product of the encounter with European astronomy. Despite this cultural combat, European methods had penetrated the thinking of Chinese scholars, and having recognised this they attempted to provide metaphysical foundations for what they were doing. The response to European methods and instruments was determined by the way they chose to see the relationship between li and qi – if the li was within the qi then they worked within the frame of European methods, but if li preceded qi then there was a negative response to European science. A variety of local contingent factors played a role in influencing the reception to European calendrical methods rather than just the efficacy of methods of calculation and scientific rationality. Pingyi Chu structures this encounter in terms of a Kuhnian crisis and the gradual stabilisation of a new paradigm, and the responses to the new paradigm as a few possible anomalies.

Examining the adaptation of modern science in South Asia, Dhruv Raina too commences by raising some questions around Kuhn's historiography of scientific change and then goes on to emphasize the role of institutions in the domestication of modern science in late colonial South Asia: he takes up three cases, that of geology textbooks, the institutionalization of psychoanalysis and of technical education. In each of the cases discussed he teases out the distance between the received historiography which is premised on the existing official record, as does Günergun in her paper, to point out how a shift in focus towards the context within which these activities were undertaken throws new light on the different outcomes of these processes. The important point which he tries to bring out is that these exchanges between Europe and Asia have in scientific terms always been two way processes, but earlier historiography was deeply anchored within the enlightenment vision of modern science as some kind of narrative of redemption. Even models of diffusion such as the centre-periphery model which are sensitive to the politics of scientific knowledge cannot in some sense overcome the limitations of the earlier diffusionist models. Shifting then the focus from the Kuhnian emphasis on science as theory building to the institutional context of the expansion of science furnishes another picture of the growth of scientific knowledge. It is on this count that the earlier part of his paper tries to draw parallels between the circulation of scientific knowledge and the sociology of religion.

The voyages of discovery have been an important subject of investigation in the history of sciences. The interesting feature of the theme has been the radical historiographic changes that have come to characterize researches in the area especially in reconstituting the theme of the expansion of European science. The concern of Harold Cook's paper is the mechanism through which the knowledge of Chinese science and medicine in particular was transmitted to Europe. Whatever the ostensible motives for the voyages of discovery, Cook points out that the trade in spices initiated by European business interests was entangled with the prospecting and trade in different goods, medicines and pharmacopoeia. However, Cook points out that early travelogues and reports are by and large oblivious, perhaps deliberately so, of the local medical practices and traditions in China and India. Cook's paper examines the early Jesuit accounts on Chinese medical practices and medical systems, paying attention to Boym's work on Chinese flora, and mathematical and medical traditions. Political reconfigurations within China and Europe played a role in shaping and reshaping the form the narrative acquired as the work passed through several hands. By the last decade of the seventeenth century a more stable account of Chinese medicine was available in Europe through the translation efforts of the Jesuits.

The introduction of modern medicine into India is currently a fast expanding area of research and Deepak Kumar's paper draws a canvas of the discussion by providing a backdrop that reviews the interactions between different systems of medicine in pre-colonial India, in particular he discusses the relationship between Ayurvedic and Unani medicines. Steeping in to the colonial period we notice that the historiography is marked by deep ambivalence around issues of empire and colonial medicine, given the complexity of the colonial situation itself involving heterogeneous actors, interests and different notions of the body and disease embedded in different knowledge forms. Kumar discusses the efforts of local rulers, doctors and healers in engaging with modern medicine and their little discussed role in the institutionalization of modern medicine. However, from a more normative position the failure of these traditional medical orders to modernize, a question naturally posed from outside the tradition, arose from the inability to forge a veritable language of dialogue between traditional and modern medicine.

Moving beyond just the interaction between Western and non-Western medicine, Akiko Ito studies the place of electrotherapeutics in Tokugawa Japan. Interestingly enough she has to explore a problematic at the double intersection of electricity and medicine on the one hand and the linkages between Western and Japanese medicine. Ito wishes to understand how as a medical technology 'electrotherapy' was culturally domesticated in Japan and the meaning system within which it came to be enveloped. The introduction of modern medicine itself provided a disciplinary frame to reinvent the explanations for nervous disease and to set the context for the emergence of electrotherapeutics. The discourse of electrotherapeutics provided a frame wherein Japanese physicians could treat a range of regional diseases within the larger frame of so called modern science and medicine. As trends in Japanese and Western medicine converged on the issue of therapeutics, the discourse on electricity and the body gradually waned. However, despite the decline of professional interest in electrotherapeutic devices, they continued to be cleverly introduced into the Japanese market through medical manuals that innovatively manipulated the Japanese notion of Genki and that of electricity. At variance with Kumar, Ito does not see such moves as proffering a derivative of Western medicine for she interrogates the notion of modernity; and argues instead that the introduction of new medical products and devices was marked by a seamless sequence wherein it is impossible to tell where the traditional ended and the modern began.

The deeper historiographic question of the identity of the different systems of medicine that historians of science often take as historic givens, as empirical data of the history of medicine or sciences is raised in Hormoz Ebrahimnejad's paper where he points out that the term Islamic medicine was never used by any Muslim historian from the medieval period except perhaps in books on scriptural healing. Ebrahimnejad reminds us that most of the so called books on Islamic medicine were written by Jews and Christians and had been translated into Arabic by non-Muslims. The term was coined by Western historians to refer to a core literature by Islamic scholars comprising translations of the works of Greek scholars into Arabic – largely elaborating upon Galenic humoral theory. Contemporary attempts by traditionalists to revive the Islamic faith have identified traditional medicine with their faith. However, in the past, the political ascent of Islam was mediated through the translation and patronage of non-Arab and non-Muslim sources and individuals. This translation project Ebrahimnejad points out was influenced by an earlier one – referred to as Zoroastrian imperial ideology – that was a project of appropriation of Greek knowledge, transmitted to the Abbasids via the Sassanids. Clearly, the appropriation of knowledge within religious traditions is an old phenomenon, but while the term appropriation connotes something negative as well – it could as well from the flip-side be seen as a kind of approval and a form of legitimating – through very 'diverse strategies'.

A number of important concerns on travelling ideas and techniques have been raised in these pages. The historical studies presented here based on encounters between Europe on the one hand and the Ottoman empire, Iran, India, China and Japan explore local knowledge and practices in their engagement with new knowledge and practices. Departing significantly from notions that, in the words of one contributor, centre 'cloning' or that recognize the limitations of center-periphery explanations, the contributions do not singularly emphasize the role of empires, religion, or state as the factors solely shaping the processes of adoption and adaptation of knowledge. One of the contributions suggests for example the possibilities of accidental discoveries and serendipitous inventions. The essays come together around the premise that even in cases where similar factors operate in shaping processes of adaptation and adoption, the outcomes differ in time, and with spatial and cultural location because of the distinct embeddings of knowledge and technological know-how. In that sense we hope the essays convey the diversity of the processes underlying the adoption and adaptation of travelling ideas and techniques.

Part I
On Technologies

Reflections on the Transmission and Transformation of Technologies: Agriculture, Printing and Gunpowder between East and West

Christopher Cullen

The period from the eighth to the twelfth century in China was marked by a number of great historical and social movements. In dynastic terms, the Tang dynasty, which came to power in 618, built on the national unity recreated by the Sui in 581 after centuries of division. The power of the central government, shaken by great rebellions in the eighth and ninth centuries, was firmly re-established by the Song in 960. This unity was to endure until the loss of north China to the Nuzhen (ancestors of the Manchus) in 1126, who established themselves under the dynastic title of Jin. In the following century both north and south were to fall under the rule of the Mongols, who named their dynasty Yuan.[1]

Socially and politically, China changed during this period from a military-aristocratic society dominated by the great north Chinese clans, with the imperial family as 'first among equals', to one in which the great families, increasingly from south China, built their power and prosperity on the high

---

C. Cullen (✉)
Needham Research Institute, Cambridge, UK; Darwin College, Cambridge, UK
e-mail: c.cullen@nri.org.uk

This paper is essentially the text of a talk delivered at Istanbul University in May 2006. It was originally intended as a series of informal reflections designed to provoke discussion, and in editing it for publication I have not tried to turn it into a formal research monograph. While I have added footnotes, these are mostly limited to suggesting further reading on topics mentioned.

[1] On the reunified empire and its history to the end of Tang see for instance J. K. Fairbank and D. C. Twitchett, *Sui and T'ang China, 589–906, Part I*. Cambridge [Eng.], New York: Cambridge University Press, 1979; in the absence of the Cambridge volume on the Song, see R. P. Hymes, C. Schirokauer et al., *Ordering the World: Approaches to State and Society in Sung Dynasty China: Conference on Sung Dynasty Statecraft in Thought and Action: Selected Papers*. Berkeley, London: University of California Press, 1993; and J. Gernet, *Daily Life in China: On the Eve of the Mongol Invasion, 1250–1276*. Stanford, Calif.: Stanford University Press, 1970; on the later regimes see J.K. Fairbank, H. Franke et al., *The Cambridge History of China, Vol. 6, Alien Regimes and Border States, 907–1368*. Cambridge: Cambridge University Press, 1994.

civil office attained by their menfolk in successive generations of service to the emperor. The Buddhist church, which in the eighth century enjoyed immense social intellectual and economic influence, lost it decisively to counter-attacks from the imperial government and Confucian scholars.

Geographically, the whole of this period saw a continuation and quickening of a long-term shift southwards of population, economic activity, and cultural strength. Economic growth went hand in hand with this southwards shift: to give one strong indicator of this, under the Song the Chinese government began to draw the majority of its income from money taxes on trade, whereas the Tang had been founded on contributions of tax grain and textiles paid in kind by farming families.[2]

The Issue of Technology

The time of the Tang and Northern Song is highly interesting to the historian of technology, science and medicine. And of course it is impossible to say much about these topics without mentioning the name of Joseph Needham. The work of Joseph Needham has certainly transformed the way in which China is seen, in East Asia as well as in the West.[3] Faced with the richness of Needham's documentation, and viewing the ingenious creations of Chinese technologists over many centuries, it is difficult not to revert to the "Needham question" in one of its many forms – all of which amount morally to a sense of puzzlement that the England of Chaucer or perhaps of Shakespeare was never visited by a well-armed Chinese fleet carrying the announcement that the British Isles had now been "discovered" and were henceforth to have the benefits of civilisation and advanced technology conferred upon them, whether they wanted or not. Why did not China get there first (wherever "there" is), long before the Europeans managed to have an Industrial Revolution for themselves?[4]

But I shall not attempt here to answer the "Needham question." What I propose instead is to take three examples of technological change in China during the period from about 700–1100 CE and attempt briefly to set them in a comparative social and historical context. I suppose there are two respectable motives for attempting to do this:

---

[2]On the finances of the Tang, see D. Twitchett, *Financial Administration Under the T'ang dynasty*. Cambridge: University Press, 1963; on the economy of the Song see M. Elvin, *The Pattern of the Chinese Past*. London: Eyre Methuen, 1973.

[3]J. Needham, *Science and Civilisation in China*. Cambridge: Cambridge University Press, 1954–.

[4]For a discussion of the history (and pre-history) of the "Needham question" and the forms it has taken, see D. Liu, "A New Survey of the 'Needham Question,' " *Studies in the History of the Natural Sciences* 自然科学史研究 XIX, 2 (2000): 293–305.

1. By thinking about the role of technical change in the larger historical process, we may hope to understand the general history of China better than before.
2. By doing our thinking comparatively we may be able to form more general trans-cultural conclusions (or at least hypotheses) about the historical role of technology.

That sort of thing is no doubt all very well, but I really wish to do something more subversive. I want to share a sense of disquiet about "doing the history of technology", which has grown on me in the course of preparing this talk. Can one really find a un-confusing object of historical discourse called "technology" or "technical progress" and see how it connects with other bits of the history of a particular culture? Is there really any such thing as a technology (such as agriculture, printing, gunpowder weapons) which can be tracked through different times and places to see what effects "it" has and what changes "it" suffers? It may be that I shall go away reassured that historians of technology are a good thing – and if so I shall be much relieved, since I am by way of being one myself. But for the moment, let me launch myself into confessional mode and ask for your help.

Agriculture in East and West

The Primacy of Agriculture

Let us begin with agriculture. It is commonly felt that the agricultural development of Europe provided an essential foundation that made it possible for the Industrial Revolution to take place there. Some would see that essential foundation as having been laid down in the Middle Ages in northern Europe as a whole.[5] Others would prefer to emphasize what happened in England in the eighteenth century. But surely agriculture was highly developed in China too? A comparison may enable us to consider more carefully just what we mean by "development".

Agricultural Development and the Industrial Revolution in Britain

For ease of comparison, I shall choose the case of eighteenth-century England. Something very important certainly happened to English agriculture in the years 1710–1790. Animals as well as people became better fed and healthier: it has been claimed that the average body-mass of oxen, calves and sheep sold at

---

[5] J. Gimpel, *The Medieval Machine: The Industrial Revolution of the Middle Ages*. Aldershot, Hants, England: Wildwood House, 1988.

Smithfield more than doubled during this period.[6] On the human level, a much improved food supply was undoubtedly a factor in the decrease in mortality levels which led to rapid population growth. What was going on?

The Size of Landholdings

In the first place, the land itself was being redefined and redistributed. At the beginning of the eighteenth century in England much land was still common land, on which the herds of various farmers mingled and their crops grew side by side on strips of common field. It was difficult for an individual farmer to improve his holding under such circumstances, and as the century progressed a series of "Enclosure Acts" passed through Parliament. Under these, the old common lands were parcelled out into private holdings divided from their neighbours by hedges or stone walls. In the decade 1700–1710 there was one such act, but in 1750–1760 there were 156, and the number continued to rise almost monotonically. As enclosure progressed, there was a parallel tendency for larger farms to grow at the expense of the small.[7]

Machinery and Management

The advantage of the larger holding over the small is critically connected with the changing nature and purpose of the management of agriculture. The key to English agricultural expansion was the dynamic management style which introduced improved husbandry techniques and new machinery on a massive scale. Great landlords went into partnership with large tenants in programmes of long-term improvement such as that which saw the rental income of Coke of Holkham rise from £2,200 per year to £20,000 per year in 40 years. Clearly the larger an estate was, the greater the benefits of skilled management and the more efficiently machinery could be employed.

Social Effects

At the top of rural society, and in the middle, the English agricultural revolution had wholly positive effects on living standards. At the bottom, however, it may have lowered them. Loss of common rights, and a decreasing demand for farm labour as machinery took over, reduced many of the rural poor to beggary. Even when they did find work, their wages were so low that they had to be topped up by social relief from local government. So when those who profited from the

---

[6]An early example of such claims is to be found in R. E. Prothero, *English Farming*. London, New York [etc.]: Longmans, Green and Co., 1917; Prothero himself expressed skepticism as to whether these figures are as reliable as they seem. But that animal husbandry made great strides in this period seems indisputable.

[7]P. Mantoux, *The Industrial Revolution in the Eighteenth Century: An Outline of the Beginnings of the Modern Factory System in England*. London: Methuen, 1964.

new surplus agricultural wealth invested it in the beginnings of industrial enterprise, they found large numbers of willing recruits to their new factories. New wealth from the land, and new poverty on the land came together to fuel the Industrial Revolution.

The Chinese Experience

So what happened in China? Was there just not enough agricultural development to start an industrial take-off? The picture is more subtle than that.

The Wet-Rice Environment[8]

In the first place, agriculture is not a single thing. We must ask what kind of agriculture there was in China throughout the middle and late imperial age, and the answer is overwhelmingly that the most important branch of agricultural activity in China during this period was wet-rice cultivation. Two thousand years ago, under the Han dynasty, the political, cultural and economic centre of gravity was in the north, in the Yellow river basin, where the predominant mode of farming was dry wheat cultivation, as it was to be in later Europe. Toward the end of the Han, there are even some signs of the growth of large centrally managed estates of the kind familiar to us elsewhere. As in eighteenth-century England, the large-scale exploitation of animal power and machinery was an important factor. But this was not to last. By 1000 CE the centres of Chinese dynamism in all spheres of human activity, including agriculture, had moved irrevocably to the wet-rice areas of the south, particularly the lower Yangzi valley.

The Primacy of Skill

Rice is an extremely attractive crop to grow, and no human group has ever reverted to other crops after having taken it up. Unlike wheat it is rendered palatable by simply steaming or boiling. Each head of rice has many more grains than a head of wheat, and so a smaller proportion has to be kept for next year's seed. A well-maintained paddy field can grow rice almost indefinitely without the need to lie fallow and without demanding heavy use of fertilisers. Of course, it is no simple matter to run a paddy field well – there are complicated requirements for seed preparation, the transplanting of seedlings into prepared ground, the careful management of water, and meticulous weeding and pest control. Each rice plant will almost certainly absorb more labour than the corresponding stalk of wheat. And this is the crucial point: so great is the return on increased human care that the most highly productive rice-lands in pre-

---

[8]On the social and physical technology of wet-rice cultivation in China, see F. Bray, *The Rice Economies : Technology and Development in Asian Societies*. Oxford: Basil Blackwell, 1986.

modern China would have had a surprisingly low-tech appearance. As the human skill input increased, the use of machinery in the fields actually decreased, so that the best farmer might use what to us seem no more than the tools of a gardener. In modern terms, we could say that software replaced hardware in large parts of Chinese agriculture.

Social Effects

The success of Chinese agriculture in the late imperial period is attested by the impressive growth in population supported without obvious social stress until some way into the eighteenth century. Wars and other disturbances apart, the admiring picture of the Chinese standard of living brought back by the first Jesuit missionaries was no doubt an accurate one. China around 1700 may well have been a better place to be a peasant than contemporary Europe. But the particularities of wet-rice agriculture meant that its success was linked with quite different social outcomes. So great is the value of dedicated human skill, and so small the value of expensive agricultural machinery, that the central management of a large rice-growing estate using hired labour was not likely to prove a success. So while the Chinese gentry were always anxious to increase their income by owning more land rented to tenant farmers, their holdings were usually in the form of many scattered parcels that could not have been centrally managed even if the owners had been interested in so doing. In some parts of China the landlord's rights were severely limited, and tenants were very difficult to displace. Unlike England, we do not find the improvement of agricultural technique going hand in hand with the creation of a large pauperised class of entirely landless labourers. And unlike England, there was no group of dynamic agricultural managers who were ready to invest their surplus in a coal mine or an iron-works in which such displaced labour might have been employed.

The Case of Gunpowder[9]

It seems, therefore, that it is not enough to be technologically innovative and technologically successful for "progress" to follow – one apparently has to be technologically successful in very particular ways, and in a particular social context. At once we begin to see the danger that confronts us: could it be that any attempt at a description of what conditions must be fulfilled by any agricultural revolution which is to serve as the basis of an industrial revolution will simply amount to demanding that the revolution should take place in the precise historical conditions of eighteenth-century England? To avoid this kind

---

[9]On the development of gunpowder and gunpowder weapons in China and their westwards spread, see J. Needham, P-Y. Ho, et al., *Science and Civilisation in China, Vol. 5, Part 7, Chemistry and Chemical Technology. Military Technology: The Gunpowder pic.* Cambridge: Cambridge University Press, 1986.

of impasse, let us try looking at an example where a Chinese priority is beyond question – the case of gunpowder.

The Chinese Invention

Anybody who reads the earliest European accounts of the invention of gunpowder will inevitably conclude that it must have been invented somewhere else. In the first place, the earliest stories, such as the account by Roger Bacon around 1267 make it plain that the first gunpowder known in Europe was already of the brisant type, which exploded with a violent report. As every school-child knows who has attempted to make gunpowder by mixing ingredients in the right proportions, there is more to the technique than just the mixture. It is extremely unlikely that brisant gunpowder could have been invented in a single step. Furthermore, the stories which attempt to give it a European origin – such as the ones about Berthold the Black, supposedly a German monk – are of the suspicious kind that start late (in the early fifteenth century) and grow more detailed as time passes.

In any case the recorded history of gunpowder in China is a long one, easily predating the references in Europe. The first explicit gunpowder recipe is given in a military manual printed in 1044 CE. Before that, we can trace the records of experiments with the rapid burning of mixtures containing saltpetre and sulphur back through several centuries. All this was not the work of military technologists, but of alchemists in search of the elixir of life. The last thing they wanted was to kill people – indeed one alchemical text of the ninth century CE cautions against heating a mixture of sulphur, saltpetre and honey on the grounds that its violent deflagration will burn down the house of the alchemist who attempts it! Gunpowder seems to have been an accidental discovery, and in part an unwanted one.

Never the less, the application to military purposes was eventually made in China – unfortunately the stories about its use being confined to fireworks are a mere canard. Initially it was hurled from catapults as incendiary parcels, or used packed into bamboo tubes which acted as flamethrowers. Around the coming of the Mongols in the thirteenth century both rockets and guns firing projectiles were in use. And it was during the age of the Mongol world empire that gunpowder found its way to our own rather backward little archipelago.

Gunpowder and Social Change in East and West

The first Western illustration of a cannon dates from 1327. From then on, the pace of development in Europe is rapid. I will not say much about what happened there as gunpowder weapons developed. It is a common-place, however, that the arrival of explosives in Europe accompanied many important socio-political changes: the vanishing of the armoured knight and the stone castle from the centre of the military stage, the accompanying growth of centralised royal power with the need for heavy capital investment in arsenals and the support of expensive specialists. If technologies do have social effects, gunpowder in Europe certainly did.

But in China, the country where the technology originated, none of this seems to have happened. In China of the thirteenth century, the age of the feudal lord in his stronghold was more than one and a half millennia in the past. China had no semi-independent military aristocracy to overthrow, and the paradigm of centralised political power governing centralised military power was already ancient when gunpowder arrived on the scene. Gunpowder was certainly recognised as an important addition to the military repertoire – as shown by the measures taken in the eleventh century to prevent its ingredients being exported through China's northern borders. But the bursting of bombs on the battlefield was not paralleled by socio-political detonations.

China was, it seems, just too advanced politically for gunpowder to be a destabilising influence. Quite the reverse was seen to be the case in Japan, where one of the early acts of the Tokugawa shoguns who forced unity on the chaos of late medieval feudalism was to ban gunpowder weapons altogether. They could see all too clearly that the new armaments were deeply inimical to the sword-wielding samurai aristocracy who were to rule in their new stable world. In China, I have not heard of any sign that gunpowder was ever seen as a social threat.

But in the long run the relative social neutrality of gunpowder in China was not to be to China's advantage. In Europe, unstable, underdeveloped and fragmented, fierce political and military competition, as well as the economic interests of those with technical skills to sell, made for rapid development and improvement of the new weapons. By the late fifteenth century, the Portuguese sailing into the Indian Ocean in search of "spices and Christians" were armed with ship-board cannon of a power inconceivable to the gunners of the Chinese Admiral Zheng He, who had sailed the same seas at the beginning of the century. His ships had mounted small antipersonnel "deck-clearers" for use before a conventional boarding attack. Vasco da Gama and his successors had long-range guns with which they could pound into submission the ports of the Malabar coast who initially refused to break the monopoly of the Muslim middle-men controlling the spice trade. And it was the sea-borne guns of successive waves of Europeans that laid the foundations of the long Western hegemony in South and East Asia, and not least in China.[10]

---

[10] See in this general area C. M. Cipolla, *Guns and Sails in the Early Phase of European Expansion, 1400–1700*. London: Collins, 1965.

The Case of Printing

The example of agriculture showed us that merely being highly developed does not in itself enable you to make what is called progress. The example of gunpowder serves to warn us that in order to gain the advantages of some new technology it is not enough to have invented it first. What is more, it may sometimes turn out that in order for the new technology to make any real impact on your society, you need to be below a certain level of political and social development. For our third and final example, we turn to a technology whose significance we can perhaps see in a new light as we compare it with other innovations in information technology which are changing our own lives today – printing.

The Invention of Printing in China

As in the case of gunpowder, there is no doubt that printing is a Chinese invention.[11] Again as in the case of gunpowder, an invention with the most pronounced consequences for secular society was invented for reasons that might seem to us to be at first sight less than practical. The key to the origins of printing seem to lie mainly in religious practice relating to texts.[12] Many Buddhist scriptures end with a short section in which are set out the benefits which will be obtained by the believer who, with a devout heart, copies out the scripture in question or teaches it to another. For the longest scriptures real devotion would have been required to undertake the task. But by the second half of the first millennium CE, a class of scriptures were circulating in East Asia which were not only shorter than others, but explicitly promised much more immediate benefits in terms of wordly prosperity and power. These were the so called *dharani* texts. Every copy of such a text was a powerful radiator of blessings for its copyist.

The technology of text multiplication was not completely undeveloped in China at that time. It had long been possible to take rubbings from inscriptions of classical texts laboriously carved on stone, and official titles with a few characters could easily be stamped repeatedly from inked seals. All that was needed for rapid multiplication of longer texts was in effect to make the seal larger, by carving the text in reverse onto a block of wood. This block was then inked, and a sheet of paper was smoothed over it, peeled off and left to dry. Once the block had been carved, the printing process was cheap and low-tech.

---

[11] J. Needham and T.-H. Tsien, *Science and Civilisation in China, Vol. 5, Part 1, Chemistry and Chemical Technology. Paper and Printing.* Cambridge [Eng.], New York: Cambridge University Press, 1985.

[12] For some of the complexity of this story in China, see T.H. Barrett, *The Woman Who Discovered Printing.* New Haven [Conn.], London: Yale University Press, 2008.

By using such a process, many copies could be produced rapidly and cheaply before the block showed signs of wear.

Literary evidence points to the use of printing in a religious context in China some time in the seventh century CE, but as it happens the earliest specimens of printing come from outside China. The first is a *dharani* found in a Korean stupa known to have been set up in 751 CE. Many specimens are also known of the million copies of a *dharani* which were commissioned by Empress Shotoku of Japan around 770, each one placed in its individual clay stupa. The printing of these short texts is relatively crude, but within a century the printed Diamond Sutra, now in the British Museum, shows a complete mastery of the technique of printing both images and text for a complete book. Movable type was in use from the eleventh century, though it never occupied a place in book production comparable to the carved woodblock. The reasons for this are, by the way, less simple than one might think.

Printing and Social Change in China

Apart from its religious role, it is indubitable that the rapid dissemination of printed texts in China had deep social and political effects. That it was expected to do so is shown by the way that the new Song government which emerged out of the disorders of the beginning of the tenth century set itself to use printing to underpin its new social order. When the Tang emperors had established their own clan as permanent leaders of the great aristocratic military families which had emerged from China's period of division, literature had not played a significant role in establishing the new polity. In a scribal age it was not in any case easy for a government to control the circulation and multiplication of texts, whose transmission was usually a matter of private relations between individuals. Under the early Song, however, when the Buddhist canon was printed in 978 it was not long before the government sponsored a complete printing of the Confucian Classics, which were to provide basic ideological texts for would-be civil servants seeking appointment through the newly regularised and broadened examination system. It would perhaps be better to say that it sponsored the carving of the blocks for this project, since that gives a less misleading picture of the pace of book production in China. Rather than a whole run of books being printed in one burst, it was much more common for blocks to be on deposit in some government office, and for small numbers of books to be printed off as required after the initial print run was exhausted. It is the temporally diffuse economics of Chinese printing, rather than the non-alphabetic script, which seems to have been decisive in marginalising movable type printing in China despite its early invention there in the eleventh century. In Korea, which also used Chinese characters for many centuries before the invention of the han-gul script, the government sponsored large runs of texts for distribution at one time, and movable type was frequently used, being broken up at the end of the print run in the usual way. When this happened in China, movable type was also used from time to time.

Whatever the precise technology used, printing in China had from the beginning a strongly official character. As already indicated, the Song government took advantage of printing to disseminate texts designed to help in the intellectual formation of imperial bureaucrats. In other spheres, such as the production of standard medical texts and historical works as well as agricultural treatises, the government used printing as a means of appropriating and controlling important fields of cultural activity. The newly extended examination system, combined with the ready availability of approved texts, enabled the government to induce all families who could afford it to subject their young men to a thorough process of ideological control as the price of entry into a civil service career. Even when printing developed a role as a commercial activity, one of the most profitable kinds of books contained model answers to examination questions. The apotheosis of the use of printed material as a means of political control might seem to be the "little red book" of Chairman Mao. But this had been anticipated as long ago as the fourteenth century, when the first Ming emperor demanded that every literate person should study his own summary of the duties of a good subject. The ability to recite the text by heart was rewarded by a reduction in the penalty inflicted for a crime.[13]

Printing and Social Change in the West[14]

In China, then, it is clear that the information revolution was a revolution in the power of the centralised state to control its subjects. The other end of the world presents a striking contrast.

The earliest examples of printing known in Europe are block-prints of the fourteenth century. The similarity of artistic form and technical execution to their Chinese equivalents is too obvious to be accidental. Clearly printing was yet another part of the package of innovation that passed from East Asia to Western Europe during the time of the Mongol Empire. Naturally, it had already passed through the Islamic states of Western Asia: printed paper money is known from Tabriz in 1294.

It is a matter of conjecture how much the innovation of movable type printing in Europe around 1450 owed to China. Movable type had already been in use in China for four centuries, and before the Mongol assault it had been used by the Uighurs in central Asia. But the technology used by the Gutenbergs was highly different from the Chinese model. Type was of die-cast metal rather than of carved wood, ink was oil rather than water based, and thicker paper was brought into contact with the type using a screw-down press

---

[13]J. K. Fairbank, F. W. Mote, et al., *The Cambridge history of China, Vol. 7, Part 1, The Ming Dynasty, 1368–1644*. Cambridge: Cambridge University Press, 1988.

[14]Much of the background for the following discussion is surveyed in E. L. Eisenstein, *The Printing Press as an Agent of Change: Communications and Cultural Transformations in Early Modern Europe*. Cambridge, Cambridge University Press, 1979.

for high pressure. As is usual with pioneers of advanced technology, the Gutenbergs soon went bankrupt. But their technology survived and prospered. Everybody is familiar with the three great earthquakes in European culture in which printing subsequently played a key role. In the Renaissance, a revival in classical studies which was already in process was re-oriented by the new technique. For the first time it was possible to guarantee the production of many identical copies of a newly edited classical text. It was thus worth while *being* an editor for its own sake. Many more books were easily available, so scholars had a much wider literary diet on which to form their taste and judgment. And of course they could now spend on reading the time previously spent on copying. In addition, the survival chances of rare books were much increased.

In the Reformation, quite apart from the effects of new scholarship on Biblical interpretation, the propagandists of new religious movements acquired a power to "preach without the voice" to remote audiences. A press in Geneva could print the works of a banned French theologian. Further, small or marginal groups that could never have made their voices heard in the age of scribal copying or oral preaching could find the means to gain cohesion and spread their ideas.

Last but not least, the scientific revolution is inconceivable without the availability of printing. Quite apart from the recovery of ancient science and its critique, we must consider, as in the case of theological innovation, the formation of international "invisible colleges" of specialists reading one another's books – and the growth of scientific periodicals to facilitate such communication. Most of all, perhaps, the possibility of claiming credit for a discovery by definitive publication led to pressure to make one's discoveries known as soon as possible rather than keeping them as a form of private intellectual property. When we move from science to technology the same story could be told with even more emphasis. Returning to China for a moment, it is significant that although there were technical publications devoted to improving practice in such fields as agriculture, they were frequently linked with official initiative and sponsorship.

Technology and Culture

I have reviewed some possible comparisons between China and Europe drawn from three very diverse fields of human activity. The message has been one of grossly different outcomes resulting from technological developments that seem at least superficially very similar.

In agriculture, we have seen that in order to provide a foundation for an industrial take-off, it may not be enough to develop a sophisticated and highly productive agricultural technology – it may be necessary to develop an agriculture of a particular type under quite specific geographical and historical conditions. If we attempt to make those preconditions more precise, it seems to me that our explanation may soon reduce to mere description of the differences between China and Europe. We may find ourselves saying that the agriculture of Europe, and not of China, formed a suitable base for industrial development, because ... well, because it was in Europe and not in China that things turned out that way. This is a kind of reductio ad absurdum of technological determinism.

The cases of gunpowder and of printing have interesting similarities and differences. Both had very marked consequences when they arrived in Europe. But while the first of these apparently had no large-scale effects on Chinese society, the second was very effective in achieving the opposite result from the European case – the massive reinforcement of the state's ideological control of its élite. It also seems to be quite possible for a technology to pass through a society without being effectively absorbed: printing, despite having passed through the Islamic world on its journey to Western Europe, does not seem to have taken significant root there. Or, as in the case of Tokugawa Japan and gunpowder, an "advanced" technology can suffer effective ejection from a society in which it has already established itself.

All this directs us in a direction which I think is helpful in confronting the Needham problem anew. Needham points repeatedly to the production of important technologies in Chinese culture, and asks why all this did not lead to the kind of developments which eventually took place in Europe. In response, I would suggest that we may need to redefine what we mean by a technology. There may be, for example, no useful unit of technological discourse that we can label "printing", in the sense of a mere material process for reproducing marks on paper. Perhaps we need to focus more broadly on the methods by which knowledge is validated, authorised transmitted and subverted. Considerations of hardware such as the printing press will certainly sometimes be important here, but what one might call "social software" will often be more important. In an age when the manipulation of information is more important than operations performed on material, this broadening of the idea of technology should give us few problems.

Indeed, I would like to take this proposal a little further. Must it always be the case that a technology involves lumps of matter or significant quantities of energy having their configurations and positions changed? Surely the obvious modern counter-example is that of the software engineer, whose work is solely concerned with the manipulation of information – although the difficulties caused by this example are highlighted by the fact that in some states of the USA it is actually illegal for the software specialist to call himself an engineer at all, since he is not formally qualified as a matter-manipulator. But once we admit the information processing expert to the ranks of technologists, it becomes hard not to make this process retrospective. Thus, why should not the elaborate information channelling and digesting systems of the imperial Chinese bureaucracy not be seen as a technological system? Once this dissolving of the boundaries of technology begins, it is very hard to stop it. To take a very different example (and I have to acknowledge Francesca Bray here[15]), what about the management of human reproduction? Once we break the link with metal-bashing or cognate activities, is there anything that cannot be treated as a technology, in the sense of a rational coordination of activities and resources designed to satisfy perceived human needs?

I think the problem is one created by the very success of historians of technology themselves. Once upon a time it was easy to draw the limits of the history of technology – they were drawn, more or less, by the refined and ignorant distaste of historians of military conflict, political struggle, or intellectual change for dealing with such dull and banausic matters as the ways people actually got their food, clothing and shelter. Technology was defined in part by its exclusion from the central discussions of "proper" historians, and indeed from the central institutions of history. This exclusion was of course always completely senseless. The retaining wall of disdain and ignorance has now largely been removed, but the consequence may well be that in its absence there turns out to be no rational way to define where technology begins and ends, unless we are prepared to insist, in the face of all pressures to the contrary, that technological activity must involve something more or less mechanical, just because the merely mechanical was once ignored and excluded.

If we are prepared to define a technology more broadly so as to include the entire cultural and social matrix within which particular manipulations of matter take place, our lives will be simpler in some ways but more interesting in others. Some problems will dissolve: for example, we will not be puzzled that China's very early lead in iron casting did not produce a Song dynasty industrial revolution.[16] On the contrary, we may well see the total phenomenon of the Song iron industry and the total phenomenon of the iron industry of late mediaeval Europe as being, in effect, quite different technologies, despite the fact that they both involve producing element number 26 of the periodic table in liquid form so that it can be run into moulds.

Further and finally, if a transplantation of a particular technical process to a different society may result in a different technology, it is clear that we shall no longer be surprised when different results are produced. Thus, for instance, what will happen to the Internet in an increasingly prosperous and technically sophisticated China? Many once talked glibly about the liberalising effects this was bound to have, but it appears that the Chinese government has other ideas. Perhaps as in the case of printing an instrument of intellectual subversion in one part of the world may function as something rather different elsewhere. The successful transplanting of a technical process may perhaps be equivalent to its re-invention and to that extent its effects may turn out to be unpredictable.

---

[15] F. Bray, *Technology and Gender: Fabrics of Power in Late Imperial China*. Berkeley, University of California Press, 1997.

[16] For a detailed survey of Chinese iron production, see D. B. Wagner and J. Needham, *Science and Civilisation in China, Vol. 5, Part 11, Chemistry and Chemical Technology. Ferrous Metallurgy*. Cambridge: Cambridge University Press, 2008.

The Ottoman Empire and the Technological Dialogue Between Europe and Asia: The Case of Military Technology and Know-How in the Gunpowder Age

Gábor Ágoston

The historiography of Muslim-Christian relations has suffered from many distortions. Clashes between Cross and Crescent have dominated the narrative, whereas intra-civilizational conflicts, that is, wars within Christianity and Islam, as well as military cooperation and acculturation between the various Muslim and Christian polities, have usually been de-emphasized. The history of Islamic warfare shows that wars within Islam and wars against non-Muslim enemies other than Christians were as important as wars against the Cross. After all, the Mongol invasion in the 1250s had far greater impact on the history of the Islamic heartlands than the Crusades. Despite rhetoric to the contrary, holy war – whether crusade or jihad – was not the only way in which relations between Europe and the Ottomans were defined either. While the Ottomans and their Christian adversaries devoted considerable resources to wars waged against one another, rivalries among the European states and the idea of a European balance of power often led to temporary alliances with the Ottomans.

The Context for Technological Dialogue

Shortly after their conquest of Byzantine Constantinople in 1453, the Ottomans became an integral part of the emerging new Renaissance diplomacy. Beginning with Venice in 1454, several European powers stationed resident ambassadors in Istanbul, who often shared information about their rivals with the Ottomans, and concluded peace and commercial treaties with the Sublime Porte. Due to their multi-layered information gathering system, the Ottomans had access to information concerning European rivalries and politics. While at the analytical level the Ottoman information gathering system lagged behind those of the leading European states, such as Venice

G. Ágoston (✉)
Department of History, Georgetown University, Washington, DC, USA
e-mail: agoston@georgetown.edu

or Spain, at the level of day-to-day politics Istanbul was quite well informed about its adversaries.[1]

The Ottomans and their European enemies constantly watched each other's military skills, learned from each other, copied and adopted their opponents' weapons, and participated equally in the early modern arms race. The sixteenth-century Ottoman Empire possessed all the characteristics that promote "brain gain" and the ability to attract foreign experts into their employ. The empire's wealth and prosperity, along with the mobility of its multi-ethnic, multi-religious and polyglot population and the relative open-mindedness, pragmatism and flexibility of its political elite made the empire, and especially its capital city Istanbul, just the sort of place where foreigners could sell their expertise, find new career opportunities, gain access to power, and advance the transmission, production and dissemination of knowledge. In addition to its capital city, the porous nature of the empire's Mediterranean, Balkan, Hungarian, Eastern European and Black Sea frontiers with European powers and Muscovy/Russia created spaces where people with knowledge about the other side of the frontier could be found, recruited and employed.

Despite prohibitions, European merchant ships, especially those belonging to France in the mid-sixteenth century and the Protestant states of England and the Dutch Republic in the seventeenth century, often brought weapons and ammunition (along with other prohibited goods) to Istanbul and Ottoman port cities. In return, the Ottomans granted – in the form of capitulations – trade, legal and religious privileges to those European states whose merchants supplied them with weapons and with such strategically important raw materials as tin and lead.[2]

Gunpowder Technology and the Ottomans

The experience of the various Islamic states with firearms varied greatly and the nature of this experience depended on historical, social, economic, geographical and cultural factors. Contrary to Eurocentric and Orientalist claims, the success or failure of these experiments had very little to do with religion. The Ottomans were especially successful with firearms, with which they became acquainted in the 1380s – that is, about 60 years after these weapons had appeared in western European sieges. While we cannot rule out the possibility that knowledge of

---

[1] G. Ágoston, "Information, Ideology, and Limits of Imperial Policy: Ottoman Grand Strategy in the Context of Ottoman-Habsburg Rivalry," in V. H. Aksan and D. Goffman, eds., *The Early Modern Ottomans: Remapping the Empire*. New York: Cambridge University Press, 2007, pp. 75–103.
[2] G. Ágoston, "Merces Prohibitae: The Anglo-Ottoman Trade in War Materials and the Dependence Theory" in *Oriente Moderno* [K. Fleet, ed., *The Ottomans and the Sea*], XX (LXXXI) n.s.1 (2001): 177–192.

firearms technology was also transmitted to the Ottomans by way of Central Asia and the Arab world, the Balkans appears to have been the region that played the most crucial role in the diffusion of gunpowder technology to the Ottoman Empire. Direct military conflicts with enemies already in the possession of firearms, including Serbian and other vassal contingents fighting alongside the Ottomans, as well as trade in weaponry all played roles in the transmission of gunpowder technology to the Ottomans.

While the relatively early Ottoman adaption of firearms is notable, the fact that the Ottomans were remarkably quick and successful in integrating gunpowder technology into their land forces and navy is even more significant. Preceding their Muslim and Christian rivals by centuries, the Ottomans set up permanent troops specialized in the manufacturing and handling of firearms from the 1390s onward, and these efforts intensified beginning in the fifteenth century: (1) artillerymen remunerated through military fiefs (timariot *topçu*s) were employed from the 1390s, salaried gunners or *topçu* from the reign of Murad II (r.1421–1444, 1444–1451); (2) armorers (*cebeci*s) perhaps from the mid-fifteenth century; (3) gun carriage drivers (*top arabacıları*) from the second half of the fifteenth century; (4) bombardiers (*humbaracı*s), perhaps from the late fifteenth century; (5) whereas the Janissaries, the sultans' elite infantry established in the 1370s, were gradually armed with hand firearms starting under Murad II. All this was in sharp contrast to most of the Ottomans' European opponents. In Europe, artillerists formed a transitory social category between craftsmen and soldiers and were usually hired by the state on a temporary basis to handle the weapons during campaigns. Their transformation from craftsmen to professional soldiers was closely related to a long evolutionary process that took place between 1500 and 1700. Thus, the Ottomans had a distinct advantage over their rivals both organizationally and in terms of the size of their artillery corps. Recent research has also shown that the Ottomans also managed to keep up with their European rivals as to the quality of their weaponry, although most of the time they were followers rather than innovators, and they lagged behind the most advanced European states with regard to the standardization of their ordnance.[3] While the Ottomans had access to a rich literature in Turkish, Arabic and Persian regarding fields such as archery and horsemanship, it seems that they had more difficulty in staying informed of the burgeoning European literature on the art of war. However, recent cataloguing of Ottoman, Arabic and Persian manuscripts, as well as specialized handbooks regarding Ottoman literature on astronomy, mathematics, geography and the

---

[3]G. Ágoston, *Guns for the Sultan: Military Power and the Weapons Industry in the Ottoman Empire*. Cambridge Studies in Islamic Civilization, New York: Cambridge University Press, 2005, pp. 15–60; G. Ágoston, "Behind the Turkish War Machine: Gunpowder Technology and War Industry in the Ottoman Empire, 1450–1700." in B. Steele and T. Dorland, eds., *The Heirs of Archimedes: Science and the Art of War through the Age of Enlightenment*. Cambridge, MA: MIT Press, 2005, pp. 101–133; S. Aydüz, *XV. ve XVI. Yüzyılda Tophane-i Amire ve Top Döküm Teknolojisi*. Ankara: TTK, 2006, pp. 12–36.

art of war[4] have brought several important treatises into light and suggest that careful study of this body of literature would help us arrive at a more informed view regarding the Ottomans' familiarity with European literature on science and the art of war. The following short overview concentrates only on works dealing with artillery and siege warfare.

The Transfer of European Military Technology to the Ottomans via Military Treatises

Although linguistic and cultural barriers significantly limited the extent to which the Ottomans could profit from the thriving European literature on the art of war, some of the best works written in Europe did reach the sultans and their advisers. It is uncertain whether Roberto Valturio's (1413–1484) *De re militari*, the first military technical treatise to be printed (1472), reached the Ottomans during the reign of Sultan Mehmed II (r.1444–1446 and 1451–1481) – to whom a manuscript copy of the work was sent by Sigismundo Malatesta of Rimini, Valturio's employer, in 1461. It is possible that the treatise was obtained only under Sultan Süleyman's reign (1520–1566), when in 1526 the sultan first entered Buda, Hungary's capital, and confiscated dozens of precious illuminated manuscripts from King Matthias Corvinus's (r.1458–1490) famous library, the Bibliotheca Corviniana, including Valturio's printed book. King Matthias had three copies of the book, and one of them later – most likely in 1526 – became the property of the Ottoman sultans.[5] However, Mehmed II managed to acquire other valuable military treatises of his time. A copy of one of the most influential fifteenth-century military books, Ser Mariano di Giacomo Vanni's ("il Taccola," 1381–ca. 1458) *Tractatus*, may have reached the sultan before the siege of Constantinople in 1453. Other European works on military technology reached the Ottomans later. Luigi Ferdinando Marsigli (1658–1730) – the Bolognese military engineer and polyhistor who fought against the Ottomans in Habsburg

---

[4]E. İhsanoğlu, R. Şeşen, C. İzgi, C. Akpınar, İ.Fazlıoğlu, *Osmanlı Astronomi Literatürü Tarihi = History of Astronomy Literature during the Ottoman Period.* 2 vols., Osmanlı Bilim Tarihi Literatürü Serisi, no.1, İstanbul: IRCICA, 1997; E. İhsanoğlu, R. Şeşen, and C. İzgi, *Osmanlı Matematik Literatürü Tarihi = History of Mathematical Literature during the Ottoman Period.* 2 vols., Osmanlı Bilim Tarihi Literatürü Serisi, no.2, İstanbul: IRCICA, 1999; E. İhsanoğlu, R. Şeşen, M.S. Bekar, G. Gündüz, A.H. Furat, *Osmanlı Coğrafya Literatürü Tarihi = History of Geographical Literature during the Ottoman Period.* 2 vols., Osmanlı Bilim Tarihi Literatürü Serisi, no.3, İstanbul: IRCICA, 2000; E. İhsanoğlu, R. Şeşen, M. S. Bekar, G. Gündüz, *Osmanlı Askerlik Literatürü Tarihi = History of Military Art and Science Literature during the Ottoman Period.* 2 vols., Osmanlı Bilim Tarihi Literatürü Serisi, no.5, İstanbul: IRCICA, 2004.

[5]This copy was found in the library of the Topkapı Palace in Istanbul in 1890. See J. Balogh, *Mátyás király és a művészet* [King Matthias and the Arts]. Budapest, 1985, pp. 197, 325. On the Bibliotheca Corviniana see also, M. Tanner, *The Raven King: Matthias Corvinus and the Fate of his Lost Library.* New Haven [Conn.]; London: Yale University Press, 2008.

service in the 1680s and 1690s – reported in his *Stato militare dell'Imperio ottomano* (1732) that Pietro Sardi's *L'Artiglieria* (Venice, 1621), one of the most celebrated books on cannons and siege warfare in seventeenth-century Europe, was translated into Ottoman Turkish and was used by the sultan's gunners in Marsigli's time.[6]

Muslims from Spain also played an important intermediary role in the transmission of military know-how and knowledge regarding the art of war to the Ottomans. One of such transmitters was the Granadan Muslim sailor and master gunner, known as Captain Ibrahim b. Ahmad (al-Ra'is Ibrahim b. Ahmad b. Ghanim b. Muhammad b. Zakariyya al-Andalusi). Born in Granada, Ibrahim settled in Seville and served in the Spanish navy. Later he escaped to Tunis, where he commanded a ship of Osman Dey, the ruler of Tunis (1594–1610). Following his capture and imprisonment in Spain, he returned to Tunis and was sent to the fort of Halk al-Wadi (La Goletta) by Yusuf Dey (1610–1637). Witnessing the incompetence of the gunners there, he decided to write a manual on gunnery. Captain Ibrahim wrote his *Manual de Artilleria* between 1630 and 1632 in his native Spanish with Arabic script, known as Aljamiado. Although the original Aljamiado work is lost, it survived in Arabic translation. The translator was the former interpreter of the Sultan of Morocco and a fellow Morisco, who, with the author's assistance, rendered the work into Arabic in 1638. Later, the translator's son made several copies of the Arabic work, one of which was dedicated and sent to the then reigning Ottoman Sultan, Murad IV (r.1623–1640). Besides his own experience, Captain Ibrahim consulted several contemporary European works on gunnery. Among these, researchers have identified Louis Collado's *Plática Manual de Artillería* (1592, Italian original from 1586), the most famous European treatise on gunnery in the late sixteenth and seventeenth centuries, and Cristoval Lechunga's *Discurso de la artillería* (1611). Captain Ibrahim was most heavily influenced by Collado's work, from which he translated entire chapters and adapted most of his 50 illustrations of cannons, mortars, hand grenades, and gunnery tools.[7]

In the seventeenth century the Ottomans continued to track European knowledge in siege warfare, and managed to further hone their skills during the long sieges of the Cretan War (1645–1669) against the Venetians, the leading experts in contemporary siege warfare. They soon put their recently acquired knowledge into good use on other fronts, as their success against the

---

[6]V. J. Parry, "Barud. IV: The Ottoman Empire," *Encyclopaedia of Islam*. New Edition, vol. 1, p. 1064.

[7]D. James, "The 'Manual de Artilleria' of al-Ra'is Ibrahim b. Ahmad al-Andalusi with Particular Reference to its Illustrations and their Sources," *Bulletin of the School of Oriental and African Studies*, University of London, XLI, 2 (1978), pp. 237–257; E. İhsanoğlu, *Büyük Cihaddan Frenk Fodulluğuna*. Istanbul, 1996, pp. 118–123. Some of the book's illustrations are reproduced in E. İhsanoğlu et al., *Osmanli Askerlik Literatürü Tarihi*. vol. 1, İstanbul: IRCICA, 2004.

Hungarian forts of Várad (1660) and Érsekújvár (Neuhausel, 1663) demonstrated. However, it seems that by the end of the seventeenth century most of the experienced Ottoman gunners, miners, and bombardiers were dead, and that Ottoman skill in siege warfare was not always up-to-date.[8] The Ottomans themselves realized this and tried to modernize the technical branches of their army. In 1735, with the help of the French renegade, Claude-Alexandre Comte de Bonneval (1675–1747), known in the Empire as Humbaracı Ahmed Pasha, the Ottomans established a new corps of bombardiers. The bombardiers and their officers were, for the first time in the history of the corps, trained in military engineering, ballistics and mathematics. Their instructor was another French renegade, Mühendis (Engineer) Selim, who had been educated in military engineering and fortress building in France. Although the corps faltered after its founder's death in 1747, it was revived under Selim III (r.1789–1807) and Mahmud II (r.1808–1839). The corps proved to be an ideal environment for acculturation and the synthesis of European and Ottoman culture: an Ottoman engineer, Mehmed Said Efendi, who taught geometry at the Corps, invented a new instrument for land surveying, by combining the European telescope and the Ottoman quadrant (*rub'u tahtası*).[9] It is thus hardly surprising that the first known Ottoman treatise on artillery, more specifically on bombs (*humbara*) and mortars (*havan*), the *Fenn-i Humbara ve Sanayi-i Ateş ba*, was written in 1736 by a certain Mustafa b. Ibrahim, one of the pupils of Humbaracı Ahmed Pasha. The work summarized European and Ottoman knowledge regarding mortars, bombs and explosives. The author and his father both served as clerks in the Bombardiers Corps (*Humbaracı Ocağı*). While writing his treatise, Mustafa b. İbrahim relied on both European works on the subject and on the experience of European and Ottoman bombardiers serving in the corps.[10]

It was around this time that the most important military treatise of Raimondo Montecuccoli (1609–1680), field marshal and commander-in-chief of the Austrian Habsburg armies in 1664–1680, was also translated into Turkish. Montecuccoli's *Memoire della Guerra* was originally published in 1704, and soon became one of the most influential military treatises in Europe, as the many foreign language editions show: Spanish (Lisbon 1708), French (Paris, 1712, 1760, Strasbourg, 1740, Amsterdam, 1752, 1770), Latin (Vienna, 1718), and German (Leipzig, 1736). The Turkish translation, titled *Fenn-i Harb* (*Fünun al-Harb*), is based on the Latin version of the book, known as *Commentarii bellici*, and was completed under Sultan Mahmud I (r.1730–1754).[11] In addition

---

[8] Ágoston, *Guns for the Sultan*, pp. 35–38.

[9] M. Kaçar and A. Bir, "Ottoman Engineer Mehmed Said Efendi and his Works on Geodesical Instrument," in E. İhsanoğlu, K. Chatzis and E. Nicolaidis, eds., *Multicultural Science in the Ottoman Empire*. Turnhout: Brepols, 2003, pp. 71–97.

[10] E. İhsanoğlu et al., *Osmanlı Askerlik Literatürü Tarihi*. vol. 1, p. 34.

[11] On the manuscript see E. İhsanoğlu et al., *Osmanlı Askerlik Literatürü Tarihi*. vol. 2, pp. 760–765.

to Montecuccoli's work, the *Fenn-i Harb* also contains three addenda (*zeyl*). Of these, the first two additions are especially important. The first is a translation of three books or chapters relating to military architecture, offense and defense, and army formation from Gaspar Schott's (1608–1666) *Cursus mathematicus*. Schott was a German Jesuit physicist and mathematician who studied in Palermo and taught mathematics in Würzburg, where his book was first published in 1661. The work saw many subsequent editions, and the three chapters (books) translated into Turkish in the *Fenn-i Harb* indeed deal with the above subjects (Liber XXII: De Architectura militaris and Liber XXIII: De polemica offensiva ac defensiva), although the last book is on tactics of the day (Liber XXIV: De tactica hodierna). It is obvious from his chapter on military architecture that Schott was familiar with, among others, the works of Pietro Sardi, also known to the Ottomans, and of Carlo Theti, whose plans of Kanizsa and Érsekújvár (Ott. Uyvar, modern Nove Zamky) and suggestions regarding Komárom and Eger were influential during the modernization of these Hungarian forts that defended Vienna and the Habsburg hereditary lands against the Ottomans.[12] According to the *Osmanlı Askerlik Literatürü Tarihi* (OALT), the second addendum consists of selections from a French book by a certain Alain Manson and titled *Fünun-i Harbiye ve Ebniye-i Harbiye Sanayii*. The author should be the famous seventeenth-century French matematician, cartographer and military engineer, Allain Manesson Mallet (1630–1706), often referred to as Allain Mallet, who served as sergeant-general in Louis XIV's artillery and later as Inspector of Fortifications. In 1671 he published his *Les Travaux de Mars, ou la Fortification Nouvelle, tant Régulière qu'irrégulière, Divisée en Trois Parties* (Paris, 1671–1672) which later saw many editions under slightly different titles.[13]

Unfortunately, the name(s) of the compiler(s) and translator(s) who produced the *Fenn-i Harb* are not indicated in the book. While some researchers have suggested that the book was compiled and translated by İbrahim Müteferrika, the Hungarian renegade and founder of the first Arabic letter printing press in the Ottoman Empire, and his unknown collaborator, others have questioned this attribution. A recently published probate inventory of İbrahim Müteferrika listed only one Latin book on warfare, named in Turkish rather vaguely as *Tertib-i Ceng*, which can be translated as Order (Organization, Method, Plan) of War(fare) or Battle. Whether the compilers of the inventory referred to Montecuccoli's *Commentarii bellici* is uncertain. Even if İbrahim Müteferrika did not possess Montecuccoli's book at the time of his death, it is

---

[12]G. Schott, *Cursus Mathematicus, Sive Absoluta Omnium Mathematicarum Disciplinarum Encyclopaedia: In Libros XXVIII. digesta. – Accesserunt in fine Theoreses Mechanicae Novae : Additis Indicibus locupletissimis.* – Bambergae; [Frankfurt, Main]: Schönwetterus, 1677, p. 486.
[13]Other titles and editions: *Les Travaux de Mars ou l'Art de la Guerre Divisé en Trois Parties*. Paris, 1684–1685; Amsterdam, 1684; Paris, 1691 etc.

still possible that he was involved in its translation into Ottoman Turkish, but without further research the question cannot be decided.[14]

The systematic translation and publishing of western military and technical books began only in connection with the military reforms of Sultan Selim III and Mahmud II. The first works were French military treatises and textbooks, often older ones such as Sébastien le Prestre de Vauban's (1633–1707) treatises on warfare, sieges and mines, which were published in Ottoman Turkish in the early 1790s. Of these, the book on mines (*Fenn-i Lağım, Sanayi-i Lağım*) was published in 1793, whereas the one on attack (*Usul-i Harbiyye, Fenn-i Harbiyye*, or *Fenn-i Muhasara*) the next year. The book on sieges (*Fenn-i Harb, Muhasara-yı Kıla, Fenn-i Muhasara*), published in 1792, is attributed to both Vauban and Bernard Forest de Bélidor (1697?–1761), who, however, used Vauban's writings in his works. All three were translated by Konstantin Ypsilanti (Konstantinos Ypsilantos) of the famous Phanariot family, son of Alexander Ypsilanti, ruler (*hospodar*) of the Ottoman vassal states of Wallachia (1774–1782, 1796–1797) and Moldavia (1786–1788), and later himself *hospodar* of Moldavia (1799–1801). All three works were printed by Raşid Mehmed Efendi, who inherited two of the four printing presses once operated by İbrahim Müteferrika.

Many of the works translated from the 1790s onward were translated for students in the newly established military technical schools: the Imperial Naval Engineering School (*Mühendishane-i Bahri-i Hümayun*), the Artillery School (*Topçu Mektebi*), and the Imperial Land Engineering School (*Mühendishane-i Berri-i Hümayun*). In addition to these, lecture notes of French and other foreign teachers at the military and naval schools were also translated and printed in either the French Embassy's Istanbul press or in one of the newly established Ottoman printing houses.[15]

Foreign Experts and "Brain Gain"

While military treatises may have played some role in the diffusion of European military technology to the Ottomans, the knowledge brought to Istanbul by European military experts was more significant. Christians, Jews, and

---

[14]A. A. Adivar (*Osmanlı Türklerinde İlim*. Istanbul, 1982, p. 184) doubts that the translator was İbrahim Müteferrika. Most recently it is posited by E. İhsanoğlu and his colleagues (see *Osmanli Askerlik Literatürü Tarihi*, vol. 1, p. LXII, and vol. 2, pp. 760–765). O. Sabev, whose book (*İbrahim Müteferrika ya da İlk Osmanli Matbaa Serüveni (1726–1746)*. Istanbul: Yeditepe, 2006) is based on new sources and a thorough utilization of all available secondary sources does not mention that İbrahim Müteferrika translated Montecuccoli's work. İbrahim Müteferrika's probate inventory is published ibid., pp. 350–364.

[15]K. Beydilli, *Türk Bilim ve Matbaacilik Tarihinde Mühendishâne, Mühendishâne Matbaasi ve Kütüphânesi, 1776–1826*. İstanbul: Eren Yayincilik ve Kitapçilik, 1995. On the authorship of the *Fenn-i Harb* see pp. 182–192.

renegades are all mentioned by sixteenth-century European sources with regard to assisting the Ottomans as military experts, working mainly in the Ottoman State Cannon Foundry (*Tophane*) and naval Arsenal (*Tersane*). Of these, perhaps the best known, the Hungarian cannon founder Orban offered his services to the sultans in 1453. Others were captured in wars and raids and then forced to work for the sultans, like Jörg of Nürnberg, who was captured in 1460 while working in Bosnia as *Büchsenmeister* or cannon founder and then served the Ottomans for 20 years. And there were those in the hundreds and thousands who were forced to work for the Ottomans, like the Christian miners, sappers, smiths, stone carvers, carpenters, masons, caulkers and ship-builders in the conquered Balkan fortresses, towns, mines, and Ottoman naval arsenals and shipbuilding centers. Most of the Ottoman *lağımcı*s, that is, the miners and sappers, were of Christian origin in the fifteenth through seventeenth centuries. In 1453 Mehmed II ordered the miners of Novo Brdo in Serbia to dig mines during his siege of Constantinople, and Ottoman Novoberda, along with other mining towns of the Balkans (e.g., Kratova, Srebreniçe or Zaplana) remained valuable recruiting areas of miners through the seventeenth century. In that century, however, Armenians (especially from Kayseri), Greeks, and Christians from Bosnia joined their ranks. Military experts and specialized craftsmen, both Muslims and Christians, also arrived to Istanbul through the *sürgün* system, that is, the Ottomans' state-organized mass resettlement policy. It seems that until the sixteenth century non-Muslim military men and artisans with war-related skills were not separated from their Muslim peers. Muslim and non-Muslim artillerymen and cannon founders, ship-builders, carpenters, caulkers and others are listed along their Muslim colleagues in Ottoman payroll and revenue registers from the Balkans and Hungary.[16]

Muslims and non-Muslims are also listed together in the first known *ehl-i hıref defteri*, that is, in the 1526 register of the names and occupations of 590 artisans in the Palace service. Of the 590 artisans, only 13 were non-Muslims, although among the *tüfekçis* or gun-makers their proportion was substantial: of the 10 *tüfekçis* only two were Muslims, while six were Jews and two were Muscovite Christians.[17] The number of artisans with specialized skills in the Palace service was substantial (1526:590 men, 1545:776 men, 1558:580 men, 1596:1451 men, 1638:708 men, 1654:718 men, 1656:690 men, 1688:312, and about 230 men in the eighteenth century), and many worked in workshops located outside the grounds of the Palace. Although the number of Christian *ehl-i hıref* was minimal, there were dozens who were Christians by birth who had converted to Islam after having been captured during raids and campaigns. In the mid-sixteenth century, for instance, there were 44 Germans, 41 Franks (general term for Western Europeans), and 34–39 Hungarians. The sharp

---

[16]On the above see G. Ágoston, *Guns for the Sultan*, pp. 39–48.
[17]İ.H. Uzunçarşılı, "Osmanlı Sarayında Ehli Hıref (Sanatkarlar) Defteri," *Belgelerle Türk Tarih Dergisi*, XI, 15 (1981–1986): 50.

increase in the number of Hungarians after Hungary's conquests in 1541 (1526:5 men, 1545:34 men, 1558:39 men) indicates that the Ottomans acquired skilled European artisans in considerable numbers following their military victories.[18] Later in the sixteenth century, there was a separate group of non-Muslim artisans in the Ottoman capital. Known as *taife-i efrenciyan*, the history of which needs further research in order to assess their role in the transmission of European (*frengi*) military skills to the Ottomans.[19]

The number of European technicians, craftsmen and military experts who served the Ottomans, however, was substantially more than the handful of *Efrenci* technicians working in the Palace. In addition to the Serbian miners, Mehmed II also employed artillerists from Serbia, Bosnia, Germany, Hungary and Italy. Ottoman pay registers from the latter part of the fifteenth and the first half of the sixteenth century list Christian smiths, stone carvers, masons, caulkers and ship-builders in Ottoman Balkan fortresses. In the sixteenth century there were Marrano, Jewish, French, Venetian, Genoese, Spanish, Sicilian, English, German, Hungarian and Slav experts working in the Ottoman cannon foundries. French and English military engineers aided the Ottomans in the seventeenth century. Italian artisans worked in the Ottoman shipyards and many of the experts aboard the Ottoman vessels were also Italians and Greeks.

Europeans travelers accused these Christians, Jews and Marranos of teaching the Ottomans how to make and use ordnance and gun carriages. Elsewhere I have cautioned against exaggerating the contribution of European technicians to the Ottoman weapons technology and arms industry, and have shown that the employment of foreign military technicians and artisans was not unique to the Ottomans. Employing foreign experts from countries that were considered to be on the cutting edge of technology was a well-established practice. From the Austrian and Spanish Habsburgs to France, and from England to Muscovy this was the major means to acquiring new technology. The Ottomans were very much a part of this transfer of early modern military technology, and I have also noted that the multi-ethnic and multi-lingual empire and its capital city was an ideal place for technological dialogue.[20]

The contribution of foreign, especially French and later English, military experts to Ottoman military technology in the eighteenth century needs further research. I have already mentioned Claude-Alexandre Comte de Bonneval, alias Humbaracı Ahmed Pasha, and his new corps of bombardiers, established in 1735. The other well-known eighteenth-century European in

---

[18] B. Yaman, *Osmanlı Saray Sanatkarları: 18. Yüzyılda Ehl-i Hiref*. İstanbul: Tarih Vakfı Yurt Yayınları, 2008, pp. 20, 32.

[19] For a Turkish-language summary of the relevant foreign-language literature and knowledge regarding the Efrencis' role in the transmission of firearm technology see S. Aydüz, "XIV–XV. Asırlarda Avrupa Ateşli Silahlar Teknolojisinin Osmanlılara Aktarılmasında Rol Oynayan Avrupalı Teknisyenler: Taife-i Efrenciyan," *Belleten*, LXII, 235 (1998): 779–830.

[20] Ágoston, *Guns for the Sultan*, p. 48.

Ottoman service, Baron François de Tott, the Hungarian-born French consul to the Crimean Tatar Khan from 1767, is more problematic. Although Tott remained on France's payroll, the Ottomans occasionally contracted him to help them with the rebuilding of the forts of the Dardanelles and to cast cannons and train artillerymen between 1770 and 1775.[21] While their role in the modernization of the Ottoman artillery and armed forces awaits further studies, one caution is warranted here. It is dubious to what extent Tott's advice could have been useful for the Ottoman gunners, given the fact that as a former cavalryman in France's famous *hussar* regiment, Tott must have had only limited knowledge in fortress building, siege warfare and cannon casting. His observations regarding the Ottoman artillery were often incorrect and superficial. However, due partly to the popularity and availability of Tott's *Memoirs*, in which he unashamedly exaggerated his role in Ottoman service, his observations have exerted disproportionate influence on later researchers, and are partly responsible for perpetuating some off the fallacies regarding the Ottoman weaponry. It is almost certain that the some 300 French officers and engineers who worked for the Ottomans in the 1780s played a more important role in the transmission of contemporary European military technology to the Ottomans than Tott.

The Ottomans' Role in the Diffusion of Gunpowder Technology in Asia

Another notable feature regarding European-Ottoman military acculturation is the Ottomans' role in the diffusion of gunpowder technology in the Middle East and Asia. Ottoman experts played roles of varying importance in the transmission of gunpowder technology to the Khanates in Turkistan, the Crimean Khanate, Abyssinia, Gujarat in India and the Sultanate of Aceh in Sumatra. As Halil İnalcık pointed out more than four decades ago, Istanbul sent cannons and hand-held firearms to the Mamluk Sultanate, Gujarat, Abyssinia and Yemen (before the latter two were incorporated into the Empire). Ottoman experts (known to us as Ali Kulu, Rumi and Mustafa) played important role in the diffusion of firearms technology and the *Rumi* methods of warfare into Babur's Mughal India. Even the Safavids, the Ottomans' main rivals in the East, acquired Ottoman artillery and muskets as a consequence of Prince Bayezid's (r.1481–1512) rebellion and escape to Iran. There were Rumlu Tofangchis, that is, Ottoman artillerists, in Shah Tahmasp's (r.1524–1576) Safavid army. The conquest of Diu in 1531 was partly the result of Muslim

---

[21]V. H. Aksan, "Breaking the Spell of the Baron de Tott: Reframing the Question of Military Reform in the Ottoman Empire, 1760–1830," *International History Review*, XXIV, 2 (2002): 253–277.

firepower over the Portuguese deployed by Mustafa Bayram, an Ottoman expert.[22]

Sher Shah ("The Lion King"), the Pashtun (Afghan) founder of the short-lived Suri dynasty that ruled over the Delhi Sultanate from 1540 through 1556 had also profited from the knowledge of Ottoman military experts. Known as Farid Khan, in 1539 he defeated the Mughal emperor Humayun, and declared himself Sultan of Hindustan (1540), and temporarily re-established Afghan rule there. In the early 1540s, he commissioned several cannons of Ottoman design from Hoca Ahmed Rumi, an Ottoman gun founder, who in 1541–1543 cast several light brass *zarb-zans* to him. Three surviving pieces found in Bengal weigh 59.79 kg, and their long cylindrical barrels are 1.346 m and the diameter of the bore at the muzzle is 3.81 cm. They also have trunnions placed in the middle or at about two-fifths of the barrel.[23] These *zarb-zans* show close similarities to the lightest Ottomans *drabzens* (also known as *zarbzen, zarbuzan*) from the 1560s: the barrel of these Ottoman light *darbzens* was 1.32–1.54 m, weighed only 54–56 kg, and fired projectiles of merely 150 g.[24] Ottoman *darbzens* were among the most popular and widely used Ottoman guns, easy to cast and transport. The knowledge to make them was available throughout the Ottoman Empire from the Balkan provinces to Egypt, Diyarbekir and Basra where dozens of *darbzens* were routinely manufactured in the mid-sixteenth century.[25] This situation is likely to have made the transmission of the necessary knowledge from the Ottoman realms to Hindustan easier. Another notable similarity between Ottoman and Hindustani *darbzens* is the fact that in both empires they were manufactured as bronze and wrought iron pieces.

Conclusion

Ottoman-European military acculturation constitutes an important aspect of Muslim-Christian relations, and should be treated alongside the narratives of military conflicts, not least because it substantially altered the outcome of military rivalries and clashes within Europe. The above short overview of the role of European technicians and works on the art of war in the transmission of firearm technology from Europe to the Ottomans demonstrates that there was no iron curtain between the Ottomans and Europe. European military treatises

---

[22]H. İnalcık, "The Socio-Political Effects of the Diffusion of Fire-arms in the Middle East" in V. J. Parry and M. E. Yapp, eds., *War, Technology and Society in the Middle East*. London, 1975, pp. 195–217; S. Özbaran, "The Ottomans' Role in the Diffusion of Fire-arms and Military Technology in Asia and Africa in the Sixteenth Century," in his *The Ottoman Response to European Expansion*. Istanbul: Isis, 1994, pp. 61–66.

[23]Iqtar Alam Khan, *Gunpowder and Firearms: Warfare in Medieval India*. New Delhi: Oxford University Press, 2004, p. 74.

[24]Ágoston, *Guns for the Sultan*, p. 83.

[25]Aydüz, *Tophane-i Amire*, p. 389.

acquired and translated by Ottoman subjects, as well as Christian captives, renegades and adventurers who served in the Ottoman military were important channels in the transmission of military technology and know-how. In addition to these, direct military conflicts between the Ottomans and their Christian adversaries in the Mediterranean, the Balkans, Hungary, and the Black Sea littoral, as well as prohibited trade in weaponry and war materials, studied elsewhere, all facilitated military acculturation.

It is also obvious that the Ottomans were more aware of some of the most influential European treatises on the art of war than previously assumed. More importantly, the acquisition, translation and adoption of these works continued after the heyday of the Ottoman armies in the fifteenth and sixteenth centuries. To what extent these works influenced Ottoman master gunners in the seventeenth and eighteenth centuries is unknown and needs further research. However, it is also clear that some of these works were written and translated exactly because the authors felt that the Ottomans were lagging behind their European opponents. This is also corroborated by recent research showing that by the eighteenth century neither the Ottoman gunners, nor their ordnance was on par with that of the European contemporaries. The Ottomans were falling behind especially in terms of standardization of their ordnance. Their weapons and munitions industry, which in previous centuries had been able to meet the needs of the Ottoman army and navy from both qualitative and quantitative points of views, also experienced difficulties from the mid-eighteenth century on, and was now unable to produce the required weapons and ammunition in acceptable quality.[26] Despite reforms in the late eighteenth century following Habsburg and French models, standardization and mobility remained major problems. However, resent research shows that the establishment of quick-firing artillery was more successful than previously thought.[27]

---

[26]Ágoston, *Guns for the Sultan*, pp. 158–163, 190–200.
[27]See K. Şakul's article in the present volume.

General Observations on the Ottoman Military Industry, 1774–1839: Problems of Organization and Standardization

Kahraman Şakul

This paper aims to analyze the gradual implementation of the Gribeauval system in the Ottoman artillery. From 1767 onwards, French General Jean Baptiste Vaquette de Gribeauval (1715–1789) succeeded in standardizing the caliber, carriages and the equipment of the artillery. The new technique of boring out the gun barrel from the solid yielded much closer tolerances in casting the metal, permitting greater range with less powder charges. The reduction of the powder load by half made possible to cast the barrel thinner and lighter, for it did not have to withstand a heavy powder charge anymore. As the risk of cracking and bursting barrels decreased, gunners doubled the rate of fire, paving the way for the Napoleonic battles. The parts of cannon were made durable, identical and hence interchangeable owing to the existence of industrial plants.[1] The introduction of the double-bracket carriage added to the manoeuvrability of the piece while the standardization of the ordnance eliminated the problem of the shortage of ammunition. Various smaller improvements in casting technique, sighting, and carriages with iron axles as well as elevating mechanisms enhanced mobility and increased the rate of fire in the artillery and this was further consolidated with a professional command structure and peculiar tactics.[2]

An analysis of the regulations concerning the standardization of the ordnance as well as of the production of projectiles, gun-carriages, and guns demonstrates that the implementation of the French Gribeauval system by the late eighteenth century was a mixed blessing in the Ottoman case. While it helped the standardization of the ordnance, the mobility and the effectiveness of the field artillery

---

K. Şakul (✉)
Department of History, İstanbul Şehir University, İstanbul, Turkey
e-mail: kahramansakul@sehir.edu.tr

[1] J. A. Lynn. "Nations in Arms 1763–1815," in G. Parker, ed., *Cambridge Illustrated History of Warfare: The Triumph of the West*. Cambridge: Cambridge University Press, 1995, p. 192; B. P. Hughes, *Firepower: Weapons Effectiveness on the Battlefield, 1630–1850*. New York: Sarpedon, 1997, p. 94.
[2] C. Duffy, *The Military Experience in the Age of Reason*. London-New York: Routledge & Kegan Paul, 1987, p. 232; Hughes, *Firepower*, pp. 41, 105–106.

remained unsatisfactory for the reformers. Organizational problems encountered in the supply of raw materials and in the production of armaments, obviously had its share in this failure. Nevertheless, more studies need to be done to understand to what extent the metallurgical problems augmented the situation. Nevertheless, late eighteenth-century military reforms meant military acculturation in which a combination of time-honored Ottoman military traditions and the western techniques resulted in an eclectic Ottoman artillery.

First experiments with the Gribeauval system in the Ottoman Empire occurred during the Russo-Ottoman War of 1768–1774 by the establishment of the quick-fire artillery corps. The Ottoman designation of the four-pounders, the new light field guns, as the quick-fire gun (*sürat topu*) has led to a terminological confusion. The Sublime Porte – as was called the Ottoman government by the seventeenth century – had believed that the quick fire of the Russian artillery devastated the Ottoman army in the pitched-battle of Kartal (Kagul, 1770). It was not a matter of technology but methodology in firing the field cannon, as the Ottomans approached the problem. They saw the solution in training a new body of gunners in quick firing with the appropriate light pieces for which Baron de Tott was commissioned.[3] Based on the Gribeauval system, the quick-fire artillery corps (*Sürat Topçuları Ocağı*) gradually became the principal Ottoman field artillery corps with its peculiar tactics and new establishments including a technical school (*Hendesehane, Ecole de Géometrie*) and a new foundry at Hasköy.[4] These early attempts at the implementation of the Gribeauval system launched an irreversible transformation of the Ottoman military technology. Immediately after the 1787–1792 War with Russia ended, Sultan Selim III (r. 1789–1807) decided to undertake a series of reforms in the empire. In 1792/93, new regulations concerning the activities of the Imperial Gun Foundry (*Tophane*) were promulgated. Accordingly, barrels of all of the howitzers and light guns such as quick-fire and *şahi* and *balyemez* of medium caliber firing projectiles of 3.85 kg and 6.41 kg (3-, 5-*çap*) were to be bored out from solid (*milsiz dökmek*) by a suspension screw (*asma burgu*; i.e., the drilling machine), while the *balyemez* of larger caliber and all mortars were to be cast hollow (*milli dökmek*) as previously.[5]

---

[3] B. de Tott, *Memoirs of Baron de Tott* (reprint of the English ed. 1785). New York: Arno Press Inc., 1973, pp. 78–81.

[4] M. Kaçar, "The Development in the Attitude of the Ottoman State Towards Science and Education and the Establishment of the Engineering Schools (Mühendishanes)", in E. İhsanoğlu, A. Djebbar, and F. Günergun, eds., *Proceedings of the International Congress of History of Science (Liège, 20–26 July 1997) volume VI, Science, Technology and Industry in the Ottoman World*. Belgium: Brepols Publisher, 2000, pp. 81–90.

[5] Ottoman Archives of the Turkish Prime Ministry (Başbakanlık Osmanlı Arşivi, BOA), Cevdet Askeriye Kataloğu (C.AS.) 13589 (23 July 1793). *Çap* means caliber but actually denotes the weight of the shot in *kıyye*. I use the equations *kantar* = 100 *lodra* = 54 kg; *lodra* = 176 *dirhem* = 0.54 kg; *kıyye/vukıyye* = 1.2828 kg = 400 *dirhems* = 2.83 English pound; *dirhem* = 3.207 g; *karış* (span) = 22 cm, see Ágoston, *Guns for the Sultans: Military Power and the Weapons Industry in the Ottoman Empire*. Cambridge: Cambridge University Press, 2005, pp. 242–247.

Standardization of the Caliber: The 1805 and the 1839 Regulations

The 1805 Regulations for the Field Army, the Fortresses, and the Navy

While we see a clear tendency to standardize the Ottoman ordnance by the late eighteenth century, the first Ottoman regulation that we could find in the Ottoman Archives concerning the standardization of the caliber is dated March 4, 1805.[6] In his memorandum Hacı İbrahim Efendi, the head of the Treasury of the New Revenues (*İrad-ı Cedid*), justified the need for the "unification of the caliber" (*çap tevhidi*) on the grounds that the great variety of guns of different caliber led to confusion in the preparation of the necessary ammunition with the result that there arose the possibility of wartime shortages. These regulations aimed to standardize all types of guns used in the field army, the fortresses, and the navy. Hüseyin Efendi[7] provided a table of the French and English ordnance regulations in a booklet which laid the groundwork for the standardization of calibers.[8] Following the French model, the 1805 regulations organized the artillery in the field army along the calibers of 9-, 7-, 5-, 3-, 1.5-*çap* (11.6, 8.98, 6.41, 3.85, 1.92 kg) and the naval artillery along the calibers of 14- (17.96 kg), 9-, 7-, 5-, and 3-*çap* (in lighter vessels).[9] Setting the types of field cannons at five, this standardization was based on the weight of the projectile, while the length of the piece was subject to change upon request.

The regulations of 1805 became implemented in the navy within 2 years. Larger guns with the caliber of 18-*çap* (23.09 kg) and the *kantar* guns –an ancient Ottoman naval gun of large caliber – were collected from the vessels and stored in the *Tophane*. Convinced of the several advantages of a galleon with heavy ordnance, the Ottomans equipped the galleon *Anka-i Bahri* in 1806 with four pieces of the *kantar* gun to gain the upper hand in the Black Sea in the face of the coming war with Russia. As a matter of fact, Hüseyin Efendi contended that the *kantar* gun was far superior to the European cannon in terms of penetrating power. According to him, the Ottoman gunners were not as well trained and competent as were the European gunners while the Ottoman gun carriages were not as handy as the European ones. He explained the

---

[6] BOA, C.AS. 39493 (15 April 1805).
[7] Probably, Hüseyin Rıfkı Tamani (d. 1807) who served as the director of the Imperial Engineering School in 1806–1817, and played a crucial role in the reforms of Muhammad Ali Pasha of Egypt after that date, see M. Kaçar. *Osmanlı Devleti'nde Bilim ve Eğitim Anlayışındaki Değişmeler ve Mühendishanelerin Kuruluşu*. Unpublished Ph.D dissertation, İstanbul Üniversitesi, 1996, pp. 138–139; F. Günergun. "Osmanlı Ölçü ve Tartılarının Eski Fransız ve Metre Sistemlerindeki Eşdeğerleri: İlk Karşılaştırmalar ve Çevirme Cetvelleri," *Osmanlı Bilimi Araştırmaları* II, 1998, pp. 23–68.
[8] BOA, C.AS. 39493.
[9] BOA, C.AS. 39493; BOA, Cevdet Bahriye Kataloğu (C.BH.) 4726 (1807) also includes the 3-*çap* caliber in the naval artillery.

absence of the *kantar* gun in the British navy by their policy of taking the enemy vessel intact rather than sinking it. His praise for the *kantar* guns and their redeployment in the navy in a couple of years after the introduction of the 1805 regulations suggest the endurance of Ottoman traditions in artillery.[10]

A list of gun-carriages sent to the front in May 1806 demonstrates that the 1805 regulations were fully implemented in the army. Out of 110 carriages sent to the army, 67 of them were for light guns of 1.5-*çap* (5 *şahi*s and 62 quick-fire guns). Others included 6 *balyemez* of 3-*çap*, 11 *balyemez* of 5-*çap*, 6 *kolomborne*s of 9-*çap*, and 20 howitzers of 7-*çap*.[11] As part of the reinforcement of the Caucasus frontier, the Sublime Porte had decided to cast locally 100 *balyemez* for the fortress of Erzurum with the calibers of 15-*çap* (19.24 kg) (30), 13-*çap* (16.68 kg) (40), and 3-*çap* (30). However, this plan appears to have been revised after the regulations of 1805 since the calibers of the guns that could be cast were those stipulated by the 1805 regulations: 17 pieces of 7-, 5-, 3-*çap*s and two howitzers of 7-*çap*.[12]

The 1839 Regulations Concerning the Field and Garrison Artillery

The 1830s witnessed another series of attempts at standardization in artillery after the abolition of the Janissaries. At the advice of the Prussian military delegation, the Sublime Porte issued the 1837 regulation which seems to have reasserted the types of caliber set by the 1805 regulation. However, it also put a greater emphasis on the use of howitzers mainly due to their light weight. The field artillery was to be composed of the quick fire gun of 1.5-*çap*, *balyemez* of 3- and 5-*çap*s and howitzers of 9-, 7-, 5-*çap*s.[13]

In the summer of 1839, a new and thorough regulation was implemented again at the advice of the Prussian delegation, aiming at standardization of the field and the long-neglected garrison guns. This regulation allowed different cannons to use

---

[10] C.BH. 4726 (13 January 1807); the Russian Black Sea fleet was founded to protect the Black Sea coasts from a possible Ottoman attack which put the priority on increasing the aggregate firepower of the fleet. Thus, Russian ships were also encumbered by heavy ordnance, J. L. McKnight, *Admiral Ushakov and the Ionian Republic. The Genesis of Russia's First Balkan Satellite*, University of Wisconsin, unpublished Ph.D dissertation, 1965, p. 26; N. Saul, *Russia and the Mediterranean 1797–1807*, The University of Chicago Press, 1970, pp. 88–89.

[11] C.AS. 27365 (12 May 1806). For the typology of the Ottoman artillery see Ágoston, pp. 61–95 and S. Aydüz, *Tophane-i Amire ve Top Döküm Teknolojisi*, Ankara: TTK, 2006, pp. 339–413. By this period *kolomborne* was no more a light gun, but meant a howitzer firing both round shell and bombshell. A document from 1805/06 describes it as a 7-span long gun firing marble balls of 14 *kıyye*, used by regular gunners, rather than the bombardiers, see C.AS. 37588. In 1809/10 the army at the front ran short of bombshells of 9-, and 7-*çap*s fired by *kolomborne*s and howitzers, respectively, see C.AS. 9929. Howitzers were sent to Aleppo instead of the demanded *kolomborne*s in the 1830s, see BOA Hatt-ı Hümayun Kataloğu (HATT) 365/20197.

[12] C.AS. 8064 (1806/07).

[13] Prussian Captain Baron Karl Vincke-Olbendorf and Captain Kockowski served as advisors for reforming the artillery beginning by late 1837, see A. Levy, *The Military Policy of Sultan Mahmud II, 1808–1839*, unpublished Ph.D dissertation, Harvard University, 1968, pp. 523, 560, 615.

the same type of ammunition interchangeably. Thus, there were only six ammunition groups for ten different types of artillery: the projectiles of 40-*çap* and 22-*çap* were to be fired both by the fortress guns and the mortars of the same calibres while the cannons and the howitzers in the caliber of 9-*çap* would use the same ammunition.

	Garrison artillery	Field artillery
Heavy guns (*top*)	40-*çap* (51.32 kg); 22-*çap* (28.22 kg)	–
Cannon (*balyemez*)	9-*çap*	5-*çap*; 3-*çap*
Mortar (*havan*)	65-*kıyye* (83.38 kg); 40-*çap*; 22-*çap*	–
Howitzer (*obüs*)	–	9-*çap*; 5-*çap*

Sources: BOA, C.AS. 35566 (5 June 1839), Levy, p. 523

The Prussian delegation suggested the replacement of the quick fire gun of 1.5-*çap* weighing 297–324 kg with the *balyemez* of 3-*çap* with a weight of 378 kg on the grounds that the former was inferior to the European cannon in terms of range and penetrating power. Nevertheless, the Council of the Sublime Porte argued that the low quality of most of the roads in the Empire necessitated the use of light artillery for easy transportation. According to the Council, the quick fire guns should not be abandoned in the event that the 3-*çap balyemez* weighed more than the projected 378 kg. It also raised objections against the abandonment of 5-*çap* in favor of 9-*çap* in the howitzer class. Despite its light weight in comparison with other types of cannons of similar caliber, the ammunition of 9-*çap* howitzer would be cumbersome for the army. Furthermore, many battles in the past had demonstrated that the Ottoman gunners excelled in firing the 5-*çap* howitzers. Thus, the Council proposed to keep 5-*çap* howitzer in the field artillery at least until the gunners gained competence in firing 9-*çap*. It also deemed the shot of the 3-*çap* howitzer too small to be an effective explosive.[14] Finally, the Ottomans decided to replace the quick fire guns with the 3-*çap balyemez*. However, the former would be used in certain military operations to support the para-military forces.[15] The Ottomans also kept 5-*çap* howitzers while dropping the howitzers of 3-*çap* from the production list, revising the original reform project proposed by the Prussian delegation.

It should be noted that these regulations under scrutiny fell short of bringing about a drastic change in Ottoman artillery in the short run. While the calibers they set had already been in use in the Ottoman army and the navy, the regulations reduced the large number of calibers into manageable numbers in

---

[14] HATT 1244/48317A (nd.); Supreme Council of Judicial Ordinances agreed with the Council of the Sublime Porte on the importance of the quick-fire guns especially in operations in the rugged terrain, see HATT 1244/48317 (nd.). These documents should be dated after March 1838 – the date of the establishment of the latter council.

[15] G. Yıldız, *Neferin Adı Yok: Zorunlu Askerliğe Geçiş Sürecinde Osmanlı Devletinde Siyaset, Ordu ve Toplum (1826–1839)*. İstanbul: Kitabevi, 2009, p.428, fn.132; HATT 307/18172.

the long run. Based on the archival sources, one can argue that the 1805 regulations largely standardized the calibers in field and naval artillery while the 1839 regulations reorganized the garrison artillery as well, completing the task of standardization launched in the 1770s.[16]

There was a clear tendency to standardize the ordnance before 1805. The calibers of 120 guns sent to the army at Babadağı (Dobrudja, Romania) in early 1771 were of 3-, 1.5-, 0.5-*çaps* (3.85 kg to 641 g) and 100-dirhem (321 g).[17] In the wake of the Ottoman wars with Russia and the Habsburgs (1787–1792) there were a total of 647 guns in the *Tophane* of which 375 pieces were light guns (*şahis* of 1.5-, 1-, 0.5- *çaps* and quick-fire guns of 1.5-*çap*), 244 guns were of medium-caliber (9-, 7-, 5-, 3-*çaps*), and 28 guns had large calibers (44- [56.4 kg] 14- and 11-*çap* [14.11 kg]).[18] Even before the 1805 regulations we see that *balyemez* guns were generally firing shots of 16, 11, 9, 7, 5, 3 *çaps*.[19] These figures are comparable to those given by Ágoston for the late eighteenth century. As he observes, small-caliber guns dominated the output of the Imperial Gun Foundry after the mid-eighteenth century.[20]

Both the 1805 and 1839 regulations concerning the standardization of the ordnance also dealt with the standardization of the projectile. The director of the projectile factory in Praveşte (near Salonica), received orders concerning the 1805 regulations that provide precious information about the Ottoman methods of standardization of caliber throughout the empire.[21] The afore-mentioned Hüseyin Efendi sent iron rings and wooden balls representing the fixed six calibers to Praveşte. They were to serve as the models for the new molds for the projectiles so that they would agree with the caliber of the guns. Other methods of standardization of caliber included the exchange between Istanbul and the periphery of a piece of circular paper cut in the same diameter with that of the muzzle or a piece of rope for that matter when standard measures were nonexistent in the empire.[22] As Günergun pointed

---

[16] Ottoman mortars were greatly diversified in calibers as late as 1797 through 1803: 36- and 22-*çap* in 1795/96 (C.AS. 1846); 65-, 36-, 22-, 14-*çap* in 1796/97 (HATT 224/12515.B); 80-, 45-, 32-, 18-, 14-*çap* in 1798/99 (C.AS. 23443); 65-, 45-, 25-, 22-*çap* in 1801/02 (C.AS. 45421); 65-, 34-, 32-, 22-, 14-, 7-*çap* in 1802/03 (C.AS. 27730).

[17] Şemdanizade (Fındıklılı Süleyman Efendi). *Mür'i't-Tevarih*. M. Aktepe, ed., İstanbul: İstanbul Üniversitesi Edebiyat Fakültesi Yayınları, 1976–1981, vol. II.B, p. 62.

[18] C.AS. 9223 (1786/87).

[19] C.AS. 52554 (19 April 1804) and HATT 1387/55156 (1789/90).

[20] Ágoston, p. 186. Various output figures are 324 in 1788–92, 182 in 1796/97, 197 in 1797–98, see table 6.5, p. 184; the proportion of light guns in this period is 82.9% in 1769–1770, 74.5% in 1771–1772, 58.4% in 1776–1779, 17.3% in 1784–1785, see table 6.6, p. 186.

[21] C.AS. 44626 (9 June 1805).

[22] BOA, BaşMuhasebe Defterleri (D.BŞM.) 6547/57 (1782/83) contains one such paper cut in the shape of the muzzle of a 16-span long iron gun using stone ball. I am indebted to Cengiz Fedakar for bringing this document to my attention. C.AS. 31042 (30 June 1836) requires the commander of the fortress of Chios to measure the diameter of the mortars of 120-, 80-, 45-*çaps* with a piece of rope and send the ropes to the *Tophane*, the Imperial Gun Foundry at Istanbul.

out, until the standardization of the Ottoman measuring rod (the *zira,* 0,757 738 m) in 1841, the measures made by the Ottoman masters in the manufacture of artillery haunted all attempts of producing guns of standard caliber.[23]

The Mobility of the Reformed Field Artillery: Guns and Gun Carriages

Unlike in Europe, the implementation of Gribeauval reforms did not increase the mobility of the Ottoman field artillery as much as the reformers had initially expected. This was partly due to the size and weight of the gun as well as the gun carriage. The Ottoman field artillery basically comprised three types of guns: *balyemez*s and *şahi*s of small caliber, and the quick-fire guns. In the aforementioned inventory of the Erzurum Foundry, there were 15 pieces of *balyemez* with a length of 13-*karış* (span). The weight of those firing shot of 3 *kıyye* ranged between 823.5 kg and 931.5 kg. The remaining, used shots of 5 *kıyye* and weighed 1.16–1.28 tons.[24] These figures are comparable to those given by Ágoston for the year 1769–70.[25]

Quick-fire guns were likely to be shorter than the *şahi*s. While *şahi*s could be of various calibers, the quick-fire guns almost always fired shots of 1.5-*çap* especially by the reign of Sultan Selim III. The superintendent of the *Tophane*, Mustafa Reşid Beğ, stated in 1793/94 that the quick-fire guns cast in 1772/73 – presumably under the supervision of Tott – were light and manoeuvrable pieces whereas those cast recently tended to be heavier and longer so as to resemble *şahi* guns.[26] In this period, *şahi*s were generally 6-, 7-, 9-, 10-, 11-span long, firing shots of 641 g to 1.92 kg (0.5-, 1-, 1.5-*çap*s). Their usual weight was roughly between 250 and 770 kg.[27]

---

[23] Günergun, "Osmanlı Ölçü ve Tartılarının", pp. 36–39.

[24] C.AS. 8064 (1806/07) two howitzers of 7-*çap* weighed 478 kg and 616 kg, respectively, and two mortars of 14 *kıyye* weighed 453 kg each.

[25] The minimum/maximum weight of 3-*çap* 13-span guns cast at the Tophane in 1769–1770 were 954 kg and 1 029 kg respectively, see Ágoston, table 58, p. 236.

[26] HATT 14553 (1793/94), first published by T. Işıksal, "III. Selim'in Türk Topçuluğuna dair bir Hatt-ı Hümayunu," *İstanbul Üniversitesi Edebiyat Fakültesi Dergisi,* 8, 11–12 (1955): 179–184.

[27] My calculation is based on the data given in Ágoston, tables 57, 58, 59, 65 in pp. 236–239; In 1790/91 Seyid Numan Efendi, the official in charge, added two spans to *şahi*s of 9- and 10-span long of 1789/90 which made them heavier and cumbersome. Upon the complaint of the head of the Gunners, the *şahi*s of 1790/91 remained in their previous size, see HATT 16056 (1790/91). However, one span increase in the length of howitzers of 5-span and 6-span in 1790–1791 improved the range considerably without sacrificing the mobility.

In his memorandum, Mustafa Reşid Beğ compared the weight and size of the Ottoman quick-fire gun with its western equivalents and pointed out that the Ottoman gun weighed 486 kg; that is, twice as much the western guns. French and the British quick-fire guns weighed only 282.2 kg while the Habsburg quick-fire gun was 295 kg with a shorter barrel than the former. He advised that the Ottoman quick-fire gun should be refashioned on the model of the French one, albeit with a thicker priming pan, which would put an extra weight of 25.5–38.5 kg to the gun.[28] The quick-fire guns, nevertheless, still weighed 486 kg a decade after he submitted his memorandum and only in the 1830s did its weight decrease to roughly 300 kg.[29] The time-honored traditions of Ottoman gunners might account for the heaviness of the quick-fire guns, as seen in Mustafa Reşid's suggestion to thicken the priming pan. The French military officer Saint-Rémy (1746–1800), a member of the French military mission to Istanbul in 1785–1787, pointed out that the head of the Bombardiers Corps had insisted on thickening the priming pan of the mortars Saint-Rémy designed following the French regulations of 1732. These two cases are clear signs of hybridization of European and Ottoman casting techniques in the *Tophane*. Apparently, the Ottomans did not merely imitate the European guns, but rather adopted them to their own system.[30]

Another factor that affected the mobility of the field artillery was the type of the gun-carriage. A crucial contribution of Gribeauval was his design of lighter gun carriages that made guns handier. In the early modern era, real development in artillery was dedicated to enhancing the mobility of carriages rather than perfecting methods of gun-making. The standard gun carriage was a double-bracket with a heavy wooden axle and a trail made of two balks of timber laying parallel to each other. In the Gribeauval system, the trail became shorter and lighter so that the gun could be traversed by a single soldier. The British General Sir William Congreve (d.1814) designed in 1792 a single pole-like structure that moved the centre of gravity forward which yielded higher rates of fire and accuracy of aiming the piece owing to the lightened trail. In addition, incorporation of the limber to the artillery piece made it possible to

---

[28] HATT 14553 (1793–1794); Resmi Mustafa Ağa, the head of the Bombardiers Corps, used copper, tin and iron in casting quick-fire guns, see C.AS. 53877.

[29] HATT 116/4668 (1804–1805); HATT 1244/48317A (1838–1839?); for a comparison of various European guns with respect to size and weight see T. Wise and R. Hook, *Artillery Equipments of the Napoleonic Wars*. China: Osprey Publishing, 2002, p. 15 and P. Haythornthwaite, *Weapons and Equipment of the Napoleonic Wars*. London: Arms & Armour, 1996, pp. 55–56.

[30] M. Kaçar, "Osmanlı Ordusunda Görevli Fransız Subayı Saint-Rémy'nin İstanbul'daki Top Döküm Çalışmaları (1785–87)", *Osmanlı Bilimi Araştırmaları* V (1), 2003, pp. 33–50; Saint-Rémy claims for himself the credit of introducing the light guns of 4 *pouce* (108 mm) in Ottoman artillery, which, before him, was composed of 8, 12, 24, and 32 *pouce* guns (1 *pouce* is 27 mm). Their insistence on thickening the priming pan might indicate that the Ottomans used heavier powder charges; the 1787 inventory of *Tophane* mentioned previously gives the powder charge as half the weight of the ball.

carry both the ammunition and the gunners which increased the mobility of the field artillery in action.[31]

As for the Ottomans, the production and the maintenance of gun carriages posed a bigger challenge than that of cannons since the wooden gun-carriage was not as durable as a metal gun despite high production costs. Ottomans acknowledged that they had to reform the design of their gun-carriages in order to have an efficient fire-power by the 1770s, if not earlier. They tried to meet this challenge by improving carriages and wagons as well as recruiting foreign experts for domestic production. The Grand Admiral Gazi Hasan Pasha (1713–1790) had bought from Britain some 63 wagons and 15 carriages for the army as well as 218 guns (iron and bronze) in addition to 91 gun carriages for the navy in 1784/85.[32] In the same year, French locksmiths (*çilingir*) were recruited for making carriages.[33]

In 1782–1789 the Porte decided to remodel the carriages of *şahi*s after the new-design carriages of quick-fire guns. By adding iron trunnion holes and capsquares, the new carriages would accommodate *şahi*s of all types. Following these changes, the documents use the terms *şahi* carriage and the quick-fire carriage interchangeably, for they looked alike.[34]

An Ottoman pasha who joined the campaign against Napoleon in Egypt was completely frustrated in the march by the heaviness of the guns and carriages while he praised the battle performance of the Ottoman field artillery. Horses had to be changed every 3–4 hours instead of eight which was the norm in Europe as they were exhausted by the cumbersome artillery. This, in turn, required additional horses to the detriment of the campaign treasury. The field artillery proved to be too heavy to be sent in to action with the cavalry forces, while the latter refused to attack the enemy without artillery support. Conversely, the European guns were so light that they could be deployed with the cavalry. Back in Istanbul, with the heads of the Gunners and the Wagon-makers, he comparatively examined the Ottoman and the European light cannon. This comparison demonstrated that the Ottoman quick-fire gun weighed twice as much, confirming Mustafa Reşid's remarks almost a decade earlier. Convinced that the carriage influenced the mobility as much as the gun's weight did, he ordered a new carriage that would carry the limber and the

---

[31] Hughes, *Firepower*, pp. 14–15, 94; Parker, *The Military Revolution: Military Innovation and the Rise of the West 1500–1800*. Cambridge: Cambridge University Press, 1988, p. 151; Black, *Eighteenth Century Europe 1700–1789*, New York: St. Martin's Press, 1999, p. 374; M. Erendil, *Topçuluk Tarihi*. Ankara: T.C. Genelkurmay Başkanlığı, 1988, pp. 34, 146.

[32] C.AS. 5939 (30 March 1787) shows the predominance of the calibers set in the 1805 regulations even earlier. Some of the wagons were sent to the *Tophane* as samples for the masters.

[33] C.AS. 13016 (3 July 1785).

[34] C.AS. 27734 (7 December 1789), 100 guns of 1-*çap* (9-span), 30 guns of 0.5-*çap* (8-span), 20 guns of 1.5 *çap* (10-span), 20 guns of 1.5-*çap* (7-span) were sent to the front; The 1839 regulations also required the construction of the carriages with interchangeable parts, C.AS. 35566.

ammunition. He proposed the formation of a mobile reserve artillery equipped with 100 guns seated on these carriages along with 20 ammunition carts. These guns were to be lighter and shorter than the quick-fire guns without compromising the range, firing shots of 0.5 kıyye.[35]

The Ottomans were aware of the carriage problem well before the Egyptian campaign. For instance, in 1792/93, the army returned 15 carriages for quick-fire guns to Wagon-makers because of their excessive weight.[36] An empty quick-fire gun carriage manufactured in the same year weighed 322.6 kg while we already saw that the piece itself weighed 486 kg which brings the total to roughly 808.6 kg.[37] They modified the carriages of the lighter type of *balyemez* guns used in the field artillery in order to improve the mobility. The axles and the hubs of the wheels of a *balyemez* carriage were quickly worn out by friction during the march since they were made of iron. The great number of axles and hubs carried in reserve meant extra load for the army baggage not mentioning long delays for repairs during the march of the army. As a solution, the Sublime Porte decreed that the hubs of the *balyemez* carriages were to be made of bronze in 1805/06. Apparently, bronze hubs and iron axles had made their first appearance in the carriages of quick-fire guns and of the howitzers several years before that date.[38] In consequence, the evidence at our disposal suggests that while the quick-fire gun set the new pattern for the other type of guns in terms of casting techniques, its carriage became the model for all types of gun-carriages used in the field artillery. Now, it is time to discuss the factors that might account for the chronic heaviness of the Ottoman guns and the carriages.

Possible Reasons for Failure

Rather than rushing into easy generalizations, it is more appropriate to speculate on possible reasons as why the reforms did not produce a field artillery as handy and mobile as the reformers had initially expected. It is reasonable to assume that the Ottomans suffered from metallurgical, technological, and organizational problems.

The mines did not yield the same quality of output in all parts of the empire. In addition to the afore-mentioned regulations, Mustafa Reşid and Saint-Rémy declared that the Kastamonu and Ergani copper was far superior to the output of the Sidrekapsi mine (Sydracopla near Salonica) and argued against the use of

---

[35] HATT 116/4668 (1804/1805?); each gun carriage would accommodate 20 shots and the required amount of gunpowder. In 1800, the commander of the fortress of Anapa (Caucasus) complained that the quick-fire guns of 1.5-*çap* (7-span) were so heavy that eight horses would be needed if they were to see action in the Kabartay region. The Sublime Porte sent him in reply that these guns were standard in size and weight, see C.AS. 38226 (15 December 1800).
[36] C.AS. 26138 (31 July 1793).
[37] C.AS. 28830 (24 January 1793).
[38] C.AS. 6048 (1805/06).

the low quality copper ores coming from the latter in casting quick-fire guns.[39] By using metals of high quality, the Sublime Porte hoped that the Ottoman guns would be as strong and polished as were the European guns.[40] Metallurgical problems, however, did not so much handicap gun making as the production of projectiles. While the *Tophane* and the Imperial Dockyards (*Tersane*) used Kastamonu, Ergani, and Gümüşhane copper, the workshops producing projectiles in Praveşte and the *Humbarahane* that manufactured bombshells in Istanbul had to make do with the low quality copper from Sidrekapsi. In 1767/68, only 60% of the copper sent from Sidrekapsi was left for making bombshell and cannon balls after the refining process.[41] This ratio was still the same in 1783 while the authorities remarked that it was as high as 85–90% in the past. It should be noted, however, that they accused the local authorities for the low quality of the metal from Sidrekapsi rather than the depletion of veins.[42] In contrast, 75.5% of the copper was left after the refining process during the afore-mentioned activities of the Erzurum foundry, which shows that the Anatolian copper mines were operating better than the Sidrekapsi mine.[43]

By the 1780s the Ottoman reformers were preoccupied with increasing the quality of the ammunition. While the British and the Saxons were entrusted with the production of bombshells in the Hasköy foundry, a Spanish noble with the name of Michel (*Mikel-beğzade*) undertook to improve the quality of the shots.[44] The British Brigadier-General George Frederick Koehler (1749–1827) who was given the task of reinforcing Gallipoli against a possible French attack in 1799, complained about the low quality of the projectiles stored in Gallipoli upon which the Sublime Porte decided to request for British masters. The Ottomans were aware of the lack of experts in the Praveşte factory and concerned with the fact that even the output of the Hasköy foundry was inferior to the British bombshells.[45] The 1839

---

[39] Kaçar, "Saint-Rémy", p. 40.
[40] C.AS. 13589 (23 July 1793).
[41] C.AS. 38303 (7 November 1767).
[42] C.AS. 20023 (13 September 1784), copper ores from Sidrekapsi were used in the production of cannon ball, bombshell, bar-shot (*plankata*) and chain-shot (*zincir*), the last two being for the navy.
[43] C.AS. 31157 (5 May 1805) and C.AS. 8064 (1806/07); The Porte ascribed the sharp decrease in the output and the revenue of the mines to the exhaustion of the richest and the most accessible veins as well as the deforestation the mine regions. Total revenue decreased by half after the mid-century, see HATT 8553 (1789/90).
[44] C.AS. 20961 (21 December 1785); HATT 1413/57668 (1790/91).
[45] HATT 129/5367; Koehler was sent along with ten quick-fire guns presumably in June 1799, see C.HRC. 2360; Great Britain sent two ships of ammunition as presents to convince the Sublime Porte to enter a coalition against France. The Porte filed an order for tin and zinc from Britain, BOA, Bab-ı Asafi Amedi Kalemi Dosyaları (A.AMD) 40/67; Hasköy foundry cast bombshells when excess copper was available in Istanbul; The superintendent of the Bombardiers Corps complained from the lack of experts on gun-carriages in the Corps, HATT 1386/55056 (30 June 1795).

regulations concerning standardization had to redress the low quality of the Ottoman shots, this time produced at the Samakocuk foundry.[46] They were cast so poorly that they fell into pieces by the time they left the muzzle, harming the gun itself. Therefore, an English engineer and master were sent to the Samakocuk works along with an Ottoman expert the same year.[47] While such cursory observations imply a technological setback in the production of arms and ammunition as well as in ore extraction and processing in the mines, they should not be treated as conclusive evidence since they need to be supported by comprehensive researches.

Organizational problems might have been more crucial in affecting the operations of military industrial plants of the empire. The guns cast in 1786/87 for the fortresses were so fragile and honeycombed (*karıncalı*) that they could not withstand the pressure when discharged with full powder load. The guncasters evaded responsibility and accused the new office of superintendent of procuring raw materials of low quality. They refused to foot the bill unless their ancient rights to materials was restored which required the abolition of the new office. Finally, Sultan Abdulhamid I (r. 1774–1789) agreed to abolish the system of superintendence as a solution.[48] Nevertheless, despite such tensions between the beneficiaries of the old system and those of the new system, the dual control embodied in the superintendent's office would emerge victorious in the last decade of the century.[49] Kaçar, based on the observations of Saint-Rémy, demonstrated how the emerging tensions between the French and the Ottoman gun manufacturers in the 1780s undermined the operations at *Tophane*. Fearful of losing their jobs to the French, the Ottoman gun casters resisted the new norms of casting, introduced by the latter and insisted on thickening the priming pan of the newly-designed mortars.[50]

The activities of the Praveşte works suggest organizational problems besides the low quality metal coming from the Sidrekapsi mine. Operating within a sort of semi-private contractual system called *muqata'a*, the target was to produce shots and bullets weighing some 255,195 kg annually, with the raw material and workers supplied by the 'service villages' in return for tax exemptions.[51] The Sublime Porte realized by 1800/01 that the quality of the shots from Praveşte was steadily decreasing after Hüseyin Ağa had become the new

---

[46] For the ongoing industrial archaeological project on the Samokocuk mines (Demirköy in Kırklareli) see, G. Danışman, "Ottoman Mining and Metal Working in the Balkans: Its Impact on Fire-Arms Technology in Southeast Europe (fifteenth–seventeenth centuries)," M. el-Gomati et al., eds., release date: May, 2007, publication ID: 701, www.muslimheritage.com/uploads/Ottoman_Mining_Metal_Working.pdf

[47] C.AS. 35566 (5 June 1839).

[48] HATT 818 (1786/87).

[49] HATT 10827 (1795/96).

[50] Kaçar, "Saint-Rémy," p. 40.

[51] C.AS. 2088 (23 February 1769).

contractor in 1797/98. The output of 1799/1800 was so faulty that the shots were conical in form with a knotted surface.[52] Strikingly, the Porte accepted these defective shots as it ran out of ammunition because of the ensuing war with France on two fronts (Egypt and the Adriatic). Defending himself against the charges directed at him, Hüseyin Ağa complained about the negligence of the service villages in supplying the necessary raw materials. The Porte increased the load of these service villages due to the armaments race going on during the Revolutionary Wars, which caused serious organizational problems. While these villages promised to procure the conventional load as previously, they claimed that it was beyond their means to supply the extra load and offered to pay back in cash equivalent. When the Porte rejected the petition because of the depletion of the storehouses, they simply refused to provide the raw materials.[53] The fact that Hüseyin was still in charge of the Praveşte works during the Russo-Ottoman war of 1806–1812 may indicate that various organizational problems due to wars gravely jeopardized the quality of Ottoman arms and ammunition, which, in turn, undermined the military performance in the war.[54]

Exorbitant costs involved in the production of arms and ammunition put a major strain on the reform efforts and may have forced the Ottomans to compromise the quality of the military equipment. In 1795/96, the masters made a sample quick-fire gun carriage in the presence of the authorities who were preoccupied with the high costs involved in the making of carriages. The carriages in question had an average cost of 969 *kuruş* 10.5 *para*, the sample cost 1,068 *kuruş*.[55] It was state policy to offer the masters lower prices than the actual market rates which should have affected the quality of the output. Upgrading of the ramrods used in the quick-fire artillery in 1793/94 is another example to the overwhelming financial costs of reforms. Upon receiving complaints from the gunners that the brush of the ramrod did not clean the barrel effectively, the authorities issued new ramrods in the Habsburg fashion with a tightly woven brush and an iron bar of good

---

[52] C.AS. 7504 (April–May 1801); C.AS. 48204 (14 December 1799); the projectiles included 10,000 shots of 9-*çap* and 5-*çap* each and 60,000 shots of 1.5-*çap*; in the war with Russia in 1806–1812, the bombshells of 7-*kıyye* and 9-*kıyye* fired by the howitzers and the *kolonborne*s respectively were in high demand, see C.AS. 9929 (1809/10).

[53] Hüseyin Ağa's debt to state accumulated over years: 124,879 kg of shots in 1797–1803 (C.AS. 36668; C.AS. 19635, C.AS. 37090), 767,960 kg of shots in 1804–1811 (C.AS. 20107).

[54] C.AS. 48204 (14 December 1799), Hüseyin Ağa was expected to cast 443.000 shots and bullets in various calibers despite the depletion of raw materials in Praveşte works, see C.AS. 20107 (13 February 1810); Sidrekapsi also failed to procure metal in the same year, see C.AS. 9929 (1809/10).

[55] C.AS. 44052 (15 April 1797). A quick-fire gun carriage cost 230,5 *kuruş* (old type carriage cost only 58,5 *kuruş*) in 1782/83 (C.AS 48516), 149 *kuruş* in 1788/89 (C.AS 27734), 577 *kuruş* in 1792–1793 (C.AS 28830). The price for 1795/96 is more realistic than the others since it represents the real costs whereas the others reflect the official prices.

quality. The prototype of the new ramrod cost 177 *para* while the old one was just 100 *para*.[56]

Conclusion

The reform of the artillery involved a high degree of eclecticism in which Ottoman techniques and tastes were "Europeanized" in certain ways. In the aftermath of the war with the Habsburgs (1787–1791), there was a clear tendency towards the Habsburg model. Besides the Habsburg style ramrods mentioned previously, the Habsburg "national pattern" in painting gun carriages also influenced the Ottomans. A drawing of a mortar of 22-*çap* with its carriage with an elevating screw possibly of bronze illustrates the carriage and the gun bed in yellow, which was the Habsburg national pattern (Fig. 1). European states began to paint their land carriages for protection against the weather in the eighteenth century. The British used greenish-grey, the French favoured olive-green, and Prussia opted for light greyish-blue carriages. The Habsburgs, on the other hand, used combinations of yellow and black or black and red. Selim III accepted the color of yellow as the standard for the carriages and the ammunition carts in the aftermath of the war with the Habsburgs. This is about the same time when the Ottomans refashioned the ramrods of the quick-fire guns in the Habsburg style, which clearly shows the Habsburg influence on the

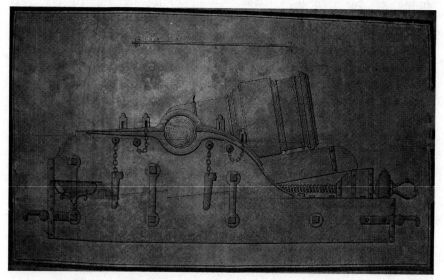

Fig. 1 The Ottoman 'national pattern' in artillery. BOA, C.AS. 24287 (20 January 1793)

---

[56] C.AS. 26142 (8 November 1793); C.AS. 11697 (27 March 1803).

Ottoman military reforms. Thus, "the Ottoman pattern" was likely to be a combination of a French style gun featuring the Turco-Islamic crescent and star, a carriage in Habsburg or French style painted in yellow following the Habsburg national pattern, and a ramrod in Habsburg style.[57]

Despite the problems of standardization and the mobility in field artillery, reforms of this period were far more successful than usually assumed. In the year of 1796/97 some detachments of quick-fire artillery had already been established in a province as far as Erzurum.[58] Sayda had received its quick-fire artillery even before Erzurum In 1784/85 at the request of its governor Cezzar Ahmed Paşa.[59] By 1793, the new technique of barrel-casting first applied to quick-fire guns served as a model for all types of light guns as did the quick-fire gun carriages for other light carriages. A large degree of trial and error was involved in these reforms. Tott's salaried bombardiers were dismissed in 1773, a year after their recruitment for their ineffectiveness in the war only to be revived in the 1780s.[60] Towards the end of his reign, Selim III discontinued with the deployment of quick-fire guns to the fortresses as they did not need light guns.[61] Despite several problems which were mainly due to wartime conditions, the process that commenced in the 1770s proved to be irreversible, dynamic and transformative of the Ottoman military.

---

[57] C.AS. 24287 (20 January 1793); for authentic drawings of the Ottoman artillery of this period see K. Beydilli and İ. Şahin, *Mahmud Raif Efendi ve Nizam-ı Cedid'e Dair Eseri*, Ankara: TTK, 2001; C.AS. 44052 (15 April 1797) mentions the "Habsburg style wagons with wooden rods." C.AS. 1846 (1795/96) lists gun carriages with the trunnion holes sited on the upper half of the bracket, following the French example (*belinden muğlulu*) besides others with the trunnion holes sited on the lower half of the bracket, which was the standard application in Europe (*dibinden muğlulu*).

[58] HATT 1533 (August–September 1796).

[59] C.AS. 9120 (1784/85); He stated that without quick-fire guns, which were drawn by horses, he could not undertake any offensive against the rebellious Druses in the mountains of Lebanon.

[60] C.AS. 1569 (1772).

[61] C.AS. 23247 (18 December 1806).

Cultural Attitudes and Horse Technologies: A View on Chariots and Stirrups from the Eastern End of the Eurasian Continent

Nanny Kim

From the Occidental perspective, the domestic horse and wheeled transport are prehistoric developments. The chariot, though an epoch-marker of great and violent transformations of ancient civilizations, did not acquire the traumatic qualities of the advent of mounted warriors roughly a millennium later, an event that gave rise to the classic dichotomy of civilization and barbarity. Again about one-and-a-half millenniums later, the introduction of the stirrup appears intimately linked with the rise of the heavily armoured mounted warriors, a new aristocratic class, which in Europe transformed society and gave rise to ideals of chivalry which still hold attraction today.

This paper explores the same developments and their cultural implications from the East Asian end of the continent. In China, the arrival of the domestic horse and the chariot is a sudden event that took place in the late Shang about 1200 BCE, in the midst of a process of early state formation. Here, as in Western Eurasia, rather than a powerful new weapon that decided the fate of states in war, it was a powerful means of representation that caused elites to go to war and to expand their polities into states. For over a millennium, the chariot remained the representative vehicle of elite locomotion, while horse-riding was an activity performed by lower ranks of soldiers and messengers. The stirrup, which first appeared in fourth-century China, may appear a non-event by comparison, a minor improvement to an already well-established culture of riding and mounted warfare. Though not unimportant for the development of ever heavier armour, it remained an almost embarrassing addition to the riding outfit. Eventually, it contributed to watering down chivalresque exclusiveness, for it enabled persons to ride, even if they did not have many years of training and experience.

N. Kim (✉)
Institute of Chinese Studies, Center of East Asian Studies, Heidelberg University, Heidelberg, Germany
e-mail: nkim@sino.uni-heidelberg.de

Why return to topics as well-ploughed as the domestication of the horse, the chariot and the stirrup? Perhaps precisely because so much effort has been expended on these issues for Western Eurasia, new insights might be gained by taking a look at them from the Eastern end of the continent. On the continental scale, the appearance of the domestic horse and the chariot in about 1200 BCE is a late event in China, while that of the stirrup around 300 CE is early. Both took place in the historical period, offering sources for a cultural perspective on horse technologies.

This exploration of cultural attitudes towards two major events in transport technologies focuses on China and Northeast Asia. The first part discusses the horse-drawn chariot from the thirteenth century BCE to its gradual demise as the only suitable form of elite locomotion in the period around the third century CE. The chariot is closely linked to the formation of state in the late Shang (ca. 1250–1050 BCE) and state systems in the Zhou periods (ca. 1050–221 BCE). I argue that horse-drawn two-wheelers, rather than a powerful new weapon that decided the fate of states in war, was a powerful means of representation, which played an important role in transforming societies, in costs and strategies of war, and in the expansion of state structures. Through these transformations, cultural patterns of elite locomotion that took shape in antiquity continued to shape attitudes. The second part investigates the stirrup, which by comparison was a non-event. In East Asia, it appears as a minor improvement to a well-established culture of horse riding. The exploration of heavy armour and new social roles of riding up to the early Tang (618–751 CE) shows that the stirrup marks a step towards a new elite culture in which riding replaced wheeled locomotion. Yet it remained an almost embarrassing addition to the riding outfit, and in the longer run contributed to watering down chivalresque exclusiveness.

On account of the thinness of the archaeological record of the Eastern Eurasian steppe, this region will initially be treated as a "black box," a transmission link that itself remains largely unknown. The investigation of East Asian experience is undertaken through a comparative perspective of conditions in Western Asia and Europe. On this basis, the paper aims at contributing an alternative perspective on the dynamics of travelling technologies, including both central civilizations and the seemingly peripheral cultures that transmitted them.

The Appearance of the Horse-Drawn Chariot in China

The two-wheeled chariot with spoked wheels together with the domestic horse as a draught animal make their dramatic appearance in the archaeological record at the Anyang sites dated to about 1250 BCE. The royal burials and sacrifices at this ritual centre of the later Shang kings contain great amounts of bronze ritual vessels together with human and animal sacrifices. Among these are a

Cultural Attitudes and Horse Technologies 59

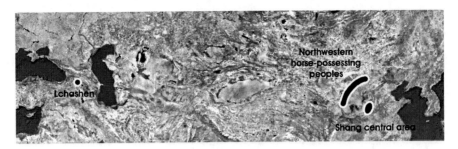

Map 1 Archaeology of early chariots: Lchashen (ca. 2000 BCE) and the late Shang (ca. 1200 BCE). Map by the author, courtesy of Google Earth

small number of chariots, with pairs of horses and charioteers. In the earliest written record, oracle bones dated to the same period, horses and chariots appear as well (Map 1).[1]

---

[1] See K. Linduff, "A Walk on the Wild Side: Horses and Horse Gear in Shang China," in C. C. Renfrew, ed., *Late Prehistoric Exploitation of the Eurasian Steppe*, Cambridge: Cambridge University Press, 2002, pp. 139–162; M. Wagner, "Wagenbestattungen im bronzezeitlichen China," in M. Fansa, ed., *Rad und Wagen: der Ursprung einer Innovation*, Mainz: von Zabern, 2004, pp. 107–122.

It is at present impossible to prove beyond doubt that the chariot was an introduced technology rather than an independent development in China, the main problem being the huge gap of firm archaeological evidence separating Sintashta and Anyang. Four arguments strongly suggest introduction. First, the late Shang were not the only ones to possess chariots, but shared this technology with their north western neighbours. Second, the great similarity of petroglyph depictions of chariots across the Eurasian steppe and Shang inscriptions suggests shared patterns, while four-wheelers, solid wheels and ox-drawn vehicles are not represented in petroglyphs east of Central Asia. On this issue, see M. Littauer, "Rock Carvings of Chariots in Transcaucasia, Central Asia and Outer Mongolia," in idem and J. Crouwel, *Selected Writings on Chariots and Other Early Vehicles, Riding and Harness*, Leiden: Brill, 2002, pp. 106–135. (Culture and history of the ancient Near East 6); Wei Si 卫斯, "Xiyu nongye kaogu ziliao xuoyin. Diyi bian: Yanhua ziliao" 西域农业考古资料索引 第一编:岩画资料 (Index of archaeological materials in agriculture in the Western regions: Part 1: Rock carvings), *Nongye kaogu*, 3 (2003); and M. Dévlet, "Felsbilder mit Wagendarstellungen in Sibirien und Zentralasien," in M. Fansa, ed., op. cit., pp. 237–246. Third, wild horses were present in the North China Plain whereas evidence of domestication prior to the Shang chariots is lacking. See K. Linduff, op. cit., pp. 142 f., R. Kolb, *Landwirtschaft im Alten China*, part 1: *Shang-Yin* (Agriculture in ancient China, part 1: The Shang period). Ruhmannsfelden: Systemata Mundi, Institut zur Erforschung fremder Denksysteme und Organisationsformen, 1994, p. 101, Chow Ben-shun, "The Domestic Horse in the Pre-Ch'in Period in China," in J. Clutton-Brock, ed., *The Walking Larder: Patterns of Domestication, Pastoralism and Predation*, London: Unwin Hyman, 1989, pp. 105–107 (One world archaeology 2). Fourth, evidence for the use of bovines as traction animals is absent, yet the Shang chariot traction system, like that of Western chariots, is evidently based on the yoke, i.e. employs the same, not entirely suitable adaptation for the horse of a traction system originally developed for bovines.

At this point, the Shang dynastic state and some of its neighbouring polities had experienced a dramatic concentration of power, population and production. These developments, most conspicuous in well-developed agriculture, urbanization and mass production, were well established at the earlier site Erlitou 二里頭, a central city which flourished between 1900 and 1550 BCE. The degree of social organization and stratification, division of labour, artisanal skill and competitive control over resources is most remarkably documented in the impressive production of cast bronze objects, which almost exclusively served ritual purposes.[2]

The Shang was a society that lived according to patterns of the agricultural year, of ritual and of war. State control may not have extended much beyond the Shang domain, but its influence reached far into the North China Plain and southwards to the Middle Changjiang (Yangzi). Lines of trade and exchange connected Central, Northeast and Southeast Asia. Differences between the Shang and its neighbours were gradual, although cultural self-imagery, certainly by the Zhou period, built on diametric contrast.[3] Millet may have been centrally important to the Shang, but they also bred cattle on a large scale. While neighbouring peoples in what is now northwestern China were more pastoral, within a broad belt of now marginal agricultural and steppe areas, settled agricultural bronze cultures existed in an age of more hospitable climatic conditions.[4] The transition to nomadic pastoralism and the formation of the steppe frontier was to set in about half a millennium after the fall of the dynasty.[5]

It is in this world that the fully developed chariot appears. Archaeology has shown the Shang chariots to have been the largest form of the vehicle throughout Eurasia with clear characteristics in construction and ornamental details.[6] The technology was not exclusive to the Shang; several polities to the

---

[2] The total weight of bronze objects found in single Shang royal burials amount to several tons. Mines were situated at a considerable distance from the manufacturing centres, attesting to the logistic capabilities and control over resources. See Liu Li and Chen Xingcan, "Cities and Towns: The Control of Natural Resources in Early States, China," *Bulletin of the Museum of Far Eastern Antiquities*, 73 (2001): 5–47.

[3] C. Kwang-chih, *Art, Myth, and Ritual: The Path to Political Authority in Ancient China*. Harvard/MA: Harvard University Press, 1983; D. Keightley, "The Shang: China's First Historical Dynasty," in M. Loewe and E. Shaughnessy, eds., *The Cambridge History of Ancient China*. Cambridge: Cambridge University Press, 1999, pp. 232–291.

[4] D. Keightley, "The Environment of Ancient China," in M. Loewe and E. Shaughnessy, eds., *The Cambridge History of Ancient China*. Cambridge: Cambridge University Press, 1999, pp. 30–36.

[5] N. Di Cosmo, "The Northern Frontier in Pre-Imperial China," in M. Loewe and E. Shaughnessy, eds., *The Cambridge History of Ancient China*. Cambridge: Cambridge University Press, 1999, pp. 885–966.

[6] The carpentry and metal fittings of early Chinese chariots have been examined in detail. See e.g. Zhang Yanhuang 张彦煌 et al., "Yinche de fuyuan yu guche zhizuo ruogan gongyi shitan" 殷车的复原与古车制作的若干工艺试探 (The reconstruction of Shang period chariots and some issues of ancient carriage building techniques), *Wenwu*, 4 (1994): 32–34, and A. Barbieri-Low "Wheeled vehicles in the Chinese Bronze Age (c. 2000–741 BCE),"

Cultural Attitudes and Horse Technologies

northwest appear in the records as possessing horses and chariots, and a few sacrificial burials of chariots with horses have been excavated.[7] The number of chariots buried as sacrificial gifts during the last two centuries of the second millennium was small, while remains of horses are slightly more numerous. Research on oracle bones has shown that gelding was known, as was the castration of bulls and pigs.[8] In some instances, charioteers and chariot owners display links to the northern steppe cultures, as they were accompanied by northern style objects.[9] Identified aspects suggest that the technological complex was adapted quickly by a culture already skilled in stabled husbandry, metal working and carpentry. In the following centuries, chariotry and horse husbandry spread throughout Northeast Asia and to the lowlands south of the Yangzi.[10]

It is not impossible that the chariot was initially the only means of wheeled transport.[11] In fact, firm evidence of the ox-drawn cart or wagon, a vehicle that would have been useful given the considerable transport volumes of the urbanized polities of ancient China, is missing throughout Central and Northeast

---

Sino-Platonic Papers, 99 (February 2000): 1–98. Because of the lack of remains in relative vicinity in space and time, however, this line of inquiry provides too few results to enable a comparative picture to be made. The earliest steppe examples from Sintashta and Lchasen are merely imprints of the lower parts of the wheels, while the vehicles' superstructure remains unknown. The next examples are the Pazyryk chariots, which are extraordinarily well-preserved but dated to about 300 BCE. The Pazyryk finds show a culture in which riding was certainly the common form of locomotion, with traces of deer husbandry preceding the introduction of the horse. The Pazyryk chariot is an extremely light, obviously ceremonial vehicle unfit for frequent use. The contrast with the solid carpentry of Shang chariots is eye-catching, but hardly useful, as the Pazyryk vehicles date from a time when the chariot was outmoded for practical use in the Northern steppe and hence cannot be used as information on vehicles in earlier periods.

[7] Di Cosmo, op. cit., pp. 902–904; Barbieri-Low, op. cit.; Keightley, "The Shang," p. 184. For a map, see Linduff, op. cit., p. 149.

[8] Kolb, op. cit., p. 104.

[9] Linduff, op. cit., p. 153.

[10] The notion that China proper is unsuitable for horses is an anachronism based on the near total expansion of intensive agriculture in the late imperial period. However, husbandry appears to have always been an activity carried out alongside, not in combination with, agriculture. This is due to the agricultural regime of millet and rice, with both main crops producing best results with intensive hoeing, weeding and fertilizing.

[11] I have excluded an often-quoted find here: The Dalitaliha site near Nomhon in Qinghai produced two hubs with holes for 16 spokes. The site is dated around 1500 BCE and the wheel remains were found at the gates of an enclosure for sheep and cattle herding. The sketchy excavation reports render dating and interpretation of this singular find too uncertain to use for a chronology. For a discussion of the site, see V. Mair, "The Horse in Late Prehistoric China: Wresting Culture and Control from the 'barbarians,' " in M. Levine, C. Renfrew and K. Boyle, eds., *Late Prehistoric Exploitation of the Eurasian Steppe*. Cambridge: Cambridge University Press, 2003, p. 173.

and Northern Asia east of the Andronovo culture.[12] In China, references to the ox cart begin to appear in written sources of the earlier Zhou period in a natural way, suggesting a familiar object.[13] The plough, the other component of the bovine traction system, makes its appearance so quietly that the dating remains hotly disputed.[14] For the time being, available evidence is too thin to support either the existence or the absence of ox-drawn vehicles. Three aspects need to be taken into consideration: First, cattle spatula for divination appearing from 2000 BCE attest to domesticated bovines and their importance in ritual. Fully congruent with this evidence, mythology accords a prominent rank to bovines and their domestication. However, reference to traction in the mythological texts is uncertain.[15] Second, the absence of ox carts from the great ritual centres – the focus of archaeological excavations – as well as from oracle bone inscriptions shows that they were not symbolically powerful objects, but does

---

[12] The question of the eastward transmission of ox-drawn vehicles is a conundrum. The wagon or cart and plough with the ox and the yoke are well established and researched for the Western Eurasian steppe from the fourth millennium onward, and the transmission to Harappan India appears certain. On the Western steppe, see E. Kuzmina, "Origins of pastoralism in the Eurasian Steppe," in M. Levine, C. Renfrew and K. Boyle, eds., op. cit., pp. 203–232. For the transmission across the steppe east of the Urals, however, evidence is scarce. Mary Littauer's investigation of rock carvings at sites along the southern rim of the steppe shows oxen-drawn wagons outnumbering two-wheelers for sites west of the Caspian Sea, while two-wheelers dominate from Kazakhstan to Mongolia. See Littauer, op. cit. The most easterly finds of petroglyphs depicting relatively primitive vehicles are in northern Xinjiang, possibly showing carts with solid wheels. These are, however, notoriously hard to date, see Wei Si, op. cit. In the same region, a solid, tripartite wheel has been excavated. The find is dated between 1400 and 800 BCE, possibly later, and thus too late to serve as a missing link. See J. Mallory and V. Mair, *The Tarim Mummies: Ancient China and the Mystery of the Earliest People from the West*. London: Thames and Hudson, 2000, pp. 142f. Few finds and large gaps permit no conclusions. The only tentative interpretation draws on cultural status: In the Pontic cultures, where the ox wagon was well established early on, it appears to have been replaced by the horse chariot about 2000 BCE as the vehicle of outstanding status that was offered as funeral sacrifice. In view of the comparable functions in Shang culture, it seems possible that both forms of wheeled locomotion travelled eastward together, with their symbolic hierarchy already fixed.

[13] E.g. the *Book of Rites*: 2 "officials of the earth" (地官司徒) with the responsibility for "horses and oxen, chariots and vehicles pulled by men" (馬牛車輦), the *Book of Songs*: Xiaoya 7, 227 with a line "my carriage, my ox" (我車我牛).

[14] F. Bray, *Science and Civilisation in China*, vol. 6, *Biology and Biological Technology*, part II, *Agriculture*. Cambridge: Cambridge University Press, 1984, pp. 141–179; Kolb, op. cit., pp. 127–136.

[15] The systematisation of mythology is thought to be a product of the Zhou dynasty, not least serving legitimation. Sequence in genealogy therefore needs to be read as a hierarchy of venerability rather than temporality. The domestication of bovines is ascribed to the pre-dynastic Shang king Hai 胲, who is said to "have made the cattle serve" (*funiu* 服牛). The phrase is mostly understood as "submit to the yoke," but the attested Shang meanings of the character *fu* 服 do not go beyond "submit," "being used" or "being subjugated." Possible readings therefore include domestication, use as a sacrificial animal, as a traction animal or as a beast of burden.

not demonstrate their non-existence. Third, attempts to push back wheeled transport to before 1500 BCE have been made on the basis of traces identified as rut marks at the city sites of Erlitou and Yanshi 偃師 (ca. 1700–1600 BCE).[16] The finds are however inconclusive, because analyses as to whether they were left by wheels or sledges have not been undertaken. Meanwhile, another feature that may indicate the presence of wheeled traffic, namely the rectangular road grids, has not been thoroughly investigated. While it seems to me that on the basis of existing materials the issue of the bovine traction complex cannot be resolved, for the horse and the chariot the uncertainty means that we cannot be sure whether we are actually observing the beginnings of wheeled transport or just the advent of a new vehicle and animal. The only firm conclusion is that horse and chariot were the only ritually important means of transport.

About 1045 BCE, the Shang were conquered by a formerly semi-barbarian statelet by the name of Zhou, which established itself as heir of the state they had conquered. Over the following centuries, Chinese classical antiquity formed across northern, central and eastern China, characterized by a vigorous expansion of the aristocracy and an agricultural economy of relatively high density controlled by urban centres. Although the real power of the Zhou soon waned, the competitive and warlike states of the "Spring and Autumn" (770–476 BCE) and the "Warring States" periods (453–221 BCE) formed a cultural universe, acknowledging the ritual centre represented by the Zhou dynasty and using a common writing system. Population increase and economic growth were marked during the period of the Warring States, when cast iron transformed productivity and war. The rise of large states ended in the formation of the first empire, the Qin (221–206 BCE). While great wars and forced resettlements during the short-lived Qin caused a population collapse, long periods of stability during the Han (206 BCE–220 CE) resulted in another age of increase in population and productivity.

Two aspects in the age of the horse-drawn chariot in China appear both remarkably specific and comparable to other cultures: the ritual ostentation and symbolic power with far-reaching catalyst functions in the state-building process, and the considerable temporal gap between the advent of a new technology and its being put to practical use.

---

[16] See Barbieri-Low, op. cit., p. 15f. Wang Xingguang 王星光, "Shilun Zhongguo niuche, mache de bentu qiyuan" 试论中国牛车、马车的本土起源 (Discussion on the origins of the ox cart and the horse carriage in China), *Zhongyuan wenwu*, 4 (2005): 28–34; Wang Xuerong 王学荣, "Shangdai zaoqi chezhe zu shuanglunche zai Zhongguo de chuxian" 商代早期车辙与双轮车在中国的出现 (Early Shang cart ruts and the appearance of the two-wheeled cart in China), *Sandai wenming yanjiu*, 1 (1999): 239–247.

Ostentation, Ritual, and Society

It has been shown for the Western Eurasian steppe, for the Near East and Egypt that rather than a military revolution, the horse-drawn chariot was a revolution in martial display.[17] First and foremost, it changed social stratification as well as the costs and requirements of warlike elites. Richard Sherratt emphasized its role as a "new language of elite ostentation" in his analysis of the transition from the ox-drawn wagon to the horse-drawn chariot in the Circum-Pontic.[18] Before the debate on the chariot as a weapon had begun, Magdalene von Dewall analysed elements of display, battle conventions and the combination of chariotry and infantry for early China.[19]

In the late Shang period, the appearance of horse and chariot coincided with the development of writing, providing sources of archaeology, oracle bones and early layers of classical literature for a reconstruction of state formation, social hierarchization and power concentration, territorial expansion, and the formation of a steppe frontier as a constitutive element to Shang identity.[20] Horse and chariot figure in all of these processes. There is no doubt that the fascination with the symbol of dominance, combining speed, control and the harnessing of strength, worked convincingly for the Shang elite.

Shang chariots played an outstanding role in ritual representation, most prominently in royal hunts and in sacrifices.[21] In the Shang and Zhou ritual system, the ruler's task as a medium was the maintenance of harmonious relations with heaven to whom he was linked through his ancestors, so that the latter would send favourable weather for agriculture. The key function of the state he presided was the control and distribution of resources, most importantly the maintenance of granaries. In burials and hunts, both involving the killing of living beings on a large scale, the king's role as the medium between the human world and heaven was enacted. In fact, the hunts were integrated into the overall agrarian goals, for they took place in the fields in order to ritually or actually rid agricultural areas of the large beasts that

---

[17] Littauer, op. cit.; A. Sherratt, "The Horse and the Wheel: The Dialectics of Change in the Circum-Pontic Region and Adjacent Areas, 4500–1500 BC," in M. Levine, C. Renfrew and K. Boyle, eds., *Prehistoric Steppe Adaptation and the Horse*. Cambridge: McDonald Institute for Archaeological Research, 2003, pp. 233–252; P. Moorey, "The Emergence of the Light, Horse-Drawn Chariot in the Near-East c. 2000–1500 B.C.", *World Archaeology*, 18, 2: *Weaponry and Warfare* (October 1986): 196–215.

[18] Sherratt, op. cit., p. 245.

[19] M. von Dewall, *Pferd und Wagen im frühen China*. Bonn: Habelt, 1964.

[20] K. Linduff, "Art and Identity: The Chinese and their 'significant others' in the Shang," in M. Gervers and W. Schlepp, eds., *Cultural Contact, History and Ethnicity in Inner Asia*, Toronto: Joint Centre for Asia Pacific Studies, 1996.

[21] von Dewall, op. cit., pp. 56–67 and 199–203; M. Fiskesjo, "Rising from Blood-stained Fields: Royal Hunting and State Formation in Shang Dynasty China," *Bulletin of the Museum of Far Eastern Antiquities*, 73 (2001): 48–192.

inhabited the savannah and open forests of the North China Plain.[22] For the most part, sacrifices consisted of alcoholic beverages – a "spiritual" form of cereals – and meat broth – sometimes involving the slaughter of a thousand heads of cattle or several hundred sheep. These offerings are interpreted as nourishment for the ancestors. Sacrifices of humans, horses and of chariots, by contrast, stand out as pure display of power and ostentation, without an inherent relation to the nourishment of ancestors or harmonising ties between the world of men and of heaven.

While in the Shang period the chariot may have exalted its rider beyond the world of common mortals, in the Western Zhou period (1046–771 BCE) chariotry became an increasingly widespread aristocratic occupation. In the oldest texts handed down in the written tradition the horse appears as the most noble of the six domestic animals. Titles of specialized royal officials are recorded, responsible for horse husbandry, shamanistic healing, and chariots. The chariot was improved, with metal nave-bands making the wheels more robust.[23] For a relatively short period in the early Zhou period, the war chariot gained military importance. Chariot warfare is recorded within a shared setting of military technology, in battles between Chinese states and in conflicts with western outsiders who also used chariots.[24] Furthermore, demonstrating the status of horsemanship, the charioteer ranked highest within the chariot teams of two to four men, above the warrior who wielded the weapons.[25] During the Spring and Autumn period, armies expanded and infantry troops increased in number, gradually reducing the role and value of chariot formations. By the Warring States period, wars between states relied on massed infantry, armed with iron weapons and the crossbow.[26] Mounted warfare in Chinese armies began in the sixth century BCE, while the increasing projectile power of composite bows and especially the crossbow from the fifth century BCE led to the rise of heavy armour.[27] Horse and chariot retained their high symbolic status through these transformations into the early empire of the Qin (221–207 BCE) and Han (206 BCE–220 CE) periods. War chariots, meanwhile, had become specialized units for specific purposes. The horse was still venerated in horse cults and the heavenly horses of Ferghana were highly prized, but a significant cultural shift

---

[22] Keightley, "The Shang," pp. 279f.

[23] J. Needham, *Science and Civilisation in China*, vol. 3, *Physics and Physical Technology*, part II, *Mechanical Engineering*. Cambridge: Cambridge University Press, 1965, pp. 73–95.

[24] Keightley, "The Shang," p. 284; di Cosmo, op. cit., pp. 904, 920; E. Shaughnessy, "Historical Perspectives on the Introduction of the Chariot into China," *Harvard Journal of Asiatic Studies*, 48,1 (1988): 208.

[25] von Dewall, op. cit., p. 184.

[26] M. Lewis, "Warring States Political History," in M. Loewe and E. Shaughnessy, eds., *The Cambridge History of Ancient China*, Cambridge: Cambridge University Press, 1999, pp. 620–632; D. Graff, *Medieval Chinese warfare, 300–900*. New York: Routledge, 2002, p. 22.

[27] C. Goodrich, "Riding Astride and the Saddle in Ancient China," *Harvard Journal of Asiatic Studies*, 44, 2 (December 1984): 297–305.

from highly exclusive and martial representation is evident. The carriage had become a vehicle for all who could afford it, subject to fashions and a wide range of general and specialized uses.[28]

Symbolic Value and Practical Use

With symbolic value established in a formative phase, horse and two-wheelers became closely associated with state and elite prestige. While their use in warfare and in ritual display imposed themselves, actual military or transport efficiency appear not to have been the primary concern. In fact, their practical usefulness as a means of speedy communication took centuries to be realized in a long process of infrastructure construction. This secondary use eventually transformed the Chinese landscape and state administration. While the situation for the Shang is unclear, there is no doubt that the Zhou, and the increasingly independent feudal states of the Zhou cultural zone, were enthusiastic road builders. Due to the nature of the written tradition, information on the road and courier networks is largely normative, making it impossible to clearly discern "realities." Yet the existence of regular grids of roads and instances of speedy overland communications show that the norms were not merely a fictive ideal.[29] The aim of road construction appears to have been twofold, the extension of the range and the enhancement of the speed of vehicles and the fixing of the agricultural landscape. The ensuing development of land transport, usually by ox cart, can hardly be overestimated. Wheeled traffic revolutionized transport and communications, providing the network upon which rested the large states of the Warring States period as well as the early empire, with monetarized economies and large trade volumes in luxury and bulk goods.

As the means of elite locomotion and of state communication, the horse chariot and carriage entered the imperial age. Han period stone reliefs demonstrate a fascination with speed, while sources on the horse cult and on horse veterinarian medicine attest to the cultural value of the animal. The courier system, which becomes visible in some detail for this period, clearly shows that roads, horses and carriages played a key role in tying the expanding empire together, both as the state communication system and as a structure representing state power.

---

[28] For horse cults, see esp. R. Sterckx, "An Ancient Chinese Horse Ritual," *Early China*, 21 (1996): 47–79.

[29] Wang Zijin 王子今, *Qin-Han jiaotong shigao* 秦漢交通史稿 (A draft history of transport in the Qin and Han periods). Beijing: Zhonggong zhongyang dangxiao chubanshe, 1994, pp. 12–14.

Horse-Riding and Social Status After the Han Dynasty

The period from the beginnings of cavalry in late antiquity to the conquest of Northern China by mounted warrior peoples between the third and sixth century CE links chariot and stirrup. Riding began in the sixth century BCE in some northern states of China. Interestingly, the first pictorial evidence survives on a laquer box from the southern state of Chu in the middle Yangzi region, south of the natural habitat of horses.[30] The recorded date for the introduction of cavalry troops by a Chinese state is the year 307 BCE. King Wuling of the northern state Zhao 趙武靈王 is said to have ordered his men to follow his example of "mounting a horse and changing his garments."[31] The shift can be regarded as a response both to the pressure by mounted warriors of the northern zone and to the growing military importance of infantry armies in Chinese wars. It seems probable that the strong cultural preference for the chariot prevented the adoption – or at least the open admission of riding – for a considerable time. Although strategists recognized the need for cavalry troops in order to insure swift movement across difficult terrain, replacing the elegant flowing robes with riding trousers and short jackets appears to have been a hindrance in the eyes of the elite. Nevertheless, the military importance and size of cavalry troops gradually increased. By the Western Han (206 BCE–9 CE) cavalry regiments were firmly established specialists, alongside regiments of crossbow-men, archers, and charioteers.[32] Elites meanwhile had increasingly turned to civilian occupations and ideals, while chariots had become carriages.[33] Riding remained a military activity of the lower ranks, fit for soldiers and messengers but hardly elegant.

The change came after the end of the Han in the early third century CE. With the demise of the imperial state, roads fell into disrepair. Climatic cooling reached its historic low point during this period. Rebellions, wars and migration left much of northern China empty. Invasions from the northern zone brought the domination of foreign peoples in the north, while massive

---

[30] Goodrich, op. cit., plate 1.

[31] Goodrich, op. cit., 280f.

[32] J. Needham and R. Yates, *Science and Civilisation in China*, vol. 5, *Chemistry and Chemical Technology*, part VI, *Military Technology: Missiles and Sieges*. Cambridge: Cambridge University Press, 1994, pp. 123–125, Li Bin 李斌, "Cong Yiwan 'Wuku yongshi sinian bingcheqi jipu' kan Handai bingzhong goucheng" 从尹湾《武库永始四年兵车器集簿》看汉代兵种构成 (The structure of Han period weaponry as reflected in the "List of armaments for soldiers and chariots of the arsenal in the Year Yongshi 4 (13 BCE)"), *Zhongguo lishi wenwu*, 5 (2002): 30–39.

[33] The Chinese term *che* 車 is originally a graph showing a two-wheeled vehicle from above and does not differentiate between chariots, carriages and carts.

southward migration caused a decisive shift in economic and cultural patterns. Riding now became a display of martial prowess, while wheeled vehicles were a civilian form of locomotion. The ox cart gained prestige as a vehicle acceptable for the transport of persons. In the South, this was probably related to the adoption of a "Southern lifestyle" in the lower Yangzi area. In this frontier area and the preferred destination of fugitives from the north, the boat gained in practical and cultural importance, while horses became rare. In the north, representations of ox carts also increase, perhaps due to the fact that the new elites did not share the classic prejudice against ox carts. The age of the chariot was past.

The Stirrup and Heavily Armoured Mounted Warriors

In this age of confusion, the stirrup is first attested in the area south of the Yangzi, but soon afterwards appears in larger numbers in Xianbei 鮮卑 graves of the Northeast (see Map 2 for the distribution of early stirrups). From here, it swiftly spread eastwards to federations and states in formation in Korea and Japan. The westward spread is less clear, but highly probable. The first archaeological evidence is a small statue, a burial gift from a grave near Changsha that is dated 302 CE. This statue, as well as a few other early examples, shows a single stirrup attached to the saddle rather high up. The device has been interpreted as a mounting help. The fact that the earliest examples of stirrups were found in the core areas of southern culture and well south of the natural habitat of horses merits attention. The use of horses in this region appears as a continuation of cultural patterns formerly established in the northern plains, while riding may have been the continuation of Han uses of escorts and couriers, perhaps also a military necessity. In this context, the development of a mounting help suggests efforts that would enable people to ride who were no longer highly skilled in horsemanship. Half a century later, the stirrup had been taken up and developed into a riding help by the Xianbei, a federation of riding peoples from the northeast, who swept across northern China. Charred remains of pairs of stirrups for real use appear as burial gifts in this culture. The Tuoba 拓跋, a branch of the Xianbei, came to dominate northern China for almost two centuries, and the formation of a culturally mixed northern culture and political system would become the basis of the second imperial age of the Sui (581–618 CE) and Tang (618–907 CE).

An exploration of the innovation in the Chinese and East Asian context shows that the use of stirrups spread together with a trend to increasingly heavy armour. Rather than a key invention associated with the rise of full body armour, however, it was a new device reinforcing an existing dynamic. The development of horse and body armour had begun early in China and in the Northern zone. A Shang bridle element has been identified as a protection of the

Cultural Attitudes and Horse Technologies

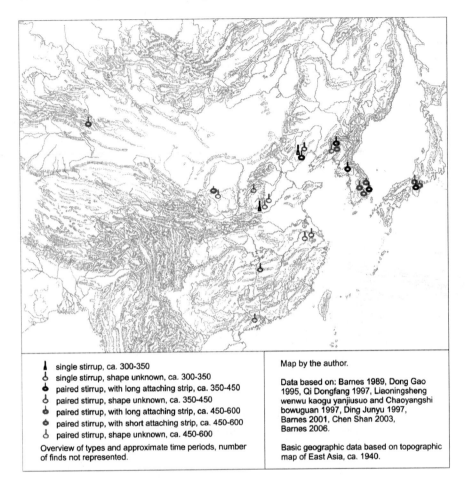

Map 2 Early stirrups in East Asia. Excavated objects (stirrups for real use, funerary goods, figurines, murals) and textual cources, ca. 300-ca. 600

horse's head; more elaborate helmets and body armour for horses and charioteers follow in the Zhou period.[34] Depictions of riders from the Northern zone cultures also show helmets. It would seem that the crossbow, a weapon with high projectile power present since the fifth century BCE, contributed decisively to an "arms race" of missile power against greater speed and better

---

[34] Tatsumi *Yoshinobu* 巽 善信, "Higashi Ajia no bachu" 東アジアの馬冑 (Horse helmets in East Asia), *Kodai bunka/Cultura Antiqua*, 47,5 (June 1995): 1–17, 61.

armour. Albert Dien has pointed out that a forerunner of the famous Qin leather (and sometimes iron lamellar) armour can be found in a late fifth century BCE Chu grave.[35]

In East and Southwest Asia, the heavy armour of mounted warriors actually preceded the stirrup. Saddles with deep groves and high backs or horns were alternative solutions that prevented armoured riders from toppling off their mounts.[36] It may be added that while specialized saddles did not help in mounting the horses, they had the advantage of avoiding the lethal danger of getting a foot caught in a stirrup in the event of a fall.[37] Nevertheless, the stirrup brought advantages to armoured warriors, for its spread coincides with the appearance of extremely heavy full body armour for horses and warriors between the fourth and sixth centuries CE. These are found from Japan to Dunhang, with the iron-armoured warriors of the Xianbei and their descendants and those of Koguryŏ the most feared.[38] While mobile frontier peoples played an outstanding role in the process, heavy armour came to prominence only when the process of state formation was well advanced, and after control over agrarian populations and metal resources had been established. It was never adopted by nomad raiders such as the early Turks.[39] The Xianbei continued to form the ruling elite of the Northern dynasties, while Koguryŏ, initially their immediate yet distinct neighbours, expanded eastward, soon controlling large parts of Korea. Over the following centuries, the stirrup became a standard part of saddles across Eurasia, integrated with existing armour and reinforcing the trend to ever heavier armour. In East Asia, full body armour for horses and riders appears to have gradually lost its importance after the end of the Tuoba Wei dynasties in the mid-sixth century CE. The gradual restoration of bureaucratic state structures during the later northern dynasties and the system of territorial soldiery established under the Sui and Tang enabled governments to raise large cavalry and infantry forces, making strategy and logistics decisive in war. By the Tang, full armour and ornate horse fittings appear as a ceremonial outfit.

---

[35] A. Dien, "A Study of Early Chinese Armor," *Artibus Asiae*, 43, 1/2 (1981): 6.

[36] M. Littauer, "Early Stirrups," *Antiquity*, 55 (July 1981): 103–106.

[37] This danger makes the leather sling, although an often suggested predecessor of the stirrup, seem rather too dangerous to have been adopted for armoured riding.

[38] Dien, op. cit., pp. 33–38; Dong Gao 董高, "Gongyuan san zhi liu shiji Murong Xianbei, Gaojuli, Chaoxian, Riben maju zhi bijiao yanjiu" 公元3至6世纪慕容鲜卑、高句丽、朝鲜、日本马具之比较研究 (A comparative study of horse outfits of the Murong Xianbei, Koguryeo, Korea and Japan from the third to the sixth centuries), *Wenwu*, 10 (1995): 34–42; G. Barnes, *State formation in Korea: Historical and archaeological perspectives*. Richmond: Curzon, 2001, pp. 125–151; Graff, op. cit., pp. 41–43.

[39] Graff, op. cit., pp. 142–144.

The Silent Adoption of a New Riding Help

In Chinese written sources, the adoption of the stirrup is hardly discernible. Mention in texts from the period between the Han and the Tang is few and far between. Altogether I have been able to locate no more than four passages.[40] In all cases the stirrup is negatively depicted. The earliest example from the *Shishuo xinyu* is a story of an event dated to 359 CE. Here the pretentious habit of a luckless general who demands to have his jade stirrup held for him even at the moment when he is forced to flee in defeat is exposed to ridicule.[41] In an episode from the Southern Qi (479–502 CE), an informant has a stirrup sent to a commander, who correctly interprets the message to the effect that an uprising is in preparation.[42] In another from the same dynasty, a prince is criticized for having even his stirrups made of silver, implying that while it may be acceptable to have other parts of a horse's outfit embellished with silver ornaments, this is ridiculous for an object as lowly as the stirrup.[43] A biography of a warrior from the late sixth century CE mentions that even when over 60 years old his feet would not touch the stirrup when mounting his horse.[44] The episodes attests to the fact that horses and horse-riding were usual elite activities in the South as well, and that stirrups had become a common part of the horse outfit and were associated with warfare. Martial prowess was enhanced by not using the stirrup.

The archaeological record of the period is especially rich in murals and terracotta figurines used as burial gifts. Depictions of stirrups are relatively rare in both forms. Although it is often impossible to make out a stirrup, the change in the posture of the riders, who now sit on their mounts with straight legs and upturned toes, documents the spread of the device.[45] It may be significant in this context that even Tang figurines mostly make very little of the stirrup, even though it evidently was part of the standard outfit by this time. Only in the depiction of particularly pronounced realist detail, such as the

---

[40] Quoted in Luo Xiaoping 骆晓平, "Tantan woguo madeng de chansheng shidai" 谈谈我国马镫的产生时代 (Remarks on the period during which the stirrup originated in China), *Wenshi zazhi*, 2 (1996): 55.

[41] Liu Yiqing 劉義慶, *Shishuo xinyu* 世說新語 (A new account of tales of the world), compiled by Yu Jiaxi 余嘉錫, collated by Zhou Zumo 周祖謨 and Yu Shuyi 余淑宜, Beijing: Zhonghua shuju, 1983, p. 570.

[42] *Nanqi shu* 南齊書 (History of the Southern Qi), compiled by Xiao Zixian 蕭子顯, 489–537, Repr. Taibei: Dingwen shuju, 1987, p. 466.

[43] *Nanqi shu*, p. 703.

[44] *Zhou shu* 周書 (History of the Zhou), compiled by Linghu Defen 令狐德棻, 583–666, Repr. Taibei: Dingwen shuju, 1987, p. 453.

[45] Dong Gao, op. cit.; B. Cooke, ed., *Imperial China: The Art of the Horse in Chinese History*, Exhibition Catalog. Lexington: Kenthucky Horse Park, 2000; J. Watt, et al., *China: Dawn of a Golden Age, 200–750 AD*. Exhibition catalogue. The Metropolitan Museum of Art, New York, New Haven: Yale University Press, 2004.

famous stone friezes of Emperor Gaozong's horses, is the stirrup shown clearly and in detail.[46] In a period of fascination with riding elegance, moreover, an uneasy awareness of a change in customs can nevertheless be found. In the section concerning regulations on court dress, the compilers of the Tang history thoughtfully comment that in the old times officials travelled by carriage rather than on horseback and their robes therefore had no need to be short so as to allow them to mount a horse.[47] We can still discern certain reservations against a form of locomotion that brought about a departure from classical garments and a posture of respect. Nevertheless, the Tang nobility, civil and military, men and women, were elegant and enthusiastic riders, as scenes of polo players most vividly show.

The attitude towards the stirrup suggests cultural attitudes comparable to those found by Mary Littauer in Near Eastern riding cultures: In both the Northeast Asian and Turkic context, the stirrup was associated with an unmanly form of riding, such as practiced by women or the infirm.[48] It seems possible that the Xianbei, who had only recently moved southward from Northern Manchuria, did not share these cultural reservations or were more disposed to put them aside in order to employ newly acquired metal resources to enhance military might. Nevertheless, cultural reservations lingered for centuries. While the stirrup swiftly spread across the continent, it would appear that similar attitudes held by cultures shaped by a long tradition of horse riding caused its adoption to be silent.

Conclusion

This brief cultural exploration of the horse, the chariot and the stirrup in China has shown the effects of cultural value, status and attitude as a factor in the promotion of and reservations against technological innovation. The example of the horse and chariot demonstrates the dynamics of an imagery of power that clearly worked in different cultural settings. To return to the under-researched Eastern steppe, there is no reason to assume that transmission required migration, and even less that it involved domination by the ethnic group first in possession of the technology. Rather, it seems that in a region of high mobility, of trade and war, the horse and the chariot as well as the ox and the cart or wagon travelled with their owners across cultures and were adopted into different settings. The finds of Pazyryk provide an example of a perfect, yet almost flimsy ceremonial chariot, attesting to its symbolic power in an Eastern steppe setting, while the wagon was the vehicle that actually provided mobility.

---

[46] See e.g. W. Watson, ed., *Art of Dynastic China,* London: Thames and Hudson, 1979, p. 35.
[47] *Jiu Tangshu* 舊唐書 (Old Tang history), compiled by Liu Xu 劉昫, 887–946, repr. Taibei: Dingwen shuju, 1980, p. 1950.
[48] Littauer, "Early Stirrups," pp. 103–104.

Furthermore, the exploration exemplifies the complex dynamics between improvements within a technology and associated cultural values. Thus, improvements to the horse traction system and the chariot, including such aspects as the quest for better horses and the development of roller bearings, illustrate the importance of prestige for technical progress. In terms of transforming society and economy, however, results that could not have been foreseen or intended by those who started the process might have made a larger impact.[49] The long process of road building and the "side-effect" of greatly facilitated road transport by ox cart is a fascinating example.

The stirrup, by contrast, is perhaps better understood as an improvement to the technological complex of riding and full body armour within the general context of social and military transformation. While the improvement reinforced an existing trend, in the longer run it contributed to the diversification of riding, making a formerly physically demanding and necessarily exclusive activity accessible to persons without lifelong horsemanship. In China, it thus appears to have contributed to the waning of martial valour associated with riding. In a comparative perspective, the story of horse, chariot and stirrup exemplifies the power of imagery travelling with technologies as well as specific cultural choices and dynamics.

---

[49] Needham, "Mechanical Engineering," p. 94.

Part II
On Maps, Astronomical Instruments, Clocks and Calendars

Patchwork – The Norm of Mapmaking Practices for Western Asia in Catholic and Protestant Europe As Well As in Istanbul Between 1550 and 1750?

Sonja Brentjes

When we wish to discuss the ways of adopting and adapting methods and techniques across different cultures, one point that deserves our attention concerns the question whether the methods and techniques scholars and other professionals talk about in their writings are indeed those they apply when producing their objects. Historians and philosophers of science in western societies have since long pointed out that there is a substantial gap between the rhetoric of legitimizing scientific or technical products and the processes that led to discoveries and inventions. Historians of Renaissance woodcuts and copper prints of paintings and maps have shown that the production process itself was imitative by its very nature relying on freehand drawing skills rather than precise constructions executed with scientific instruments and that the production costs induced the acquisition of copperplates owned by a passed away printer and map publisher by their heirs or competitors who then would slightly modify plates rather then order completely new ones. The result of these constraints imposed by production and commerce was a substantial gap between the ideal of scientific map production and the factual creation of the maps we find in atlases, geographical handbooks and single sheet collections. Such observations beg the question whether maps found in manuscripts do indeed differ substantially with regard to their scientific character on the level of production methods from these early modern and modern printed maps. They also imply to ask whether conflicts between ideal and practice can also be found on a different level, i.e. the process of creating a new map by a geographer, cosmographer or other professional involved in such a kind of innovative procedure. In this paper I will briefly survey some of my findings to the first question and then devote most of its pages to second one.

S. Brentjes (✉)
Department of Philosophy and Logic, University of Seville, Seville, Spain
e-mail: brentjes@us.es

Observations on Practices of Mapmaking Before Printing

There is very little discussion in writings on history of cartography about the practical methods applied for producing a preserved cartographic artefact. While I did not do an extensive study of this question I have analysed a number of portolan charts extant in the Bibliothèque nationale de France when searching for their possible usage of pictorial elements of Asian provenance. I found confirmed what some earlier historians of cartography had already pointed out for charts held in other European libraries. Most of the specimens kept at the BnF were not made following the method seen as the basis construction tool of portolan charts – compass roses and the network of rhumb lines connecting their partitions with each other. The sequence of colours found on the maps rather indicates that the person who painted the map started first with coastal lines that separated the Mediterranean Sea, the Atlantic or the Black Sea from the main land. Then he added the names of the ports, rivers, landmarks, towns and villages along those coastal lines. If interested in displaying some information about inland geography, since producing for instance a princely gift, he continued the process with filling out the interior of the *terra firma*. Only after major parts of the maps had been drawn, and occasionally may be even only at the end of the entire process, did the producer finally add the compass roses and rhumb lines. Examples for my claim are Angelino Dulcert's portolan chart from 1339, the Catalan Atlas presumably produced in 1375 and the so-called Catalan world map ascribed to the period stretching from the late fourteenth to the middle of the fifteenth century.[1] Similar observations can be made when analysing the Byzantine maps illustrating Ptolemy's (fl. ca. 125) *Geography* made in the thirteenth and fourteenth centuries.[2] Here too, the grid was not used as a constructional device, but drawn freehand and added in all likelihood after the contours of the landmasses had been put on the paper.

When we turn to Arabic, Persian or Ottoman Turkish maps made between 1100 and 1700 we find a more mixed pattern. Some of the maps I have seen in manuscripts in Istanbul, Paris, Berlin, Oxford and other libraries imply that their painters did not use any kind of instrument, but produced each and every element of the map through freehand drawing skills. Examples are the map of Sicily in the newly discovered Fatimid geography now at the Bodleian Library in Oxford or the world

---

[1] A. Dulcert, Paris, BnF, Cartes et Plans, Rés GE b?; Catalan Atlas, Paris, BnF, Esp.30; Catalan World Map, Modena, Biblioteca Estense, C.G.A.1.

[2] MSS Rome, Vat. Urb. Gr. 82, ff 60b–61a; Istanbul, Topkapı Palace Library, Gl 57, ff 73b–74a. F. Sezgin, *Mathematische Geographie und Kartographie im Islam und ihr Fortleben im Abendland; Geschichte des arabischen Schrifttums, bis ca. 430H.*, vol. 10–12, Frankfurt am Main: Institute for the History of Arab-Islamic Science, 2000, vol. 12, Plate 28.

map in a 1553-copy of an Ottoman Turkish translation of Zakariyā' Qazvīnī's (d. 682/1283) *'Ajā'ib al-makhlūqāt* preserved in the Library of Congress.[3]

If circles are part of the map, as in the case of many world maps, a certain number of maps show holes and hence suggest that a compass was used for drawing the circle. Straight lines also are an occasion for applying instrumental techniques. A good number of them appear to have been drawn with a ruler. Examples are world maps in Gotha MSS A 1514 of Pseudo-Ibn al-Wardī's cosmography (composed 522/1419) and A 2157 of al-Nawajī's (d. 859/1455) thematically ordered collection of poems and prose and regional maps from manuscripts of al-Iṣṭakhrī (fourth/eleventh century) and Ibn Ḥawqal (d. ca. 367/977).[4]

But even when a compass and a ruler were used for producing basic mathematical features, as is the case with the image of the sphere in Topkapı Palace Library, R 1622, this does not mean necessarily that the figure was constructed (Fig. 1).[5] The application of instruments can merely reflect a mechanical procedure. This effect can be easily observed in world maps attached to very different kinds of manuscripts ranging from the mathematical sciences to history. Examples are maps in a copy of Niẓām al-Dīn Nīsābūrī's (d. 730/1329) commentary on Naṣīr al-Dīn Ṭūsī's (597–672/1201–1274) astronomical textbook *Tadhkira*, several copies of Ḥafeẓ-e Abrū's (d. 833/1429) geographical part of his *Kitāb-e Tārīkh* or Sirāj al-Dīn al-Sajawandī's (d. 607/1210) text on so-called folk astronomy.[6]

So far I encountered only one Arabic map that suggests that in its case the sequence of working steps which I have described so far as dominant practice was not followed. The producer of the map seems indeed to have started with the construction of the mathematical elements, i.e. the circle, its diameters and the grid. This map is found in a fourteenth-century manuscript of Ibn Faḍlallāh al-'Umarī's (d. 749/1349) encyclopaedia

---

[3]MSS Oxford, Bodleian Library, Arabic c.90, ff 32b–33a; Washington, D.C., Library of Congress, LC 25-6 (9/53); E. Edson and E. Savage-Smith, *Medieval Views of the Cosmos*, with a foreword by T. Jones. *Picturing the Universe in the Christian and Islamic Middle Ages*, Oxford: Bodleian Library, Oxford University, 2004, figure 44, p. 90.

[4]MSS Gotha, Forschungsbibliothek, Schloß Friedenstein, A 1514, ff 1b–2a; A 2157, f 164a. Orientalische Buchkunst in Gotha. Ausstellung zum 350jährigen Jubiläum der Forschungs- und Landesbibliothek Gotha, Spiegelsaal, 11 September 1997 bis 14. Dezember 1997, Forschungs- und Landesbibliothek Gotha, 1997, pp. 165–167.

[5]MS Istanbul, Topkapı Palace Library, R 1622, f 15b.

[6]I thank J. Ragep, McGill University, Montreal, for providing me with a copy of the map found in one of the copies of Nīsābūrī's commentary. D. A. King, "A world map in the tradition of al-Bīrūnī (ca.1040) and al-Khāzinī (ca. 1120) presented by Sirāj al-Dīn al Sajwandī (1210)," in Frank Dalemans, J.-M. Duvosquel, R. Halleux, D. Juste, eds., *Mélanges offerts à Hossam Elkhadem par ses amis et ses élèves*, Archives et Bibliothèque de Belgique, Archief- en Bibliotheekwezen in België 83, Bruxelles 2007, 131–160, Figure 1, p. 137.

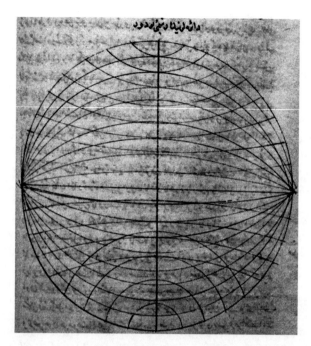

Fig. 1 MS Istanbul, Topkapı Palace Library, R 1622, f 15b.
Courtesy Topkapı Palace Library

Masālik al-abṣār preserved in Istanbul.[7] If one has a closer look on the ways the painter of this particular map uses the grid it seems, however, clear that it was not a mathematical, but rather a painterly auxiliary device. None of the place names is accompanied by a coordinate point, while mountains, lakes and rivers end either on lines or cross-points of the grid.

As in the case of portolan charts, a study of the relationship between the colours on Ottoman maps helps to find information about working techniques. In general, the sequence of colours indicates whether elements such as the equator or lines for the climates were added before or after the world map had been filled with continents, islands and oceans. While in the case of portolan charts, rivers and mountains often seem to have been added and larger areas of colour filled in only after the network of rhumb lines was drawn over the *terra firma*, in the case of the Ottoman maps it is not easy to decide whether the climate lines or the grid were incorporated in the course of the painting procedure or at its very end. A very good example of the latter order of steps can be found in one of the world maps added in the late sixteenth century to the Turkish translation

---

[7] F. Sezgin, *Mathematische Geographie und Kartographie im Islam*, vol. 12, plate 1a.

Patchwork – The Norm of Mapmaking Practices 81

of Meḥmed b. 'Alī Sipāhīzādeh's (d. 997/1589) geography *Awḍaḥ al-masālik ilā ma'rifat al-mamālik*.[8]

Manuscript copies of Ḥajjī Khalīfa's (d. 1067/1657) and Abū Bakr al-Dimashqī's (d. 1102/1691) geographical works possess the same features as Catalan and Italian portolan charts and Arabic, Persian and Ottoman Turkish world maps made between 1100 and 1600. Most of the specimens found in Ḥajjī Khalīfa's and Abū Bakr's extant manuscripts in Istanbul, Paris or Berlin were not constructed with scientific instruments, but drawn freehand. An example is the hemispheric world map in the so-called fine copy of Abū Bakr al-Dimashqī's rendition of the Willem and Joan Blaeu's *Atlas Maior*.[9]

Those drawn with mathematical instruments were not necessarily also constructed, i.e. those who drew them did not use necessarily mathematical knowledge. Examples are the maps of Europe, Asia, Africa and the North Pole in two manuscript copies of Ḥajjī Khalīfa's *Cihān-nümā* which were most likely produced in different workshops. They follow undeniably similar, but different styles and copy different ancestors.[10] The second world map and all continental maps of Topkapı R 1632 show a similar representation of the *terra firma* and the islands as that used in Hamidiyye 988.[11] If not for their straight partitioning lines in form of a simple wind rose, their occasional ornamentation and their names they would be of the same type as the map of Europe in Hamidiye 988 which does not contain a grid or any other kind of division.[12] The rhumb lines in these maps were drawn with a ruler. But not in every case are the angles of each pair of lines indeed rectangular.[13] The frames around the maps of Europe and Asia betray insecurities of the painter since in each map at least one line of the frame is paired with an unfinished parallel suggesting that the painter abandoned an unsuccessful first attempt.[14] Both the rhumb lines and the frames were added only after the pastel coloured contours of the *terra firma* and the islands had been drawn.

Hamidiye 988's map of Africa also betrays the fact that the grid was composed only after the contours of the African landmass had been painted in pinkish colour (Fig. 2). Its rivers were added afterwards and the equator and tropics followed suit. This map is also one of the rare specimens that leave no doubt that at least in this case the islands and hence probably all of the contours were outlined before being coloured.[15]

---

[8]MS Istanbul, Süleymaniye Library, Ismihan Sultan 298, f II.
[9]Ibid., Topkapı Palace Library, B 325, ff 11b–12a.
[10]Ibid., R 1632, ff 13a, 15a, 25a, 30a; Süleymaniye Library, Hamidiye 988, ff 39a, 41a, 42b, 51a.
[11]MS Istanbul, Topkapı Palace Library, R 1632, ff 12a, 13a, 15a, 25a.
[12]MS Istanbul, Süleymaniye Library, Hamidiye 988, f 39a.
[13]MS Istanbul, Topkapı Palace Library, R 1632, f 15a.
[14]Ibid., R 1632, ff 15a, 25a.
[15]MS Istanbul, Süleymaniye Library, Hamidiye 988, f 41a.

Fig. 2 MS Istanbul, Süleymaniye Library, Hamidiye 988, f 41a.
Courtesy Süleymaniye Library

The spot with the two remaining, uncoloured, hence apparently wrongly drawn islands suggests that the grid was added only after these two mistakes occurred. Hamidiyye 988's map of Asia is an obviously unfinished map. It shows a complete map of Asia, but no grid. Neither has it latitude and longitude values in its frame.[16] This is a very strong confirmation that these mathematical elements of the maps were added very late, often only when everything else was finished. The maps of the North Pole suggest that the painter of Hamidiye 988 used one or more auxiliary tools for drawing the various circles and the eighteen straight lines that partition the central circles, since the sectors are of equal size and all

---

[16]Ibid., f 42b.

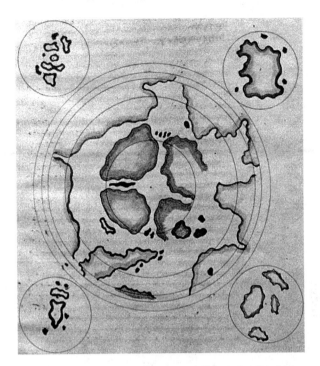

Fig. 3 MS Istanbul, Topkapı Palace Library, R 1632, f 30a.
Courtesy Topkapı Palace Library

lines are drawn without deviations, while the painter of Topkapı R 1632 used a compass for drawing the circles, but omitted the sectors (Fig. 3).[17]

Finally, maps of the same type and illustrating the same text are not necessarily the result of one and the same procedure. Examples are the hemispheric world map in two manuscript copies of Ḥajjī Khalīfa's *Cihān-nümā*. The colour sequence in BnF, Suppl Turc 215 indicates that the grid and all other mathematical curves except the encompassing circles were in all likelihood drawn only at the very end.[18] In contrast, the colour sequence in Süleymaniye, Nuruosmaniye 3275 seems to say that the grid was drawn after the islands, but before the contours of the *terra firma*. The islands were coloured in, while the equator, the tropics and the polar circles were added at the very end.[19] Colours, drawings and artistic style, however, suggest that the two manuscripts originated from

---

[17] MSS Istanbul, Topkapı Palace Library, R 1632, f 30a; Süleymaniye Library, Hamidiye 988, f 51a.
[18] MS Paris, BnF, Supplément Turc 215, f 33a.
[19] MS Istanbul, Süleymaniye Library, Nuruosmaniye 3275, f 26a.

one and the same workshop. These observations mean that the techniques for producing maps as finished products were not standardized yet, not even within one workshop and one artistic style. The implications of such a result for the degree of professionalism achieved in eighteenth-century Istanbul are impossible to ignore. Map production for sale was not in the hand of specialized cartographic experts, but in the hands of manuscript workshops with their calligraphers and miniature painters as I have already argued elsewhere.[20] The fluidity of working steps within one workshop suggests that there was no agreed upon hierarchy of such steps. The visibility of the grid and the other mathematical curves in a finished map was obviously more important than the map's scientific construction and the grid's usage as an auxiliary tool for drawing.

One way to question these observations is to assume that what is visible on the finished products does not provide a direct access to the actually performed process of production. Unfinished, uncoloured maps in Ḥajjī Khalīfa's autographs offer support as well as challenge for my interpretation of the later coloured variants. The hemispheric map, for instance, in Topkapı R 1624 shows unambiguously that the diameters of the circles were drawn only after the contours of the continents had been placed on paper (Fig. 4).[21] It is even possible to speculate that the circles too were drawn only second to the land contours. The map of Asia, on the other hand, leaves no doubt that the grid was drawn first.[22]

Taking all my observations together, I suggest as a conclusion that the process of translating Gerard Mercator's (1512–1594) and Henricus Hondius' (1597–1651) *Atlas Minor* and Willem J. (1571–1638) and Joan Blaeu's (1596–1673) *Atlas Maior* did not incorporate the transfer and subsequent application of the ideal methods of cartographic science, i.e. the usage of a number of scientific instruments, the application of mathematical methods such as projections, the consecutive execution of a grid and the placement of locations according to their geographical coordinates. This does not mean that the knowledge of such methods was not available or was not discussed in the process of translating the Latin books. It means that other methods were applied in the actual process of producing the various kinds of manuscripts: the autographs of the translations, the new texts compiled on the basis of these translations and most of the copies of these new texts. Hence,

---

[20]S. Brentjes, "Mapmaking in Ottoman Istanbul between 1650 and 1750: A Domain of Painters, Calligraphers or Cartographers?", in C. Imber, K. Kiyotaki, R. Murphy, eds., *Frontiers of Ottoman Studies: State, Province and the West*, London, New York: I.B. Tauris, 2005, vol. II, pp. 125–156.
[21]MS Istanbul, Topkapı Palace Library, R 1624, f 21a.
[22]Ibid., f 25a.

Patchwork – The Norm of Mapmaking Practices

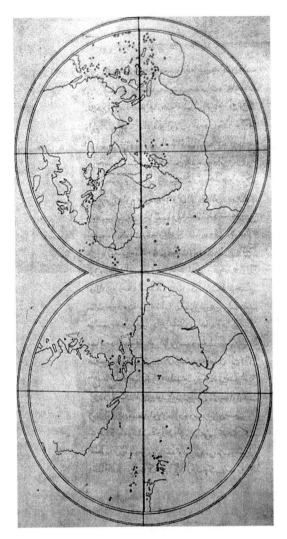

Fig. 4 MS Istanbul, Topkapı Palace Library, R 1624, f 25a.
Courtesy Topkapı Palace Library

neither adoption nor adaptation of mathematical methods occurred in this specific context nor did the translations and their considerable cultural success contribute immediately to a greater standardization of the process of producing maps for sale and better technical and mathematical skills among those who drew and coloured them.

On Methods and Techniques Used for Creating Early Modern Maps in Venice and Istanbul

In addition to the question whether mapmakers produced their maps by applying scientific instruments or cooperated with painters whose education included the training of skills in freehand drawing of straight and curved lines, we need to ask ourselves what we can discover about the methods which the original creator of a specific map had used for determining position and size of the map's individual elements, i.e. places, rivers, mountains, boundaries, coastal lines, lakes and deserts. The study of two maps made by Giacomo Gastaldi (d. 1566) in Venice in 1548 and 1564 (reprints 1566, 1570 and undated exemplars) and their comparison with maps in the translations and editions of Gerard Mercator's and Henricus Hondius' *Atlas Minor* and Willem and Joan Blaeu's *Atlas Maior* in Ḥajjī Khalīfa's and Abū Bakr's manuscripts will demonstrate that the three men whose path to mathematical, cartographic and geographical knowledge differed substantially used a variety of methods for producing their maps, least important among them was the application of mathematical constructions and astronomical observations.

After Ptolemy's *Geography* had been translated into Latin in Florence between 1406 and 1409 and maps were added to the translation in about 1427, at least three processes evolved that aimed to harmonize the newly acquired ancient geographical knowledge with the previously available mix of geographical information and its visual representation. One process consisted in re-introducing ancient geographical terminology into contemporary maps that did not belong to the Ptolemaic tradition, for instance portolan charts. Another process consisted in binding Ptolemy's *Geography* in print editions together with medieval texts. The third process of harmonization introduced new maps into Ptolemy's work itself. These new maps were of two kinds. One kind of maps was non-Ptolemaic in content and form and either of medieval provenance or fifteenth- and sixteenth-century products. The other kind was non-Ptolemaic in form and mixed, Ptolemaic and non-Ptolemaic, in content. Both kinds were presented as modern or novel. This process is usually seen by historians of cartography as a critique at the outdated character of Ptolemy's cartographic and geographical knowledge and as an important step towards modern, mathematically based cartography. This focus on the revolutionary critique overlooks, however, the fact that certain components of Ptolemy's *Geography* shaped substantially the form and content of the new maps. This is the case in particular for maps of western Asia and northern Africa or, as mapmakers of the early modern period labelled the maps, of *Natolia*, *Persia*, *Imperium Turcicum* and *Imperium Persarum* or *Sophi*.

The major figure of early modern mapmaking whose work influenced maps of Anatolia and Iran for more than 200 years was Giacomo Gastaldi from Piedmont. Gastaldi arrived in Venice early in his life where he began as an author of an introduction into astronomy and a builder of one of Venice's new churches. In the 1540s, he started making maps. At first, he mapped parts of Catholic Europe, but soon he also turned to mapping other parts of the world. His maps were very successful. The Venetian Doge ordered him to paint wall maps of Asia, Africa and the New World in his palace integrating the new information collected by the Portuguese. Repainted in the eighteen century, Gastaldi's wall maps are lost. There is no information about how Gastaldi had represented the territories of western Asia and northern Africa.

Parallel to his painting a room in the palace of the Doge, Gastaldi illustrated in 1548 the first Italian translation of Ptolemy's *Geography* made by the physician Pietro Andrea Mattioli (1501–1587).[23] This Italian translation possesses three features that pertain to the question of Gastaldi's methods in mapmaking. The first feature is that Gastaldi added names from other sources to Ptolemy's ancient names of places, mountains, rivers and seas. He was not the originator of this procedure that posed as identifying and modernizing Ptolemy's names. The Latin translation which Mattioli had chosen to translate was not the one produced in Florence in the early fifteenth century, but the one published by Sebastian Münster (1488–1552) in 1540 in Basel or one of its subsequent editions. Münster had taken over most of the contemporary as well as ancient names added to those given by Ptolemy, mostly chosen from ancient geographers, medieval portolan charts, the Bible and some Jewish sources, that Willibald Pirckheimer (1470–1530) in his fresh translation of Ptolemy's Greek text published in Strassburg in 1525 and Michael Villanuova, alias Servetus (ca. 1511–1553) in his edition made in 1535 in Lyon had joined to the text in order to correct, elucidate and modernize it.[24]

---

[23]Ptolemeo. *La geografia di Clavdio Ptolemeo Alessandrino, Con alcuni comenti & aggiunte fatteui da Sebastiano munstero Alamanno, Con le tauole non solamente antiche & moderne solite di sta<m>parsi, ma alter nuoue aggiunteui di Meser Iacopo Gastaldo Piamo<n>tese cosmographo, ridotta in uolgare Italiano da M. Pietor Andrea Mattiolo Senese medico Eccelle<n>tißimo. Con l'aggivnta d'infinit nomi moderni, die Città, Prouincie, Castella, et altri luoghi, fatta co<n> grandissima diligenza da esso Meser Iacopo Gastaldo, il che in nissun altro Ptolemeo si ritroua. Opera ueramente non meno utile che necessaria. In Venetia, per Gioa<n> Baptista Pedrezano. Co'l priuilegio dell"Illustriß. Senato Veneto per Anni.x. M.D.XLVIII.*

[24]S. Münster, *Geographia universalis, vetus et nova, complectens.* Basel 1540. See http://www.maphist.com/artman/publish/munster_1540.html, Map 17 and Claudius Ptolemaeus. S. Münster, ed., *Geographia.* Basle 1540. With an Introduction by R. A. Skelton. Theatrum Orbis Terrarum Ltd., Amsterdam MXMLXVI. Third series – volume V, p. vii; Ptolemaeus, *Geographicae enarrationis libri octo.* B. Pirckheymero interprete. Annotationes Ioannis de Regio Monte in errores commissos a Jac. Angelo in translatione sua. Straßburg, J. Grüninger for J. Koberger, 30 March 1525.

According to archival documents in Venice, Gastaldi tried to form in 1550 a partnership with one of the Signoria's interpreters for Oriental languages, Michele Membré (1505–1595), in order to produce a new map of Asia.[25] Producing a list of geographical names that equated ancient with contemporary knowledge was meant to be part of this venture. No list or map of this 1550-venture has, however, survived. The partnership indicates one method that Gastaldi considered particularly valuable for his approach to generate modern maps – the acquisition of contemporary local knowledge via oral sources.

When we study the new maps of Asia that Gastaldi drew since 1555, we discover that this approach must have been indeed a central method used by the cosmographer of the Signoria. His map of Anatolia made in 1564 is densely covered by names, many of which have not only a distinct Turkish flavour such as *Isnich*, *Acsar*, *Mugla*, *Culeisar* and *Mentese*, but reflect indeed Turkish geographical and political features of Anatolia at his time. When we compare this map with Gastaldi's first printed map of Anatolia, added to the Italian translation of Ptolemy's *Geography*, we find a substantial change, both in terms of quantity and quality of the inscribed names (Fig. 5). The map of 1548 also appears densely covered, but in comparative terms it is mountains that rule this map, not names as in the 1566-reprint of the 1564-map (Fig. 6). The names displayed on the 1548-map come predominantly from ancient geography. Additionally, the map contains elements from Christian church history, portolan charts, Venetian or Genoese trade and possibly western travel accounts. These names as well as some of the ancient names appear in Venetian and other vernacular dialects or transcribe Arabic and Turkish words in forms found in earlier portolan charts or slightly later world maps such as those produced by Fra Mauro (d. 1460) and Andrea Bianco (fifteenth century) in the San Michele monastery at Murano. The names on the 1564-map, however, are an intriguing mix of Ptolemaic and other ancient names, names taken from letters and reports of Venetian merchants and envoys to various Muslim rulers and names of undeniably Ottoman Turkish provenance. They are more homogeneous in their linguistic appearance and political outlook than those on the map of 1548. Latin has disappeared completely. Italian forms are used for seas, gulfs and other elements of physical geography. The territory of Anatolia is now structured by Turkish dynasties and local tribes.

---

[25] R. Almagià, *Monumenta cartographica Vaticana*. Città del Vaticano, Biblioteca Apostolica Vaticana, vol. II, 1948, p. 34 (Tavola VIII); G. Bellingeri, "Un prospetto geografico di Michele Membrè (1581)," in *"Turcica et Islamica" Napoli Studi in memoria di Aldo Gallotta*, a cura di U. Marazzi; Venezia, Università L'Orientale, 2003, pp. 15–36.

Patchwork – The Norm of Mapmaking Practices

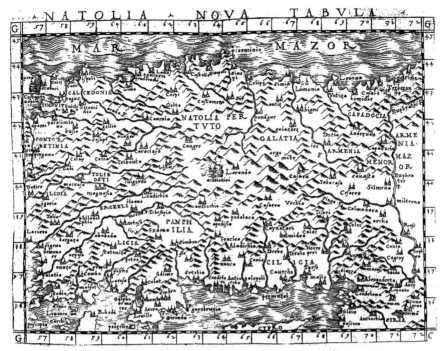

Fig. 5 Giacomo Gastaldi, Map of Anatolia, 1548.
Courtesy The Map Collection, Harvard University

Despite spelling errors and Italianized transformations, the transliteration of most names in the 1564-map and their later copies is of an extraordinarily high quality. The relationship between the Turkish names and their Italian transliteration in Gastaldi's map surpasses all forms of transliterated Turkish, Arabic and Persian geographical names found in travel accounts, letters and reports by Venetian diplomats, merchants, gentlemen travellers, soldiers and adventurers of the period. This also holds true for transliterations in travel accounts, letters and reports by travellers from other Catholic and Protestant nations or towns in Europe during the sixteenth and seventeenth centuries. The source of such an excellent linguistic understanding must be sought in a person with proper understanding of Italian, Turkish and perhaps even Arabic. None of the letters or accounts written by Venetian envoys or interpreters shows this high quality of transliteration. None of the Ottoman Turkish *futuhnames* (letters sent to announce a conquest)

Table 1 Turkish names in Gastaldi's map of Anatolia of 1564 (reprint 1566)

1564 map of Anatolia (reprint 1566)	Identification
Aldinelli reg.	Aydın-eli
Germian reg.	Germiyan
Sacrvm reg.	Saruhan
Pegian reg.	Besyan (Kurdish tribe)

extant in the Venetian State Archives contains enough names to be Gastaldi's source nor does the transliteration of these documents come close to the quality of Gastaldi's names. Hence the most likely conclusion is that Gastaldi cooperated with a highly educated and skilled multilingual visitor of Venice, perhaps a former prisoner of war or one of the many merchants from the Ottoman and Safavid empires.[26] One such informant was Muhammad Khwājā, a merchant from Tabas in Gilan.[27] But while he probably was a good source for Iran and some regions east of it which he knew thanks to his trading business he may not have been equally well informed about the towns, villages and rivers of Anatolia. Above all he may not have mastered Ottoman Turkish and Italian well enough to serve as Gastaldi's only source. The difference in transliteration between the latter's map of Anatolia and his depiction of Iran in his new map of *Asia* (First Part) made in 1559 strengthen the assumption that Gastaldi cooperated with more than one local informant.

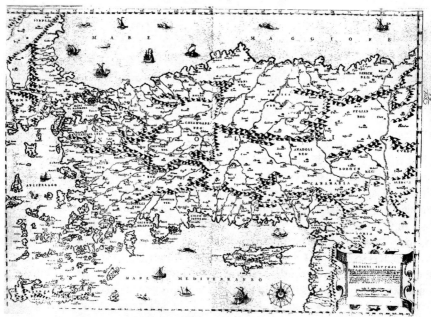

Fig. 6 Giacomo Gastaldi, Map of Anatolia, 1564 (copy 1566). Courtesy The Map Collection, Harvard University

---

[26] For a more detailed discussion of Gastaldi's maps of Anatolia and the problems involved in their interpretation see S. Brentjes, "Giacomo Gastaldi's maps of Anatolia: The evolution of a shared Venetian-Ottoman cultural space?" in C. Norton, A. Chong, eds., *Cultural Encounters: Europe, the Ottomans, and the Mediterranean World*, Boston: Periscope Press and the Isabella Stewart Gardner Museum, forthcoming.

[27] *Navigationi et Viaggi*, Venice 1563–1606, R. A. Skelton, George B. Parks, eds., Second Volume, Amsterdam: Theatrvm Orbis Terrarvm Ltd., 1968, f 16b.

The second feature that is important to the quest for Gastaldi's methods is the material shape of Anatolia and its placement within the set of geographical coordinates. Gastaldi used at least two different kinds of overall contours, one for the first map of Asia within the Ptolemaic text and one for Anatolia with only minor differences between the 1548-map and the copies of the 1564-map. The first contour shows clear similarities with the first map of Asia printed in the edition of Ptolemy's *Geography* in 1507 (Rome 1507). The second contour Gastaldi appropriated most likely from portolan charts.[28] A configuration close to the 1566-reprint of his 1564-map appears in portolan charts in the second half of the fifteenth century, differing slightly from earlier forms. When compared to a current depiction of Anatolia's shape, Benincasa's form seems to be slightly closer to physical reality than the Ptolemaic form of 1507. This does not mean, however, that Benincasa, Gastaldi or their shared source(s) used necessarily astronomical observations. Even the use of tables of geographical coordinates does not seem to have ranked high among Gastaldi's working methods, at least as far as western Asia is concerned. The longitudinal and latitudinal framing of *Natolia* and *Caramania*, for instance, has extended westwards and eastwards by approximately 1° in each direction and northwards and southwards by approximately 2° in each direction in the 1566-copy of his 1564-map of Anatolia in comparison to his 1548-map. This extension caused some surprising results. Greek place names taken over from the 1548-map often moved 1° to 2° degrees either in latitude or in longitude. *Laranda*, for instance, has moved almost 2° to the south. It is now situated at the beginning of a river not present in the earlier map. A similar observation can be made with regard to the newly incorporated knowledge of Ottoman Turkish provenance. Almost all names of provinces are placed too far to the east and in a number of cases to the south. The provincial capitals are often situated outside the true territorial borders. The relative position of towns is rarely correct and occasionally even totally wrong as in the case of *Gerede* and *Bolli*. In a few exceptional cases, such relative positions come close to reality as in the case of *Sinopi* (Sinop), *Osmangiuch* (Osmancık) and *Amasia*. Gastaldi obviously had no clear understanding of the physical geography of Anatolia and no qualified help for checking errors. Moreover, Gastaldi apparently worked at least with two sets of sources: one for the Turkish names of towns, villages and rivers and their relative positions, another one for the names of provinces, territories and people.

The third feature of relevance to the analysis of Gastaldi's working methods is the presence of mountains and rivers in his two maps. In the 1548-map, almost the entire territory of Anatolia is covered by mountains and all four boundaries of Anatolia are marked by several larger and smaller rivers in fairly regular distances from each other. The purpose of this generous

---

[28]Almagià suggested already in 1948 that Gastaldi had followed portolan charts when drawing the contours of Anatolia. Almagià, *Monumenta*, vol. II, p. 34.

endowment of Anatolia with physical features that do not exist in reality may have been to set this *Natolia Nova Tabvla* apart from his version of Ptolemy's *Tabvla Asiae I*. This map, namely, has much less mountains and rivers. In the 1566-copy of the 1564-map, the regularity that marked the *Nova Tabvla* of 1548 is gone and the river system has become more complex, covering also the interior. As a look into any current map of Anatolia teaches us, this change is not an expression of Gastaldi's better understanding of Anatolia's physical features. Hence, the reasons for this change need to be sought elsewhere, probably in aesthetic principles that determined what was permissible and beautiful.

Summarizing the findings I wish to emphasize that Gastaldi worked with written and oral sources, mostly with regard to names. He did not, however, utilize the most easily available prints such as the collection of travel accounts compiled by his friend Giovanni Battista Ramusio (1485–1557) in the 1550s and 60s, but rather privileged prints from the first half of the sixteenth century. This may indicate that he started to work on his map of Anatolia some time before 1550. He paid little attention to geographical coordinates available in scholarly works and manuscript collections. He chose outlines of Anatolia in accordance with the type of map he wished to produce, i.e. for mapping the ancient territories in Ptolemy's *Geography* he adopted a map from a Latin print edition of the work, while for the so-called new maps he preferred portolan chart patterns. Gastaldi and perhaps his patrons and customers apparently considered visual knowledge enshrined in portolan charts, not Ptolemaic maps as more fitting for contemporary cartographic purposes. New verbal knowledge, on the other hand, was appropriated predominantly from contemporary local sources from the mapped region, whether written or oral. Gastaldi had no problems fusing this new verbal knowledge with knowledge taken from several other types of written sources from his own culture, whether ancient, medieval or almost contemporary. As for the physical features of interior Anatolia, in their totality they are neither taken from Ptolemy's work nor from any other specific source. Most of them do not exist in reality. They reflect possibly a perception of visitors according to which the parts of Anatolia they had seen were covered by high mountains and hills and nourished by numerous streams and rivers. Gastaldi may have taken this perception as valid for all of Anatolia. Aesthetic considerations were possibly an important additional factor that governed the shifts and changes from one map to the other.

Perusing the maps of Anatolia produced by subsequent generations of mapmakers in Catholic and Protestant Europe until the second half of the seventeenth century indicates that not much changed in these 100 to 150 years. It is unknown when the first maps of Anatolia with a mathematical grid were produced in Catholic or Protestant Europe. The first with a complete grid known to me is the map in the *Atlas Maior* which is a copy of Willem Blaeu's map produced in 1635. When we compare this map with those by Abraham Ortelius (1527–1598) made in 1570 and Gerard Mercator/Henricus Hondius printed in 1609, we see that no substantial shifts took place between these three maps and that of their ancestor, one of the reprints of Gastaldi's map of 1564. The only major differences are the map's overall position, i.e. vertical as with Ortelius or horizontal as with the other mapmakers, the range of the coordinates and the presence of mountains and the Euphrates or their absence. Ortelius and Mercator/Hondius obviously disliked mountains. None of Gastaldi's four successors discussed here believed overly much in the need for accuracy and the power of geographical coordinates, since all of them vary the numerical frame of their maps between $1°$ to $10°$ for portraying the same physical space. Mercator and Hondius omitted the Euphrates despite of its physical presence and size, its historical importance and the fact that it's geographical coordinates in Gastaldi's and Ortelius' maps were part of Mercator's and Hondius' numerical frame. All this ascribes minor relevance to values such as precision, accuracy, reliability and truth. The most important goal apparently was to enable the viewer to recognize the mapped territory through shape, naming and decorative features.

Ḥajjī Khalīfa's and Abū Bakr's Adaptations of Mercator/Hondius' and Willem and Joan Blaeu's Maps

Ḥajjī Khalīfa, as is well known, translated together with Mehmet Ikhlāṣī, a convert of probably French origin, Henricus Hondius' reprint of Gerard Mercator's *Atlas Minor* as well as extracts from Philipp Clüver's (1580–1622) *Introductio in universam geographicam tam veteram quam novam* (Leiden, 1624) and Abraham Ortelius' *Theatrum Orbis Terrarum* (Antverp, 1570 and following years). This translation, called *Levāmiʿ al-nūr*, is extant in what Hagen described as an autograph copy in Süleymaniye, Nuruosmaniye 2998.[29] Ḥajjī Khalīfa used this translation as a basis for the second, substantially reworked version of his geographical masterpiece *Cihān-nümā*. An autograph copy is extant in Topkapı R 1624.[30] According to Hagen, nine copies of the *Levāmiʿ* exist in libraries in Istanbul. In a footnote, however, he repeats Taeschner's warning formulated in 1926 that it is often unclear whether the manuscript contains indeed a copy of the *Levāmiʿ* or rather one of *Cihān-nümā* 2.[31] Of the *Cihān-nümā* (version 2), Hagen enlists 23 complete copies, four fragments, seven doubtful copies and one unclear version.[32] The difference between the maps in those of the later copies that are illustrated and the autograph of version 2 is substantial with regard to the finesse of their execution and the type of artistic

---

[29]G. Hagen, *Ein osmanischer Geograph bei der Arbeit. Entstehung und Gedankenwelt von Kātib Čelebis Čihānnümā*, Berlin: Klaus Schwarz Verlag, 2003, p. 186.
[30]Ibid., p. 187.
[31]Ibid., p. 186, footnote 10.
[32]Ibid., pp. 421–426.

design. The maps in the autograph are crude, black freehand drawings, often unfinished, with ochre colouring of coastal lines and red coloured titles, islands and names of seas. The maps in later copies range from elegant, pastel coloured, miniature-like images over darker coloured and less finely painted specimens to carelessly executed exemplars. Several late copies also contain printed maps that were hand coloured, possibly by the buyer.

We have even less precise information about the relationship between the various copies and paraphrases of Abū Bakr's translation of Willem and Joan Blaeu's *Atlas Maior*. There are a six-volume, a nine-volume, a two-volume and at least two one-volume versions. The six-volume version is seen by Halasi-Kun and following him Hagen as the original translation, while the nine-volume edition, prepared for Sultan Mehmed IV, is seen by Ihsanoglu and others as the original translation.[33] The shorter versions pose as a rule as epitomes. One of the two one-volume versions is an Arabic summary of Abū Bakr's views on geography. None of these versions has been systematically analysed so far, neither their texts nor their maps. Halasi-Kun, in his study of the map of *Yeni Felemenk maa Ingiliz*, has shown that this map does not come from the *Atlas Maior*.[34] This is an important insight in the working style of Abū Bakr and his collaborators. Maps were apparently not seen as components that were untouchable and immovable.

As for the maps of the Ottoman and Safavid empires and their parts, the manuscripts in Istanbul which I had access to, i.e. those in Topkapı Palace and in the Süleymaniye Library, show a fairly stable style in respect to design. They follow principally their source, i.e. Willem and Joan Blaeu's *Atlas Maior*. In regard to artistic execution, two major tastes seem to have governed the workshops of Istanbul, one that chose to colour the contours only; the other preferred fully coloured images. The first style worked either with bright colours, gold and silver or with light colours.[35] The second focused on bright and often darker tones.[36] Mixing styles was by no means frowned upon, whether in terms of colouring or when joining maps of the Abū Bakr tradition to manuscripts of Ḥājjī Khalīfa or vice versa.[37] The examples suggest that the differently painted maps were executed by different painters or perhaps even

---

[33] Ibid., p. 257, footnote 148; G. J. Halasi-Kun, "The Map of "şekl-i yeni felemenk maa ingiliz" in Ebubekir Dimişki' "tercüme-i atlas mayor"," *Archivum Ottomanicum* 11 (1986), 51–70, p. 53; E. İhsanoğlu, 'The Introduction of Western Science to the Ottoman World: A Case Study of Modern Astronomy (1660–1860),' *Science, Technology and Learning in the Ottoman Empire. Western Influence, Local Institutions, and the Transfer of Knowledge*, Ashgate, Variorum CS773, 2004, II, p. 11.

[34] Halasi-Kun, op.cit., pp. 51–70.

[35] MSS Istanbul, Süleymaniye Library, Köprülü 173, Nuruosmaniye 2995; Topkapı Palace Library, R 1629, R 1634.

[36] MS Istanbul, Topkapı Palace Library, R 1636.

[37] MSS Istanbul, Süleymaniye Library, Nuruosmaniye 2995; Topkapı Palace Library, R 1629, R 1634.

Patchwork – The Norm of Mapmaking Practices 95

bought in another workshop. In these two ways – introducing maps from different western sources and mixing styles –, patchwork or eclecticism is a visible aspect of the working practice of workshops that produced maps for sale in eighteenth-century Istanbul. It remains unclear, however, who defined the choices – the producer or the customer.

Finally, the question needs to be addressed which other ways for joining elements belonging to different traditions, sources, techniques or tastes were followed by scholars or painters. Halasi-Kun's observation that maps from a different source could be joined even to the fine copy of Abū Bakr's work dedicated to the sultan can be confirmed for eighteenth-century copies of Ḥajjī Khalīfa's as well as Abū Bakr's works. Within a few years after Guillaume Delisle (1675–1726) had published in 1721 and 1723 new maps of the Caspian Sea derived from a map sent by Peter I from Moscow to Paris for the Academy of the Sciences to evaluate its novelty and reliability, manuscripts of Ḥajjī Khalīfa's *Cihān-nümā* (version 2) included a new version of the unfinished and rather crude, but remarkable map of the lake in Ḥajjī Khalīfa's autograph, remarkable because it deviated substantially from any of its forms available to Ḥajjī Khalīfa and Mehmet Ikhlāṣī in the Latin books they worked with. This new version reflects closely the fundamental shift achieved in the new Russian and French maps.[38] The fact that this was indeed a cartographic innovation was not lost to Istanbul's scholars as indicated in the collection of single-sheet maps preserved in Süleymaniye, Atıf Efendi of which I unfortunately do not own a copy: *resm-e jedīd-e baḥr-e khazar*.[39] This collection was assembled in 1141/1728, i.e. a year before BnF, Supplement Turque 215 of Ḥajjī Khalīfa's *Cihān-nümā* (version 2) with its map of the Caspian Sea was produced. Its first owner was the scholar Atıf Efendi who founded 14 years later, i.e. in 1154/1742, the *madrasa* with its attached library where the collection was preserved. The scholars and mapmakers did not, however, learn of this new form of the Caspian Sea directly from Delisle's maps. A look on the map of the *Persian Empire* by the German mapmaker Baptist Homann (1664–1724) indicates that it may have been his slightly modified form of the Caspian Sea that came to the attention of the workshops in Istanbul as well as Ibrāhīm Müteferriqa (d. 1157/1744) when he printed the map of Iran in 1142/1729.[40]

The relationship between several manuscripts of Ḥajjī Khalīfa's *Cihān-nümā* (version 2) and Müteferriqa's print of 1145/1732 deserves more attention than it has found so far. The maps in Müteferriqa's print constitute a mixture of maps of the *Cihān-nümā* (version 2) tradition in the style of the three manuscripts produced around 1141–1142/1728–1729 in Istanbul and maps of three

---

[38] Guillaume Delisle 1722, London, British Library,; MSS Istanbul, Süleymaniye Library, Hamidiye 988, f 236b; Nuruosmaniye 3275, f 260a; Nuruosmaniye 2996, inserted after f 274b.
[39] MS Istanbul, Süleymaniye Library, Atıf Efendi 1693, fourth map.
[40] Joann Baptist Homann 1723, *Imperii Persici... Tabula Geographica...*, The Harvard Map Collection, Mt 18.50 (113).

manuscripts of the Abu Bakr tradition.[41] A detailed comparison indicates, however, that while the relationships in content and style are very close between those maps taken from the *Cihān-nümā* (version 2) and the three manuscripts Hamidiye 988, Nuruosmaniye 3275 and BnF Supplément Turc 215, none of them was directly copied by the men working for Ibrahim Müteferriqa. Occasional deviations in naming, borderline drawing and the placement of mountain chains are the most visible differences. The same applies to maps closely related in content and style to the three manuscripts Nuruosmaniye 2996, Köprülü 173 and Topkapı R 1634. The colours used for lakes, borderlines and contours in the two sets of manuscripts were also used in some of the printed copies available for this comparison. But as in the case of borders, names and mountains they were not left unchanged. The silver preferred in Köprülü 173 for colouring lakes and occasionally rivers found the approval of the people who hand-coloured maps of Iran in the printed version, while the maps of the Arabian Peninsula and Anatolia show a preference for the pastel colours of Topkapı R 1634. The colour scheme of bright yellow, pink, blue or green of Hamidiye 988 and BnF Supplément Turc 215 was supplemented by a washed-out grey and blue. On the level of technical skills, the maps in Müteferriqa's print also show clear parallels to the manuscripts. In the maps of Asia, Africa and America certain lines of the grids are not continuous. They either miss their end point or consist of two lines that run for a short while parallel to each other. The engraver's hand may have wavered or the ruler he used may have moved. Other details such as the spaces left for writing names in the map of Africa confirm that the grid was added to the map as the very last element.

These observations as well as the substantial similarity of the mapped territories, in particular their contours, between the maps in Müteferriqa's print and those in the manuscripts suggest that the maps for Müteferriqa's book were produced by the same kind of professionals, i.e. painters, who produced the maps in the manuscripts. The introduction of printing did not change the professional landscape of map production in early eighteenth-century Istanbul. Müteferriqa and his collaborators rather appropriated the skills, tastes and visual styles of two (or three) workshops, combined their products within the printed book and took the liberty to alter some of their visual and rhetorical features without, though, thoroughly unifying them. This kind of eclecticism may have contributed to the introduction of maps from the workshop where Hamidiye 988 and BnF Supplément Turc 215 were produced into manuscripts of the Abū Bakr tradition.

The scale of patchwork in Müteferriqa's print is not exhausted by the features listed so far. It also included printing contradictory maps that showed, for instance, different forms of the Caspian Sea (Ptolemaic and the new form given in variants by van Verden, Deslisle and Homann), of

---

[41] MSS Istanbul, Süleymaniye Library, Hamidiye 988, Nuruosmaniye 3275; Paris, BnF, Supplément Turc 215; Istanbul, Süleymaniye Library, Köprülü 173, Nuruosmaniye 2996; Topkapı Palace Library, R 1634.

California (as a peninsula and as an island) and of the east coast of North America. Another form of syncretism, also found in several manuscripts, concerned the choice of cartographic representation. Some maps present a grid and latitude as well as longitude coordinates. Some maps present a scale. Several maps have a compass rose with or without rhumb lines and possess in the same time a frame with latitude, but no longitude values. These conflicting elements underline the fact that the printed book was produced to a substantial degree by the same kind of practices as the manuscripts. It did not involve the application of new mathematically based techniques nor did it undergo a more rigorous control of the geographical content its maps meant to teach and illustrate.

A further variant of eclecticism in the manuscript maps can be discovered when we study major singular features of individual maps such as their central meridian or the position and form of certain geographical units in East and South East Asia, Northern Europe or the Americas, for instance *Terra Australis*, Japan, Korea, Greenland or the imaginary island of *Friesland*. The central meridian in the map of Europe in Nuruosmaniye 2998 (*Levāmi'*, autograph) and in Topkapı R 1624 (*Cihān-nümā* (version 2), autograph), for example, is marked as 50°, while the same map in Topkapı R 1622 (copy *Cihān-nümā* (version 2)) gives 40° for this meridian.[42] Comparing this difference with Gerard Mercator's *Atlas* published in 1595 and Hondius' 1621-edition of the *Atlas Minor* documents that the map of Europe in the atlas of 1595 also has its central meridian at 50°, while the map of Europe in the 1621-edition of the *Atlas Minor* has its central meridian at 40°. Without going into more details, what appears to be probable is that the producers of the individual manuscripts did not always work from a single ancestor, but used several western and Ottoman sources for their work.

Another form of eclecticism appears in copies of maps of Abū Bakr's work. Here, transliterations of names from Latin, Spanish, other Indo-European languages and indigenous names from the Americas and the Pacific region can be found side by side with Ottoman Turkish and Arabic geographical nomenclature as well as with mixed forms.

In certain manuscripts of Abū Bakr's epitomes a change in naming can be observed. Territories which were designated earlier by either *saltanat* or *mamlakat* are now also called *iqlim* such as *Iqlim-e Çin, Iqlim-e Ma wara' al-nahr* or *Iqlim-e Iran*. Other copies of the same maps continue, however, using the older terminology. Hence, this process of change took place within the workshop or in agreement with the customer, not, however, as a process of scholarly adaptation. In general, in the maps illustrating Abū Bakr's Turkish versions of the *Atlas Maior* adoption and adaptation of geographical terminology seem to

---

[42] MSS Istanbul, Süleymaniye Library, Nuruosmaniye 2998, f 17b; Topkapı Palace Library R 1624, f 32a; R 1622, f 41a. The map of Europe in Topkapı R 1622, however, was affixed later to the manuscript and pasted over the original map which is now invisible except for some shadowy contours on the back of the folio.

have taken place concurrently as well as in minor subsequent steps of change. The flexibility of terminology indicates a lack of standardized disciplinary language to which producers and customers alike felt bound to adhere.

The maps in Abū Bakr's Arabic synopsis of modern geography are predominantly bilingual, i.e. use Arabic and Ottoman Turkish terminology, in those regions where it was appropriate and possible. In territories for which no Arabic or Ottoman Turkish terminology was available, Abū Bakr continued to work with transliterations of names found in his western sources. The most important feature in comparison with the maps in the Turkish manuscripts of his works is the undeniable effort he made to assemble a detailed, contemporary set of names for regions and cities in regions outside the classical Islamic world. This effect is particularly visible for South, East and South East Asia.[43] The effort to modernize focused on terminology. As for the positioning of the names, Abū Bakr did not invest a comparable care. While coordinates are inscribed in the margins of the map, the grid is largely incomplete. Moreover,

Table 2 Types of geographical names in maps of Abu Bakr's epitomes

Name	Indo European	Turkish/ Arabic	Indigenous	Mixed
Hibernia	x			
Dania	x			
Litu'aniya	x			
Dalmasiya	x			
Nova Zemla	x			
Jaza'ir al-khalidat		x		
Bahr-e atlas		x		
Qubrus		x		
Anadole		x		
Qara Deniz		x		
Bahr-e Faris		x		
Qalmuq		x		
Yuqatan			x	
Yeni Holland				x
Bahr-e daqaliduniya (transliterates: Deucaledonius)				x
duqiyanus-e shamal (error for Oceanus)				x
Java saghir				x
Nurd deryasi				x
Zur deryasi				x

---

[43]MS Istanbul, Süleymaniye Library, Köprülü 176, f 28b.

several names are visibly misplaced. This also applies to larger geographical units such as islands in Abū Bakr's map of Europe.[44] Hence, precision with regard to location was not Abū Bakr's aim, while modernity and accuracy with regard to naming ranked high in his priorities.

As for the physical and political properties displayed in the maps, similar observations can be made. While the general scheme of contours, boundaries, geographical symbolism and distribution of physical units did not change significantly between the fine copy made for Sultan Mehmet IV (1642–1693) and the various manuscripts of the Turkish epitomes, there are numerous tiny changes in detail. Mountains, islands, rivers and boundaries obviously were not yet stabilized in the views of Ottoman Turkish mapmakers and their clients. Much more research is needed, however, in order to establish whether this flexibility results from intentions to change or reflects rather properties of handicraft. Since the maps made in the seventeenth and eighteenth centuries in various Catholic and Protestant countries of Europe seem to possess similar properties as far as terminology, contours, boundaries, mountains, rivers and other geographical units in Asia are concerned, comparison with newly arrived maps and atlases may have contributed to this relative instability in maps of copies of Abū Bakr's works.

Conclusion

The analysis of maps in manuscript copies of Ḥajjī Khalīfa's and Abū Bakr b. Bahram al-Dimashqī's translations and subsequent geographical works shows a confusing manifold of differences in detail on several levels – techniques, coordinates, terminology, units of physical or political geography and dependence on securely definable ancestors. This manifold indicates that processes of adoption and adaptation did not lead to a stable standard of mapmaking in Istanbul in the second half of the seventeenth and during the eighteenth centuries. A closer study of the differences reveals that most of them are best understood when we consider eclecticism as the main work ethic that producers of manuscript maps subscribed to in this period. In this sense, Ottoman scholars and craftsmen did not deviate substantially from the approach that dominated map making in Catholic and Protestant Europe between the 1550s and 1750s, at least with regard to maps of western Asia, Anatolia and Persia.

The analysis of the various acts of adoption (translation, integration into new compilations, production of epitomes and summaries, inclusion of translations of new maps from Catholic and Protestant Europe, reproductions of translations, new compilations, epitomes and summaries) and adaptation (change of what is mapped, decisions on how regions are portrayed in a map, alterations of contours, geographical content and names, inclusion of

---

[44]Ibid., f 30a.

translations of new maps from Catholic and Protestant Europe) uncovers, on the other hand, a substantive difference between the atlases printed in Italy, France, The Spanish Netherlands, The United Provinces, Germany or England and the geographical manuscripts adorned with maps produced in workshops of Ottoman Istanbul. In Catholic and Protestant countries the production of printed maps was a complex procedure that began with a cosmographer or geographer and ended with the owner of a publishing house. In several cases the cosmographer or geographer published his own maps either alone or in cooperation with an already established owner of a print shop. Many of the leading cosmographers or geographers of the sixteenth and seventeenth centuries produced their maps together with other male relatives, mostly sons, sons-in-law and nephews. Owners of print shops also relied on the work done by their family members, both male and female. But they also paid engravers, instrument makers and other craftsmen for producing new variants of established maps and for modifying acquired copper plates of maps. The production process of printed maps thus included several groups of experts, some of whom specialized in mathematical methods for creating a map, while others had learned the skills of a painter, a copper plate engraver, a wood cutter and a type setter.

In the map producing workshops of Istanbul only one of these groups worked – craftsmen with particular skills in painting and writing. They collaborated with the workshop owner/s in completing the maps of the autographs and altering them into the highly sophisticated art works produced in some workshops and to the cheap, carelessly made and unskilfully coloured specimens produced in others. They compared them with other maps of European provenance available in Istanbul and adapted their archetypes to such newly acquired objects as well as to Ottoman geographical terminology. Mastering mathematical projections and other constructions, though, did not belong to their standard repertoire. Hence they did not develop standards of technicality based in the mathematical sciences for producing and reproducing maps.

The adoption and adaptation of the *Atlas Minor* and *Atlas Maior* into the culture of the Ottoman capital did not lead to a spread of mathematical and technical skills among those who participated most actively in these processes – the manuscript workshops and their craftsmen. The maps rather testify to the domination of these processes by Ottoman standards of producing illuminated literary and historical manuscripts. Local skills and tastes as well as the monetary means of the customers and workshop owners determined the overarching framework into which the maps and the knowledge they represented were integrated – the arts of the eye and the ear, i.e. images and stories. Only one foreign framework made the transition into the host culture successfully – the geographical knowledge and its verbal presentation. Foreign artistic decoration was not even included in the translation process, i.e. was no part of the activities that led to the maps' adoption among Ottoman producers and consumers of maps. The artistic character of the Ottoman copies of these maps was the result of the activities of adaptation undertaken in the workshops and dominated by the available local standards, skills, tastes and means. Foreign scientific and instrumental skills, while acknowledged and rhetorically appreciated as the main reasons for Christian scientific progress and military as well as commercial success in Asia, did not pass beyond the level of the word. They remained a stable part of the texts and their glosses, but only exceptionally made their way into the reproduction of the maps appropriated from Latin, French or other European sources.

The maps joined to the autographs, in particular of Ḥajjī Khalīfa's and Mehmet Ikhlāṣī's works, were unfinished sketches that either were mostly empty or contained numerous unintelligible transliterations. These features set too loose boundaries for mathematical and instrumental skills to emerge in the process of their reproduction as needed, indispensable elements for map reproductions. They also forced those who wished to reproduce them for sale to complete their contours and content and thus engendered the various options for adaptation that were taken up in Istanbul's workshops.

The Ottoman Ambassador's Curiosity Coffer: Eclipse Prediction with De La Hire's "Machine" Crafted by Bion of Paris

Feza Günergun

The relations between Europe and the Ottoman Empire founded by Turks of Central Asian origin have been subject to numerous studies focusing mainly on military conquests or political negotiations. While economical and social aspects of this centuries-long relation started to be discussed some decades ago, the historiography of cultural interactions has mostly been directed to the mutual impact in arts and literature. Relations in the field of technology and science have come to be studied rather recently and mainly within the modernization of Ottoman educational institutions and the renovation of the Ottoman military. Histories of the military schools of engineering and medicine, biographies of their staff were among the earliest works published by researchers in history of science and medicine. These histories mostly emphasize the initiatives taken by the Ottoman administration to transmit European technical and scientific knowledge through newly created institutions and by recruiting European experts who were instrumental in conveying new knowledge even after Ottomans set sail to Europe to be trained in European universities or military plants. Few studies deal with individual efforts deployed in the transmission process or discuss the encounter between the newly transmitted knowledge and the local practices. The present paper focuses on an eclipse calculator brought from Paris to Istanbul in the eighteenth century and on the translation of its manual to Ottoman Turkish. The additional explanations offered by the translator are witnesses to the calculations he made in order to adapt this instrument for predicting the date of eclipses to the Islamic calendar in use in Ottoman society.

Diplomats Pave the Way

The expanding circulation of diplomatic envoys between the Ottoman and European capitals during the eighteenth century was considerably influential in the introduction of new scientific and technical knowledge and practices into

F. Günergun (✉)
Department of the History of Science, Faculty of Letters, Istanbul University,
Istanbul, Turkey
e-mail: gunerfez@istanbul.edu.tr

Turkey. Books, atlases and instruments were either presented as gifts to the Ottoman court by European delegations, or they were purchased in Europe by Ottoman diplomats who brought them back to Istanbul. While some books likely remained untouched, others were translated into Ottoman Turkish and the knowledge they contained could be disseminated, albeit on a limited scale. Crucial in the transmission of new knowledge and practices, the translation of European books that had begun in the seventeenth century was carried on in eighteenth century due to the collective work of Ottoman scholars, bureaucrats, interpreters, and European residents in the Empire.[1]

The twelve month tenure (1720–1721) of the embassy of Yirmisekiz Mehmed Çelebi (Chelebi, d.1732) in France was crucial for the introduction not only of French cultural life but also of scientific institutions and industrial workshops to the Ottoman palace and elite class. The ambassador's report[2] describing his voyage from Istanbul to Paris, and the various institutions he and his suite visited in the French capital triggered a long-lasting admiration of French art works and technology among the Ottoman intellectuals. Mehmed Çelebi's son Mehmed Said Efendi's (d.1761) discovery of scientific cabinets during his embassy in Paris in 1741–1742 stimulated the import of an eclipse calculator to Istanbul and the translation of its manual into Ottoman Turkish. The present case study of this import and translation process will try to expose the role of Ottoman high officials or elite bureaucrats in the introduction of European technical knowledge and its adaptation to Ottoman context. In particular, it will point to the translator Sıdkı Efendi's introduction of additional operations on the instrument, in order to enable its use according to the Islamic calendar.

The Ottoman Ambassador Sets Eyes on the *Eclipsarium*

Mehmed Çelebi was amazed and fascinated by the large number of astronomical instruments he saw at the Paris Observatory which he visited twice during his mission. He seems to have appreciated the significance of these instruments: "Someone who has little knowledge of the science of stars can became a master

---

[1] F. Günergun, "Ottoman Encounters with European science: sixteenth and seventeenth-century translations into Turkish," in P. Burke and R. Po-chia Hsia, eds., *Cultural Translation in Early Modern Europe*. Cambridge: Cambridge University Press, 2007, pp. 192–211.

[2] J. Galland, *Relation de l'Ambassade de Méhémet Efendi à la Cour de France en 1721, écrite par lui même et traduite du Turc*. Paris 1757; *Relation de l'Ambassade de Mohammed Effendi*. Paris: Typographie Firmin Didot Frères, 1841; *Tacryr ou Relation de Mohammed Effendi*. Ali Süavi, ed., Paris: Imprimerie de Victor Goupy, 1872, 48 pp.; F.R. Unat, *Osmanlı Sefirleri ve Sefaretnameleri* (Ottoman ambassadors and their reports). 2nd ed., Ankara: Türk Tarih Kurumu, 1987 (1st ed. 1968); Mehmed Efendi, *Le Paradis des Infidèles. Un ambassadeur Ottoman en France sous la Régence*. G. Veinstein (Commentaires), J.-C. Galland (Traduction), Paris: Librairie François Maspero, Collection La Découverte, 1981; F. M. Göçek, *East Encounters West – France and the Ottoman Empire in the Eighteenth Century*. New York, Oxford: Oxford University Press, 1987.

thanks to these instruments". The term he used for the observatory -*müneccimhane* (the house of the astrologer), but not *rasadhane* (house for observations) or *muvakkithane* (timekeeper's office or horology room)- is indicative of his perception of the utility of astronomical instruments. It is, however, difficult to decide whether he used the term *müneccim* to denote an "astrologer" or a "person who observes the stars".[3] The term *rasad* had already been used by the Ottomans to denote astronomical observations, since the observatory built during the reign of Sultan Murad III (r.1574–1595) by the astronomer Taqi al-din (d.1585) was called *Darü'l-rasadü'l-cedid* (the house for new observations).

Çelebi inventoried the following instruments that he observed: an instrument to observe the stars, an instrument related to mechanics, the vacuum instrument, instruments to raise water, many other devices related to "curious arts", large concave mirrors, gigantic globes and many geometrical and astronomical instruments. Two instruments captured his attention the most: The telescope through which he observed the surface of the Moon, and the instrument which allowed predictions of solar and lunar eclipses. Çelebi gave a brief description for the eclipse calculator which fascinated him, especially after he learned that the movement of its circular plates could aid the prediction of eclipses:

> There is an instrument newly invented to know [predict] the solar and the lunar eclipses. It is made of a few circles with numerals, the Moon and the Sun marked round the edge. When these circles rotate, a pivotal needle similar to the hour hand and having a rounded end like a coin, points either to the Moon or to the Sun. When pointed to the Moon and according to whether the needle covers it partially or totally, one declares that a Moon's eclipse will occur on such and such month and it will be of such and such magnitude. It is the same for the Sun. It is a strange/marvelous thing that a circle helps one know that a solar eclipse of such and such magnitude will occur at such and such year, month and day. They [the French] show great care to this instrument which was created thanks to the favor of the Old King [Louis XIV] and no other state possesses such instrument.[4]

The Ottoman ambassador's statements and his enthusiastic description quoted above imply that he considered the *eclipsarium* as a novelty. Mechanisms designed to predict eclipses were however used by astronomers since medieval times and earlier.

---

[3] The term *müneccim* is generally used to denote an astrologer and is derived from the Arabic word *necm* (a star).

[4] This passage is translated from the Turkish text given in *Yirmisekiz Çelebi Mehmed Efendi'-nin Fransa Sefaretnamesi*. Beynun Akyavaş, ed., Ankara: Türk Kültürünü Araştırma Enstitüsü, 1993, pp. 48–49. This version of the embassy report is based on the Ottoman Turkish text published in 1283 (1866) in Istanbul; The text in Ottoman is given as an appendix in the 1993 edition. See also, Mehmed Efendi, *Le Paradis des Infidèles*, pp.149–150.

Predicting Eclipses at a Glance: Ingenious Mechanisms Come to the Rescue

The quest to predict eclipses varied throughout ages and cultures. Mesopotamians predicted eclipses for divination purposes.[5] Chinese astronomers devoted much attention to the prediction of eclipses: Han people had developed a 135 months period during which 23 eclipses took place while in the third century CE methods of eclipse prediction recognized the nodes and assessed the angle of the path with the ecliptic as 6 degrees approximately. By seventh century CE, times of the first and last contacts were given together with the extent of partiality.[6] The two main reasons presumably led the Muslim astronomers to make eclipse measurements: (i) to test the reliability of contemporary eclipse calculations and (ii) to determine the difference between the longitudes of two locations on the Earth's surface.[7] According to al-Biruni (973–1048), one needed to predict eclipses to make the necessary arrangements beforehand: i.e. to hire persons who could accurately measure the times of the various durations of the phenomenon in the towns where one planned to measure the longitudes.[8]

The prediction of eclipses, either lunar or solar, required tedious calculations of the positions of the Sun as well as a sound understanding of Moon's orbit and its speed.[9] The design of mechanisms to predict eclipses can be traced back to prehistory. Stonehenge, dating back to 2500 BCE, is claimed to have been used to predict eclipses. This was done either by the 56 evenly spaced "Aubrey holes" arranged in a circle surrounding the Stonehenge monument or the builders used a simpler way.[10] The pattern of glyphs marked on the Antikythera mechanism built around the second century BCE, that match the dates of commencement of one hundred eclipses that occurred during the final four centuries BCE, point out to its usage also as an eclipse calculator.[11] Inscriptions on its spiral dial in the lower back indicated the months during which lunar and solar eclipses were

---

[5] http://articles.adsabs.harvard.edu//full/1992ASPC...33..205R/0000205.000.html

[6] J. Needham, *Science and Civilisation in China*, vol. III. Cambridge: Cambridge University Press, 1995, pp. 420–422.

[7] F. R. Stephenson and S. S. Said, "Precision of Medieval Islamic Eclipse Measurements," *Journal of History of Astronomy*, 22, 69/3(August 1991): 195–207.

[8] Ibid.

[9] P. Duffett-Smith, *Practical Astronomy with your Calculator*. Cambridge: Cambridge University Press, 1979, p. 157 (lunar eclipse), p. 161 (solar eclipse).

[10] R. Colton and R. Martin, "Eclipse cycles and eclipses at Stonehenge," *Nature* 213 (04 February 1976): 476–478. Bill Kramer, "Stonehenge Eclipse Calculator," http://www.eclipse-chasers.com/tseStonehenge.html

[11] For the various publications mentioning the use of the Antikythera as an eclipse calculator see 'The Antikythera Mechanism Research Project' at http://www.antikythera-mechanism.gr/

expected to occur.[12] Greek astronomers such as Ptolemy were able to predict eclipses approximately. Archimedes who lived in the third century BCE is reported to have built two planetariums accounting for the motions and the relative positions of the Sun, Moon and the planets. The tradition of designing and manufacturing mechanical devices was carried on by Muslim scholars from the ninth century onwards. The *equatorium*, a mechanical device for computing the past or future positions of the planets and the Moon was largely used in medieval Islam and Europe. While many *equatoria* made of parchment or paper, survive on the pages of manuscripts and printed books, only a few brass *equatoria*, or fragments thereof, are preserved.[13] Early examples are given by Abu Gafar Muhammad b. al-Husain al-Hazin (10th c. CE),[14] Abulqasim of Granada (11th c. CE) and by al-Zarqali of Toledo (11th c. CE).[15] The instrument was further improved both in the Latin West in the early fourteenth century and in the East.[16] The Plate of Zones (*Tabaq al-Manateq*) designed by Jamshid al-Kashi (1393–1449), a distinguished scholar at Ulugh Beg's Observatory, was an improved *equatorium* used to find the true positions of longitude of the Sun and the Moon; the longitudes, distances and equation of the planets.[17] Made of brass or wood or other suitable material, this *equatorium* consisted of a circular plate with an alidade like an astrolabe pivoted at its center. The plate was surrounded by a graduated circular ring, and in addition to the alidade there was a ruler, of length equal to the diameter of the plate, but not attached to it. The same instrument was also used as a lunar eclipse computer and presented an alternative to lengthy numerical computations.[18]

---

[12] T. Freeth, "Decoding an ancient computer," *Scientific American*, December 2009, pp. 52–59.

[13] *Equatoria* were never as common as astrolabes, playing no part in the curricula of the schools. They became outdated as the parameters of the planetary theory were revised. As a purely calculating device (in contrast to the astrolabe which was used both as an observing instrument and as a computer) the equatorium did not require great durability. See J. D. North, "A post-Copernican equatorium," *The Universal Frame: Historical Essays in Astronomy, Natural Philosophy, and Scientific Method*. London: The Hambledon Press, 1989, pp. 145–184. The reason may be that astrolabes were used to determine the positions of the sun and the stars, the *equatoria* those of the moon and the planets.

[14] F. Sezgin and E. Neubauer, *Science et Technique en Islam*, Tome II (Astronomie). Frankfurt am Main: Institut für Geschichte der Arabisch-Islamischen Wissenschaften, 2004, p. 177.

[15] D. J. Price, *The Equatorie of the Planetis*. Cambridge: University Press, 1955.

[16] Ibid.

[17] E. S. Kennedy, "A Fifteenth-Century Planetary Computer: al-Kashi's "Tabaq al-Manateq" I. Motion of the Sun and Moon in Longitude," *Isis*, 41, 2 (July 1950): 180–183; E. S. Kennedy, "A Fifteenth-Century Planetary Computer: al-Kashi's "Tabaq al-Manateq" II.Longitudes, Distances and Equations of the Planets," *Isis*, 43, 1 (April 1952): 42–50; E. S. Kennedy, "An Islamic Computer for Planerary Lattitudes," *Journal of American Oriental Society*, 71, 1 (January–March 1951): 13–21.

[18] E. S. Kennedy, "A Fifteenth Century Lunar Eclipse Computer," *Scripta Mathematica*, 17 (1951): 91–97.

Al-Kashi also gave the description of another instrument, namely the Plate of Conjunctions (*Lauh al-Ittısalat*) to determine the time of an anticipated conjunction.[19]

With the development of printing in the fifteenth century in Europe, the *volvelles* (wheel charts), devices consisting of movable discs, were designed and incorporated into books. They were used to determine the positions of celestial bodies, especially of the Moon and the Sun. Some would show the phases of the moon and were used as a lunar clock to determine the time at night. Others could solve mathematical equations or other astronomical problems. Made of paper and generally incorporated into books, *volvelles* were more easily accessible than expensive instruments made of brass. They could be used both for scientific and astrological purposes. Made of wood or brass, the *nocturnal* which employed stars to determine the time at night, was mostly used in navigation.[20]

Towards the end of the seventeenth century, eclipse calculators were designed and manufactured mostly in Paris.[21] Despite the more precise results obtained based on the observation of Jupiter's satellites, astronomers would still rely on the observation of lunar eclipses to determine the longitude. The interest of academicians for finding the longitude accurately was linked with the desire to create more precise maps. In January 1681, the Danish astronomer Ole Römer (1644–1710) who worked at the Paris Observatory between 1671 and 1681 presented his "Planisphère pour les éclipses" to the French Academy of Sciences.[22] Jean-Dominique Cassini (1625–1712), a member of the Academy, endorsed Römer's *eclipsarium* which was manufactured by Isaac Thuret (1630–1706) in his workshop located at the Galeries du Louvre. This instrument enabled the prediction of eclipses until the end of the nineteenth century with an error of a day.

---

[19] E. S. Kennedy, "Al-Kashi's "Plate of Conjunctions"", *Isis*, 38, 1/2 (November 1947): pp. 56–59; See also *The Planetary equatorium of Jamshīd Ghiyāth al-Dīn al-Kāshī (d. 1429): an edition of the anonymous Persian manuscript 75 <44b> in the Garrett Collection at Princeton University; being a description of two computing instruments, the plate of heavens and the plate of conjunctions*. With translation and commentary by E. S. Kennedy, Princeton: Princeton University Press, 1960. For the replicas of al-Kashi's and other equatoria see F. Sezgin and E. Neubauer, op. cit, pp. 173–201.

[20] http://www.sciencemuseum.org.uk/broughttolife/objects/display.aspx?id=1746; http://www.mhs.ox.ac.uk/students/98to99/Inst/Instpgs/8.tinyVOLVELLEhoraeDiei.html

[21] Existing samples are rather rare. A copy of Römer's eclipsarium is kept in Rosenburg Castle (Kopenhagen) and Bibliothèque Nationale de France (Département des Cartes et Plans, Paris). For the copy at Kopenhagen see http://www.rosenborgslot.dk/v1/genstand.asp?GenstandID=170&countryID=2. An eclipseometrum which is similar to de La Hire's "machine a eclipses" is kept in Adler Planetarium and Astronomy Museum (See Fig. 3).

[22] "Planisphère pour les éclipses inventé par M.Roemer," in Gallon, *Machines et Inventions Approuvées par l'Académie Royale des Sciences*, Paris, 1735, pp. 85–89.

In 1704, an eclipse calculator designed by Philippe de La Hire (1640–1718) was placed in the pavilions located in the garden of the Château Marly together with the two globes (celestial and terrestrial) ordered from Vincenzo Coronelli by Cardinal d'Estrée and dedicated to Louis XIV.[23] De La Hire gave the description and usage of this instrument in his *Tables astronomiques* and mentioned that he had invented it "il y a déjà fort long-temps." Composed of three circular superimposed plates made of copper or cardboard, this instrument would predict "assez exactement" the date and the time of the eclipse. According to De La Hire, it found great favor among public.[24] Authors such as Nicolas Bion (1652–1733) and Jacques Ozanam (1640–1717) introduced the calculator in their works. Other kind of machines to predict eclipses were constructed in England and Italy throughout eighteenth century.[25]

Which eclipse calculator did Mehmed Çelebi observe in 1721 when he visited the Paris observatory? Bearing in mind that both Römer's instruments were presented to Jacques II Stuart (1633–1701) during his visit to the observatory in 1690,[26] and that "Monsieur De La Hire's *eclipsarion*" remained in Marly until 1722, leads us to surmise that it was Römer's eclipsarium that Mehmed Çelebi inspected.[27] Although he gave a brief description of the *eclipsarium*, he must have been impressed by this instrument of about one meter in height and which combined astronomical knowledge with fine craftsmanship. This was a remarkable instrument that had also captivated the attention of Louis XIV when

---

[23] "Une machine des éclipses de La Hire où l'on voyait d'un coup d'œil les éclipses de chaque année et par le mouvement de quelques platines de cette machine on apercevait les éclipses d'une autre année et ainsi de suite pour les années *à* venir et passées." In 1722, these instruments were transported to the Library of Louis XV. See L. Paris, *Essai Historique sur la Bibliothèque du Roi: Aujourd'hui Bibliothèque Impériale*. Paris 1856, p. 117.

[24] Philippe de La Hire, *Tables astronomiques*. 3rd edition, Paris, Chez Montalant, 1735, pp. xiv–xv (first published in Latin in 1702).

[25] Among them were Johann Andreas von Segner's (1704–1777) "Macchina ad Eclipses" (1741), James Ferguson's (1710–1776) "Piece of Mechanism contrived for exhibiting the time, duration and quantity of solar eclipse" (1754), and Veneziani's 'Macchina pel cui mezzo si predice l'avenimento di ecclissi del Sole e della Luna (1807). See M. L. Todd, *Total Eclipses of the Sun*. Boston: Robert Brothers, rev. ed. 1900 (rep. 2008), p. 6.

[26] "Sa majesté [britannique] monta ensuite à la Salle des Machines [de l'Observatoire] où elle admira principalement celle des Eclipses inventée par M. Roemer, & exécutée par le Sieur Thuret d'une manière toute particulière." *Histoire de l'Académie des Sciences*, Tome II (Depuis 1686 jusqu'à son renouvellement en 1699), Paris 1733, p. 63.

[27] G. Veinstein (*Le Paradis des Infidèles*, p. 149, n. 222), guided by M.-J-P. Verdet also assumes that Mehmed Çelebi alluded to Römer's eclipsarium. Interestingly, Römer's instruments are not deposited at the Paris Observatory nowadays (see the catalogue at http://patrimoine.obspm.fr/Instruments/Instruments/Instruments.html), but at BNF as mentioned above. According to the testimony of the members of the Académie des Sciences (Paris) these instruments served for about 50 years to confirm astronomical calculations, then stored at the Academy. Following the French Revolution, they were deposited at BNF (J.-Y. Sarazin, "Belles et obsolètes: deux 'machines' astronomiques," *Revue de la Bibliothèque nationale de France*, 14 (2003): 46–47). These instruments, after being presented to academy members, have been moved to the Paris Observatory for calculations.

visiting the Royal Academy of Sciences in 1681.[28] Sets of *eclipsarium* and *planetarium* sent to the Siamese King and the Chinese Emperor in 1685 together with other instruments and a group of French mathematicians, had also received high esteem.[29]

No information is available on whether Mehmed Çelebi or his retinue purchased any astronomical instruments in Paris and took them to Istanbul. Before leaving for Turkey, the ambassador was given, among other valuable goods, six clocks and six watches, as examples of fine French craftsmanship. On his return home to Istanbul, some optical instruments were commissioned from Istanbul to Paris: The list of items ordered included besides textiles, bulbs and parrots, a number of field glasses or telescopes (*lunettes* or *longue vues*), convex lenses (*verres ardents*), convex mirrors (*miroirs ardents*), a few Roman telescopes (*lunettes de Rome*) and microscopes.[30] Did these devices arrive in Istanbul? If yes, for what purpose were they used? Were they utilized in the field for military ends or for celestial observations or were they objects of entertainment or curiosity in the imperial palace? These are open questions to be investigated further. The fact that all the instruments ordered were optical, may point both for a need and absence of such items in Turkey in the early eighteenth century.

An Ottoman Text Introducing De La Hire's Eclipse Calculator

Some 30 years after Çelebi gave a description of Römer's *eclipsarium* in his embassy account, Mustafa Sıdkı Efendi, an Ottoman functionary and mathematician was commissioned to translate into Ottoman language,[31] a text accounting for the construction and use of an instrument that predicts lunar and solar eclipses. The translation, presumably carried out in Istanbul, is dated

---

[28] "Sa Majesté passa ensuite dans la Salle des Assemblées ordinaires de l'Académie.... M.Cassini expliqua ensuite la construction & l'usage des deux machines astronomiques de M. Römer auxquelles le Roi s'arrêta assez longtemps. L'une sert au calcul des Eclipses & l'autre représente toute la Théorie des Planètes." *Histoire de l'Académie des Sciences*, Tome I (Depuis son établissement en 1666 jusqu'à 1686), Paris 1733, pp. 319–320. A year earlier, in 1680, Römer had demonstrated to the Academy in Paris, a machine showing the orbits of the planets (a planetarium) and a machine for calculating the Moon's eclipses (an eclipsarium): *Histoire de l'Académie des Sciences*, Tome I, p. 206.

[29] I. Landry-Deron, "Les Mathématiciens envoyés en Chine par Louis XIV en 1685," *Arch. Hist. Exact Sci.*, 55 (2001): 423–463 (p. 434).

[30] For a list dated 15 Juillet 1722 of items ordered from the Ottoman court to France see "RF Affaires Étrangères. Cp Turquie, vol. 64, s.99RV in M. Kaçar, Osmanlı Devleti'nde Mühendishanelerin Kuruluşu, unpublished PhD dissertation, Istanbul University 1996, p. 217.

[31] A multicultural written language formed as a result of the merging of Arabic and Persian words with the vocabulary, syntax and grammar of Turkish. Present day Turkish linguists term this composite language "Ottoman Turkish", the language employed in literary and scientific works as well as Ottoman State's official correspondence.

3 *Cemaziyelevvel* 1161 after Islamic/Hijri Calendar, which corresponds to 1 May 1748. In the introductory paragraph which may be regarded as an extended title, the translator mentions that the text deals with the "cycles of conjunction and opposition invented by Monsieur De La Hire" and that the translation was made from "a treatise written by Nicolas Bion".[32] The King's engineer for mathematical instruments (globes, sundials, mathematical and mechanical instruments), N. Bion was well known for the instruments he manufactured in his Parisian shop, and for his several writings, edited many times.

The Commissioner: An Ambassador Fond of Technical Novelties

The translation was prepared at the behest of a former ambassador and deputy Grand Vizier, Mehmed Said Efendi (d.1761).[33] He was the son of Mehmed Çelebi, who had admired the eclipse calculator in the Paris Observatory. Said Efendi had visited Parisian libraries housing many printed books in 1721, in his father's retinue. Back in Istanbul, he opened in 1726, the first Ottoman printing house producing books with Arabic characters, together with his partner Ibrahim Müteferrika (1674–1747). In the footsteps of his father, Said Efendi acted as the Ottoman ambassador in Paris in 1741–42. This gave him the opportunity to visit the libraries, cabinets and workshops in the pursuit of technical novelties.[34]

Said Efendi was very interested in the natural history collection displayed in the cabinet of curiosities of Chevalier de Julienne (d.1766), director of the Gobelins tapestry factory. He spent four hours in the cabinet of Sieur Pagny, who used to teach experimental physics for many years in Paris. Professor of physics of the Queen and demonstrator at the University, Pagny had a collection of more than 400 machines. Said Efendi admired the arrangement of the cabinet, the experiments performed and Sieur Pagny's intelligible teaching.

---

[32] The manuscript in Turkish is catalogued under the title "Devair-i ictima ve istikbalin resm ve istimali" (The construction and usage of circles of conjunction and opposition) and kept in Selimiye Library (Selimiye Yazma Eser Kütüphanesi, Edirne, Turkey) at Nr.560/4 (ff. 64–72). It is bound together with a treatise in Persian on calendar computing and two others in Arabic on mathematics dated to the seventeenth and eighteenth centuries. The owner of the corpus is unknown but it was donated to the library in 1797 by Çelebi Mustafa Pasha (d.1811). A high military official, Mustafa Pasha donated his private collection of manuscripts to the library he had founded next to the Selimiye Mosque during his mission in Edirne. This corpus might have once belonged to the Pasha's private library.

[33] In 1748 (date inscribed on the translation), the translator refers to Mehmed Said Efendi as "former deputy grand vizier". Said Efendi was appointed to this post in 1746. Apparently he was holding a different, lower position when he commissioned the translation.

[34] T. Timur, in his article "Said Mehmed Efendi" (*Toplumsal Tarih*, Nr.128 (Ağustos 2004): 55–61) besides analyzing Çelebi's diplomatic mission gives a summary of Çelebi's visits to the cabinets as based on information published in *Mercure de France* of Juin 1742.

Furthermore, he examined the *estampes* – especially those depicting plants and animals – at the Royal Library.[35]

Said Efendi spent a whole day in the cabinet and laboratory of the Count of Onsenbray's at Bercy where the Count (Louis Léon Pajot, 1678–1754) conducted his experiments, kept his rich collection of natural history and physics and often received prominent science enthusiasts.[36] Said Efendi's father Mehmed Çelebi had already visited this cabinet in June 1721. Çelebi had been shown different experiments (*jeux de la nature*), various mechanical devices, drawings of plants and animals, wax anatomical models and burning mirrors. He had also visited the small menagerie where animals imported from East Indies were kept.[37] His son Said Efendi was presented several experiments with the burning mirror, such as the combustion of a piece of wood under water and the melting of a diamond. He asked to take some specimens to present to the Sultan. He also visited the sections on chemistry, physics, natural history and mathematics and inquired about the many devices displayed. Those who were present during his visit, made a note of the questions he asked about the machines and recognized his interest for the exact sciences. His *noble inclinaison et goût décidé pour les sciences utiles et les arts*, was however considered atypical, since such interest was regarded as exceptional among Ottoman dignitaries Said Efendi represented.[38]

Had Said Efendi accompanied his father on his visit to Count of Onsenbray's cabinet some 20 years ago? Were visits to the cabinets arranged upon Said Efendi's request or did they follow a standard tour offered by the French government to official visitors? Whatever the case, both ambassadors witnessed new experiments and inventions. The presence of a microscope invented by M. Lebas[39] were among the gifts given to Said Efendi by the King,[40] may be a indication of the acknowledgement by the French authorities of the

---

[35] *Mercure de France*, Juin 1742, pp. 984, 985, 987.

[36] Honorary member of the Académie des Sciences, Count Onsenbray invented many mechanical devices, among them an anemometer. See B. Jacomy, "L'Anémomètre de Pajot d'Ons-en-Bray," *La Revue* (CNAM), No.30, Juin 2000.

[37] Mehmed Çelebi does not mention his visit to d'Onsenbray's cabinet in his embassy report. An account is given in *Mercure de France*, Juin-Juillet 1721. See Mehmed Efendi, *Le Paradis des Infidèles*, pp. 198–199.

[38] *Mercure de France*, Juin 1742, pp. 1016–1017, 1023.

[39] In the last decades of the seventeenth century, small microscopes of 6 or 7 in. high with a silver foot and screw and composed of three lenses were being sold in Paris. C. Huygens wrote of them that they were "polished by the expert process of which the little widow Lebas jealously guards the secret. "Optician to the French king, Philippe Claude Lebas (1637–1677) developed his own personal method of polishing lenses. Following his death, his wife and son carried on the workshop which remained in production at least until 1721. Lebas, later on his wife and son were the suppliers of lenses to the Paris observatory. M. Daumas, *Scientific Instruments of the 17th & 18th Centuries and their Makers*. Trans. by M. Holbrook, London: Portman Books, 1972, pp. 73–74.

[40] *Mercure de France*, Juin 1742, p. 1028.

Ambassador's curiosity for scientific instruments. A lexicon of botanical and zoological terms he compiled in Istanbul in 1753 seems to attest that his interest lay beyond the mathematical sciences.

A few days after his arrival in Paris, Said Efendi visited, together with his son,[41] the shop of Sieur Nicolas Bion[42] then at the Quai de l'Horloge (Île de la Cité). There he studied every instrument with great attention and deliberated on the selection of the most curious, useful and perfect instruments to be brought to Istanbul. It was agreed that the selected instruments would be manufactured by Sieur Bion himself, so that they were of the maximum accuracy. The manufacture of the instruments took more than three months. During this period, Sieur Bion frequently conferred with Said Efendi in order to comply with his taste and opinions. Said Efendi was highly satisfied with the instruments and displayed them to distinguished persons with special care and delight. He would keep them as precious jewels in a box and the keys to the box were in his custody.[43]

Which instruments did Said Efendi order and purchase from Bion's atelier? Was there an eclipse calculator among them? Did Sieur Bion provide an operating manual together with the calculator? This is not unlikely, bearing in mind that Said Efendi asked for the translation of a text for its construction and usage some years after his return to Istanbul. A calculator of De La Hire type, made of brass, wood or pasteboard was a handy and portable instrument,[44] looking very much like an astrolabe that Ottoman timekeepers used extensively for centuries. Its function as an instrument of prediction, as well as its physical shape may have attracted Said Efendi's attention.

The Translator: A Bureaucrat Competent in Mathematics

The translator Hadji[45] Mustafa Sıdkı Efendi (d.1183/1769–70) was the son of a high-ranking military official at the Corps of Armorers, attached to the Janissaries. After being trained in the administrative division (*kethüda kalemi*) of the corps in Istanbul, he left for Egypt on an unknown mission.[46] During his 20-year stay in Cairo, he copied, studied and edited many mathematical texts. More than fifty, these texts attest to his deep interest in studying classical texts

---

[41] Said Efendi is reported to have a son named Süleyman Tevfik (d.1767–68) who was a madrasa teacher (*Sicill-i Osmani*, vol. 3, pp. 29–30). *Mercure de France* mention his name as Meksous Bey.

[42] At the time of Said Efendi's mission, the French instrument maker Nicolas Bion (1652–1733) had passed away and his atelier was run by his son Jean-Baptiste Nicolas Bion who was also among the King's engineers.

[43] *Mercure de France*, Juin 1742, pp. 1022–1023.

[44] See the figure (pl. 3, fig. 9) with a ring and ribbon given in J. Ozanam's *Recreations in Mathematics and Natural Philosophy* (vol. 3, London 1814).

[45] The Hadji denomination testifies his pilgrimage to Mecca.

[46] Mehmed Süreyya, *Sicill-i Osmani*, vol. 3, [Istanbul]: Matbaa-i Amire 1311 (1893), p. 225.

of geometry and astronomy.[47] He collected the texts in two volumes comprising the Greek works translated to Arabic in the ninth century and edited by Islamic mathematicians subsequently. Among them, there were the works of Autolykos (4th century BCE), Hypsicles (2nd century BCE), Theodosius of Bythinia (2nd century BCE), Euclides (ca. 300 BCE), Archimedes (3rd century BCE) and Aristarchos (3rd century BCE) translated by Thabit ibn Qurra (836–901) in Baghdad and elaborated by Nasir al-Din al-Tusi (1201–1274). Mustafa Sıdkı's collection included several works of the geometer Abu Sahl al-Quhi (fl.10th century), a treatise on geometry of Al-Biruni (d.11th century) and some anonymous works on geometry. Mustafa Sıdkı, probably considering the astronomical work of Taqi al-Din (d.1585) as significant, incorporated a copy of his treatise accounting for the astronomical instruments of the Istanbul observatory into his collection of texts. The only reference to eighteenth century mathematics was the work of the Ottoman mathematician Esad Efendi of Yanya (Janina, d.1729) on the squaring of a circle, a problem which goes back to Antiquity.

Mustafa Sıdkı seems to have worked in Egypt between 1727 and 1747, being these the earliest and the latest copying dates of the texts he brought together in Cairo. Since the translation he made from Nicolas Bion is dated 1748, he must have left Cairo in 1747 or 1748. The ambassador Said Efendi might have met Sıdkı Efendi when on mission in Cairo in 1744–1745 and arranged an occupation for him in Istanbul. In the course of his Istanbul years, Sıdkı Efendi was conferred several posts in the Ottoman military and state institutions: He was appointed, in turn, secretary to the steward of the Grand Vizier, the paymaster-general of the Imperial Dockyards and Navy, the director of the Imperial Mint, the accountant of the administration of Holy Shrines (Mecca and Medina), chief-secretary in the Finance Department. These were posts requiring a sound knowledge of mathematics. Heavy administrative duties, however, seem to have prevented Sıdkı Efendi from working on mathematical texts in Istanbul.[48] It is however unclear where he learned French. While he clearly mentions himself as "the translator" (*li-mütercim el-fakir*) in the manuscript, he may have collaborated with other persons for translating the text.

The Translation into Ottoman Turkish: "On the Cycles of Conjunction and Opposition"

According to the introductory note, the text accounting for the eclipse calculator was a translation from a report (*resm ü takrir*) written by Nicolas Bion.

---

[47] C. İzgi, *Osmanlı Medreselerinde İlim*, vol. 1, Istanbul: İz Yayıncılık, 1997, pp. 302–306.

[48] He is said to have invented a surveying instrument to measure the area of lots (Bursalı Mehmed Tahir, *Osmanlı Müellifleri*, vol. 3, Istanbul 1975, p. 287).

The *Traité de la construction et des principaux usages des instruments de mathématique* of Bion comprises a chapter introducing De La Hire's eclipse calculator with the title "De la construction et usage d'une machine qui montre les éclipses, tant du soleil que de la lune, les mois et les années lunaires avec les épactes." Bion's *Traité* had first been published in 1709 and subsequent editions saw the press throughout the eighteenth century. Thus, either the father (Mehmed Çelebi) or his son (Said Efendi) could have brought the book from Paris to Istanbul in 1721 or 1742. On the other hand, in his *Traité*, Bion mentions that he manufactured this calculator and had a pamphlet or manual published on his usage:

> J'ai divisé & fait graver des planches d'une bonne grandeur, pour monter cet instrument en cartons. J'ai fait aussi imprimer séparément un petit livre pour expliquer son usage.[49]

It is highly probable that Said Efendi bought the instrument together with the booklet when visiting Bion's shop in 1742. The fact that the Turkish translation includes a "table for the epochs of the lunar years corresponding to the solar years for the meridian of Paris," which is not extant in the *Traité*, corroborates this assumption. One may assume that the texts accounting for the eclipse calculator comprised similar information, whether in the form of a booklet or as a chapter in the *Traité*. In case the translation was made from the booklet bought in Paris in 1742, it is not easy to ascertain, at the moment, if the text is a full or a shortened translation. The comparison of Chapter V of the *Traité* with the Ottoman manuscript confirmed that the translation was an abridged version of the chapter.[50]

The Ottoman text, in accordance with the Chapter V of the *Traité*, is divided into three parts followed by a conclusion. However, the translator added an appendix in which he expressed his own assessment of the instrument. The first two parts offering a description of the three circular plates and the method of making divisions on them, are translated without omission. The third part which describes the usage of the instrument is, however, translated only partially: The first paragraphs explaining the determination of the first new moon,

---

[49] N. Bion, *Traité de la construction et des principaux usages des instruments de mathématique*. Paris: Chez Jombert et Nion fils, 1752. For the Chapter V on eclipse calculator see pp. 231–236. The reference to the pamphlet is on p. 236. I have not been able to consult earlier editions (1709, 1723, 1725) of the *Traité*, but I assume that same statements are included.

[50] E. İhsanoğlu et al., in their work entitled *Osmanlı Astronomi Literatürü Tarihi – History of Astronomy Literature during the Ottoman Period* (vol. 2, Istanbul: IRCICA, 1997, p. 466) argued that the Turkish text is the translation of N.Bion's book titled *L'Usage des astrolabes tant universels que particuliers accompagné d'un traité qu'en explique la construction*. There is, however, no account of an eclipse calculator in *L'Usage des astrolabes*. This statement implies these authors assumed that the translation dealt with an astrolabe and not with an eclipse calculator. This misleading information is echoed in Mustafa Sıdkı's biographies published from 1997 on.

the full moons and the eclipses of the year 1703 are fully translated. The following explanations about the determination of the first new moon of 1704 are omitted, as well as Bion's warning that circular plates and the ruler should be adjusted very precisely in order to prevent errors, and that tables for the epochs of lunar years should be used to rectify the usage of the instrument. The translator also skipped the explanations given on the calculations of the tables of the epochs, and the paragraph accounting why astronomers, in order to facilitate the calculations, prefer to use the mean movements of the Moon and the Sun. The latter information, known to all astronomers, was probably deemed unnecessary for the functioning/use of the calculator and thus was not translated. It is possible that these paragraphs were not included in the operating manual which was published by Bion and sold together with the instrument.

Arabic and Persian technical terms reflecting the translator's acquaintance in medieval astronomical Islamic texts, abound in the text. An Arabic equivalent is provided for every French term, and no European term is borrowed. A phrase such as "*machine qui montre les éclipses tant du soleil que de la lune*" is translated as *devair-i ictima ve istikbal* (circles of conjunction and opposition) terms used by Islamic astronomers to denote solar and lunar eclipses. No new word is coined for *machine*. Instead, the translator preferred to use *devair*, the Arabic word for circles, based on the physical appearance of the device.

The Ottoman text does include the tables of "époques des années lunaires pour le meridian de Paris" which are necessary for the use of the calculator, but lacks the figure of the calculator itself. The translator, either having the instruments in his hands or its figure present in the French text, probably deemed it unnecessary to draw it. Another explanation may be that the booklet accompanying the calculator was devoid of figure.[51]

The Eclipse Calculator "De La Hire – Bion"

The calculator consists of three circular plates (*platines*) made either of metal or cardboard (Figs. 1 and 2). The upper and the smallest plate is divided into 12 lunar months of 29 days 12 hours and 44 minutes each. It bears an index attached to its border and a ruler or alidade fixed to its center. The upper plate has also two circular belts with holes. The holes on the superior belt represent the new moons and when colored black, they denote solar eclipses.

---

[51] The most precise of the figures is published by De La Hire in his *Tables astronomiques*. The figure given by Bion is the *Traité* is less precise. The one in Ozanam's *Récréations mathématiques* is the less technical of all three figures.

The Ottoman Ambassador's Curiosity Coffer 117

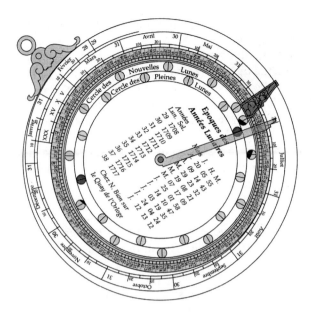

Fig. 1 The eclipse calculator De La Hire – Bion with its three plates superposed

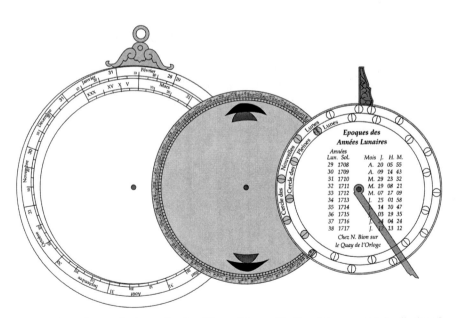

Fig. 2 The eclipse calculator De La Hire - Bion with its plates separately displayed. Figures 1 and 2 are reproduced after the information and the drawings of the calculator given in N. Bion's *Traité* (Paris, 1752) and P. de la Hire's *Tabulae astronomicae* (Paris, 1702)

The holes on the inferior belt represent the full moons and when colored red, they mark lunar eclipses.

The plate in the middle represents the lunar years. According to the text of Bion's *Traité* and the figure given by De La Hire in *Tabulae astronomicae*, the border of the middle plate is divided into 179 equal parts.[52] 179 is the maximum number of lunar years of 354 days 9 hours each that one can perfectly fit into 360 degrees. Thus the instrument can be used for a duration of 179 years. In the figure given in the *Traité*, the middle plate bears 45 divisions, while the text speaks about dividing the plate into 179 parts. In all likelihood, the middle plate of the calculator crafted by Bion had 45 divisions, a simplification which probably facilitated its manufacturing: dividing a circle into 45 should be easier than dividing it into 179 parts. The red and black mound-shaped forms depicted on the middle plate represent respectively the nodes where lunar and solar eclipses occur (Fig. 2). On the border of the lower and the largest plate, the days and the 12 months of the solar/civil calendar are inscribed.

Eclipses are predicted by revolving the two upper circular plates, the index and the ruler according to given instructions and by having recourse to a table of epochs. The latter lists the epochs of the lunar years corresponding to the civil years, i.e. the dates of the first new moons that occur in each civil year. This table is calculated for the years 1680–1854 and for the meridian of Paris but it can be set for cities located on other meridians.

To find out if any eclipse would occur in a civil year (i.e. 1703) one should first read on the epochs' table which lunar year corresponds to 1703: it is the 24th. Then, the index attached to the border of the upper plate is taken to the 24th division of the middle plate which is divided into 179 equal parts. By consulting again the table of epochs, one finds out the date and the time of the first new moon of the 24th lunar year: it is 14 June 1703, 9 h 52 mn.

The two upper plates are then rotated together so as the index denotes 14 June 1703, 9 h 52 mn on the lower and largest plate. Subsequently, without changing the position of the three plates, the movable ruler is turned till it pass through the middle of the hole of the first full moon. The edge of the ruler would point to the time of the first full moon: 29th of June, 4 h 15 mn. The red color which appears in the hole of the first full moon marks the occurrence of a lunar eclipse on that date and time. No lunar eclipse would occur on dates corresponding to colorless holes. Other lunar eclipses occurring in 1703 are read by the coloration in red of other apertures.

Similarly, on the belt of new moons, one may notice that a hole is partially colored in black. When the ruler is turned till it passes through the middle of this half-back hole, its edge points to the 14th of July and 3 hours, the time when a partial solar eclipse would occur.

---

[52] See *Traité*, p. 231. For the figure given by De La Hire see *Tabulae astronomicae*, Joannes Boudot, Parisiis 1702, http://gallica.bnf.fr/ark:/12148/btv1b26001153.item.r=Philippe+de+la+Hire.f3.legendes.langFR

Fig. 3 An *eclipseometrum* made of wood, pasteboard, paper and brass, bearing Latin inscriptions. Adler Planetarium Collection.
The division of the middle plate into 179 lunar years and the table of epochs given on the back for the civil years 1680–1854 make it a typical De La Hire eclipse calculator. Its provenance and manufacturer are unknown. 38 x 35.7 x 2.4 cm.
(B.Stephenson, M.Bolt, A.F.Friedman, *The Universe Unveiled*, Cambridge University Press & Adler Planetarium and Astronomy Museum, 2000, p. 33)

The Translator's Evaluation of the Eclipse Calculator

Sıdkı Efendi's personal evaluation of the calculator is not altogether favorable. After having used the instrument according to the rules given by Bion in the manual, he probably tried to determine whether the machine is compatible with Islamic (Hijri) calendar used by the Ottomans. To determine the hijri dates of the eclipses he needed to read on the calculator, the beginning of the hijri (lunar) years corresponding to the first new moon of civil years given in the table of epochs. He found out that the date of the first new moon given for a civil year did not mark the beginning of the corresponding lunar/hijri year. For example, 17 February 1681 (first new moon of 1681) which should correspond to the beginning of the hijri year 1092 (1st *Muharrem* 1092) erroneously coincided with the second month (*Safer*) of the hijri year 1092. Sıdkı Efendi proposed the following operations to correct this discrepancy of one-lunar-month.

Take the 1st lunar year (1681) recorded in the table of epoch and move the index attached to the upper plate to the 1st division of the middle plate. Then rotate both plates until the index points to the 17 February 1681, the first new moon of the year 1681. Retrograde the index 4 divisions[53] on the middle plate and fix it on the 179th division. Place the ruler on the hole preceding the one below the index. The ruler will point January 21th 1681. This is the beginning of the hijri year, 1st Muharrem 1092. Once the beginning of the hijri years is found on the instrument, it is possible to calculate the hijri dates of the eclipses. Therefore, the instrument was less practical for those using the Hijri/Muslim calendar and needed additional operations to calculate the date of the eclipses. Sıdkı Efendi noted that the device helped predict only the eclipses to occur between 1680 and 1854, and if one wished to predict eclipses beyond that period, one would need to set a new table of epochs. His second remark was that the calculator was not a precise instrument giving the exact hour and minute of an eclipse, and one could only make approximate predictions, a disadvantage already mentioned by Bion. The translator also noted the lack of information in Bion's text regarding the use of the extra 30 days preceding the month of March and marked with Roman numerals on the lower plate. It is possible that the presence of these days representing the epacts (period added to harmonize the lunar with the solar calendar) was not elucidated in the manual accompanying the instrument. This remark of Sıdkı Effendi makes clear that he had the instrument in his hands when translating its manual.

After reading his end note, one can argue that the translator Sıdkı Efendi lacked the enthusiasm or admiration of Mehmed Çelebi who came to know Römer's *eclipsarium* in the Paris Observatory. Astrolabes (*usturlab*), the astrolabe quadrants (*rub-i mukantara*) and the sine quadrants (*rub-i müceyyeb*) were the most frequently used instruments by Ottoman astronomers and timekeepers. The two former were used to determine the positions of celestial bodies (the Sun or a star), while sine quadrants helped to solve trigonometric functions. In daily practice, they also helped calculate the hours of prayer. Instruments for determining the direction of Mecca, i.e. qibla compasses (*kıblenüma*) were also widely used. These were designed after the medieval Islamic astronomical instruments.[54] The *equatoria* used for calculating the positions of the moon and the planets were known in the Ottoman Empire in the fifteenth century: An anonymous Persian manuscript (Nr.75, Garrett Collection, Princeton University Library) accounting for al-Kashi's equatorium was dedicated to the Ottoman Sultan Bayezid II (1450–1512).[55] This dedication, however, doesn't imply a common use of the instrument in the Ottoman world. Nevertheless, Mustafa Sıdkı, knowledgeable in mathematics

---

[53] Each lunar year is represented by 4 divisions on the middle plate.

[54] D.A. King, *Islamic Astronomical Instruments*, London: Variorum 1987; *World maps for finding the direction and distance to Mecca*, Leiden: Brill 1999; *In Synchrony with the Heavens – Studies in Astronomical Timekeeping and Instrumentation in Medieval Islamic Civilisation*, vol. 2 (Instruments of Mass Calculation), Leiden: Brill 2005.

[55] Kennedy, "Al-Kashi's Plate of Conjunctions", p. 57.

and astronomy of medieval Islam, did not consider Bion's eclipse calculator as a novelty as one can deduce from the evaluation he appended to the translation.

Concluding Remarks

The text in Turkish accounting for the eclipse calculator crafted by N. Bion after De La Hire's design is among the earliest works introducing an astronomical instrument from Europe in the Ottoman Empire. Sıdki Efendi's work was not a simple translation. Besides translating Bion's text into Ottoman Turkish, he tried to adopt the use of the calculator to the Islamic (Hijri) calendar and correct the incompatibility he noticed in the instrument. The additional operations he suggested were helpful in finding the beginning of hijri years on the instrument and then calculating the dates of the eclipses in the Islamic calendar.

Mustafa Sıdkı's translation form French coincided with a Turkish translation of an Arabic text accounting for lunar and solar eclipses. The latter translation was made by Abbas Vesim Efendi (d.1760)[56] from Al-Birjandi's (d. after 1529) *Haşiya ala Şarh al-Mulahhas fil-Hay'a.*[57] Thus the year 1748 saw the emergence of two texts in Turkish dealing with eclipses: one being a text explaining the construction and use of a late seventeenth century device elaborated by a French astronomer and the other a sixteenth century text accounting for the eclipse phenomena. Since only single copies are available in Turkish libraries, their influence seems to have been confined to the year of their creation.

Bearing in mind the relatively small number of texts dwelling on eclipses in the Ottoman literature, one wonders about the coincidence of two translations in the same year. A glimpse at the solar eclipses observed from the Ottoman territories in the mid-eighteenth century may provide the key. An annular solar eclipse was observed on 25 July 1748 throughout Asia Minor and the Middle East. This solar eclipse, in all likelihood, put the calculator and the manual purchased by Said Efendi in Paris in 1742 on the agenda. It may explain why the translation was made in 1748 and the 6 years of oblivion that elapsed after they

---

[56] Abbas Vesim Efendi (d.1760) was a Persian scholar who moved to Turkey in early sixteenth century. He is well known for his two-volume medical encyclopedia combining both the Islamic and European medical knowledge. He was a strong supporter of iatrochemistry, called "new medicine" by Ottoman physicians. His encyclopedia also introduced the practices of sixteenth- and seventeenth-century alchemists such as J. B. van Helmont, Daniel Sennert and many others. See F. Günergun, "Science in the Ottoman World," in *Imperialism and Science* by G. N. Vlahakis et al., Santa Barbara California: ABC-Clio, 2006, pp. 87–89.

[57] Al-Birjandi's work is an annotation of Kadızade-i Rumi's commentary on al-Caghmini's (d.1221) work *al-Mulahhas fil-Hey'e*. Birjandi's annotation became popular in the Ottoman world and was taught in the madrasa). See Cevat İzgi, *Osmanlı Medreselerinde İlim*, vol. 1, Istanbul: Iz Yayıncılık, 1997, p. 381.

had been purchased. The translation was completed on May 1st, 1748, approximately three months before the solar eclipse occurred. Ottoman astronomers/astrologers had certainly determined the date of the eclipse by calculation before Sıdkı Efendi undertook the translation. The work was translated either to test whether the calculator designed by a European was precise, or it was the result of Said Efendi's curiosity for devices coupled with the coming of an annular eclipse in 1748. Giving less accurate dates then those obtained by calculations, the translation does not seem to be related with timekeepers' demands of setting a precise calendar or astrological concerns or religious requirements. The translation might have been undertaken in order to check the reliability of predictions made with an instrument imported from Europe. Its lack of precision hindered its popularity among Ottoman astronomers who preferred to make predictions by calculations, a method far more precise. Although the instrument could appeal to amateurs, there is no indication that the eclipse calculator was manufactured by the Ottomans and released to the Ottoman market.

The translation of the manual on the eclipse calculator, points out to the fact that the translation of monographs and pamphlets preceded the translation of textbooks from Europe. On the other hand, Bion's writings and possibly his instruments reached the Ottoman capital. Nevertheless, the texts introducing European astronomical instruments were scarce in the first half of the century. Ottoman timekeepers and astronomers/astrologers of the eighteenth century stuck to the traditional instruments such as the astrolabe and quadrants and Ottoman craftsmen pursued this tradition. As Bion's eclipse calculator was designed for Gregorian calendar, it was not of any help in timekeeping or mathematical calculations, it remained an object of curiosity and astronomical exercise.

Despite the eclipse calculator did not become popular among Ottoman astronomers, the ordering, the purchase of this instrument by an Ottoman ambassador in Paris, its traveling to Istanbul and the translation of its user's manual in the year of the 1748 solar eclipse is evidence that the transmission of knowledge into the Ottoman sphere happened also outside the educational institutions and military facilities, and should direct our attention to individual enterprises.

Acknowledgements I am grateful to the following persons who kindly sent me material and shared information during my research: Paul Gagnaire who, in 2006, kindly sent me a copy of the "Chapter V" together with the figure of the eclipse calculator from Nicolas Bion's *Traité de Construction* (Paris, 1752). Thanks to this material, I was able to compare the French text and the Turkish translation. My cordial thanks go to Atilla Bir and Mustafa Kaçar who have kindly contributed to the technical drawing of the eclipse calculator "De La Hire – Bion" and to Gaye Danışan who diligently assisted in accounting for the astronomical principles involved in its design and functioning. I am also grateful to David A. King who affirmed that *equatoria* were known to the Ottomans; to Marvin Bolt (Adler Planetarium, Chicago) for sending me information on and a high resolution image of the *eclipseometrum* kept in Adler Museum's collection; to Peter Kristiansen (Royal Danish Collections, Rosenborg Castle,

Kopenhagen) for providing me with Claus Thykier's article on O. Römer's instruments; to M. Kaçar who kindly fulfilled my request and made a photocopy of an issue of the *Mercure de France* (issue Juin 1742) at the Bibliothèque Nationale de France (BNF, Paris); to Sara Yontan and Nilgün Paner from BNF for sending me J.-Y. Sarazin's article; to Christophe Benoit (Nice Observatory) for directing my attention to the *Histoire de l'Académie Royale des Sciences*; to Nilüfer Gökçe (Trakya University, Edirne) and Musa Öncel (The Public Library of Edirne) for providing me with information on the provenance of the manuscript containing Sıdkı Efendi's translation kept at the Selimiye Library, Edirne; to Darina Martykanova for reading the final manuscript and her remarks. Last but not least, I am much indebted to Şeref Etker for his inspiring and useful comments.

The Clockmaker Family Meyer and Their Watch Keeping the *alla turca* Time

Atilla Bir, Şinasi Acar, and Mustafa Kaçar

Praying five-times-a-day is one of the most important religious obligations for Muslims, and prayers should be performed during certain periods defined by the position of the Sun at a given location. Prior to the invention of the mechanical clock that enabled the precise measurement of time, and the diffusion of these clocks among Muslims, instruments such as sundials, astrolabes and quadrants were widely used for timekeeping. Astronomical instruments measuring the local apparent time, i.e. the time indicated by the Sun at a particular location were widely used in Antiquity and the Middle-Ages. Solar time, with sunset taken as reference, with its unequal hours, proved a major difficulty in reckoning with the time of religious obligations and the mechanical clocks had to be set daily. Before the full adoption of the international standard time in Turkey in the twentieth century, the mean solar time, popularly called 'European time' (*alla franca* time) started to be used in the major cities of the Ottoman empire, mainly in communication and transportation networks connecting the empire with European countries. Thus both the solar mean time and the local apparent time (Turkish time or the *alla turca* time) were concurrently used in the Ottoman empire in the latter part of the nineteenth century. The mechanical clocks produced in the empire and those manufactured in Europe for the Ottoman market bear elements of this co-existence.

A. Bir (✉)
Faculty of Electrics and Electronics, Istanbul Technical University, Istanbul, Turkey
e-mail: atilabir@gmail.com

Ş. Acar (✉)
Faculty of Arts, Anadolu University, Eskişehir, Turkey

M. Kaçar (✉)
Department of the History of Science, Faculty of Letters, Istanbul University, Istanbul, Turkey
e-mail: mkacar@istanbul.edu.tr

The *ezânî* or *alla turca* Time

Prior to 1927, the enforcement date of the laws making the use of the universal time (UT) and the Gregorian calendar compulsory in Turkey, days were reckoned from sunset to sunset. This practice was inspired by the fact that an 'Islamic day' begins at sunset, because the Islamic calendar is lunar and the months begin with the sighting of the crescent shortly after sunset.[1] It was convened that the time of sunset was 12 o'clock or 00 hours. Thus a day would end and a new day would begin at 12 o'clock. The day was divided into two parts, each consisting of 12 equal hours: the first part would span the period from sunset to sunrise; the second part from sunrise to sunset. This system of designating time is known as *alla turca* time (Turkish time) or *ezanî* time (call for prayer time).

The convention of starting the day at sunset had the following drawback: Clocks keeping the Turkish time needed to be adjusted by a few minutes every day. The reason was that the time of sunset varied daily in each locality according to the seasons and the present latitude. Thus the beginning of the day changed daily from one locality to another and coordination remained relative. This meant that in a determined place, the clocks had to be adjusted every day at 12 o'clock – at the hour of sunset. In order to allow townspeople to adjust their clocks and watches daily, as well as help them to obtain the real time, horology rooms (*muvakkithane*) managed by timekeepers (*muvakkit*) were established in the busiest parts of Ottoman towns.

Alla turca Clock Mechanisms

Mechanical clocks were first introduced to the Ottoman Empire in the fifteenth century. They were offered as presents to the members of the imperial palace or to the senior officials of the State by European emissaries. Repair shops were gradually opened in Istanbul by European craftsmen and their Ottoman apprentices learned the art of repairing and manufacturing mechanical clocks. The Ottoman imperial astronomer Taqi al-Din (d.1585), after studying the collection of mechanical clocks of the governor of Egypt Ali Pasha (d.1565), compiled a pioneering treatise that accounted for the construction of 'clocks functioning with springs and weights' entitled *Al-kavakib al-durriya fi vaz al-bangamat al-davriyya* (The brightest stars for the construction of mechanical clocks, 1558–1559). This was a unique treatise written on European clocks in the Middle East.[2] Well

---

[1] D. A. King, "Astronomy and Islamic Society: Qibla, gnomonics and timekeeping," in R. Rashed, ed., *Encyclopaedia of the History of Arabic science*, vol. 1 (Astronomy). London – New York: Routledge 1996, pp. 128–184 (p. 182, Timekeeping in Ottoman Turkey).

[2] O. Kurz, *European Clocks and Watches in the Near East*. London: Warburg Institute, University of London, 1975, pp. 49–51; F. Günergun, "Science in the Ottoman World," in *Imperialism and Science* by G.N. Vlahakis et al., Santa Barbara California: ABC Clio, 2006, pp. 78–79 in pp. 71–118.

informed on medieval Islamic literature on mechanics, Taqi al-Din also manufactured a mechanical clock that he used in his astronomical observations.[3]

Clocks were mostly manufactured by a guild of horologists in the Galata district of Istanbul between the years 1630–1700. According to Otto Kurz, the products of the Galata clockmakers were as good as those produced by their Swiss and English masters. The making of clocks flourished especially among the Mevlevî order of dervishes. Apart from these, there were clockmakers trained at the Sultan's palace.[4] The work done in the shops consisted mainly of assembling pieces from clock kits imported from Europe, and parts such as hands, dials, and case were produced after the Ottoman aesthetic values. The guild at Galata died out in the beginning of the eighteenth century as a result of European restrictions set on providing material to the Ottomans out of military concerns. The Swiss and English horologists, taking advantage of the circumstances and considering Ottomans' interest in mechanical clocks began the manufacture of clocks specially designed for the Ottoman market. Subsequently, European companies which provided clock kits to Ottoman horologists throughout seventeenth century refrained from delivering material due to mercantilist policies.[5]

Mechanical clocks are ideal devices for time measuring and determining equal hours. In order to use mechanical clocks – much more precise than sundials, water clocks or quadrants – for measuring unequal hours, the Turkish or *alla turca* time system was invented. The Turkish time system is a compromise between the division of the day and night separately into 12 parts and the use of semi-variable hour duration. In this system the beginning of a new day is defined by the sunset which has to be daily corrected on the clock. Thanks to this system, the time of the evening prayer and the beginning of the new day could keep pace with the variable hour concept, and thus, the mechanical clocks of precision can be used conveniently. This system disregards the rising of the Sun at 12 o'clock. The length of the day and the length of working hours became thus equal and independent from seasons. It is not exactly known when the *alla turca* time came to be implemented. One may argue that it was in use in early seventeenth century when people started to possess clocks, and horology rooms (*muvakkithane*) became annexed to mosques. As the time difference between the

---

[3] A. Bir and M. Kaçar, "Takiyüddin bin Maruf's Astronomical clock," in *Visible Faces of the Time – Timepieces*. Istanbul: Yapı Kredi Yayınları, 2009, pp. 67–82.

[4] R. Hildebrand, "Isaac Rousseau à Péra: heurs et malheurs d'un compagnon horloger," in P. Dumont, R. Hildebrand, and R. Montandon, eds., *L'Horloger du Sérail: aux sources du fantasme oriental chez Jean-Jacques Rousseau*, Paris: IFEA, Maisonneuve & Larose, 2005, pp. 28–47.

[5] R. Murphey, "The Ottoman Attitude Towards the Adoption of Western Technology: The Role of the efrenci Technicians in Civil and Military Applications," in J. L. Bacqué-Grammont and P. Dumont, eds., *Contributions à l'histoire économique et sociale de l'Empire Ottoman*, Collection Turcica III, Louvain, 1983, pp. 287–298; F. Çakmut, "Turkish Clocks and Watches in the Topkapı Palace Museum Clock Collection," in *Visible Faces of the Time – Timepieces*, Istanbul: Yapı Kredi Yayınları, 2009, pp. 91–103.

alla turca time kept by the clock and the true prayer time is negligible for noon and afternoon prayers, and almost non-existent for the night prayer, mechanical clocks and *alla turca* time became popular in the Ottoman society up to twentieth century.

Johann Meyer's Manually Adjusted *alla turca* Watch

In the mid-nineteenth century, the Longines company (founded 1838) constructed a pocket watch for the Ottoman market, showing the both the Turkish time (*alla turca*) and the European time (*alla franca*) on the same dial (Fig. 1a). In this watch both hour- and minute-hands took the movement from the same spring mechanism. Therefore, at the sunset or at the end of each day, one had to manually regulate the hands of the watch to adjust the Turkish time.

Johann Meyer (1843–1920), inspired by Longines, manufactured such watches in Istanbul with kits he imported from Europe and added to them *alla turca* dials and adjusting mechanisms (Fig. 1b). They were named *Hamidiye* after the Sultan Abdulhamid II (r.1876-1909). Johann Meyer, the founder of Meyer company (est.1878) in Istanbul, also served as clockmaker in the Ottoman Palace for many years.

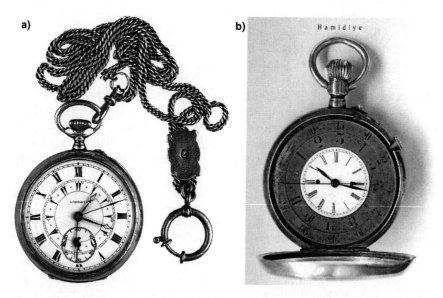

Fig. 1 19th-century pocket watches with manual adjustments for keeping both Turkish and European times. **(a)** A *Longines* watch manufactured for the Ottoman market (Ş. Acar Collection). **(b)** the *Hamidiye* watch designed and crafted by Johann Meyer in Istanbul (Ş. Acar Collection)

On the clock face of the *Hamidiye* watch crafted by Johann Meyer, there are three dials. The central dial with Roman numerals is fixed and shows the European time. The two adjustable dials surrounding the central European dial represent the *alla turca* time. The inner one indicates the Turkish hours and the external one, the minutes. Each day at sunset, the inner dial for hours is adjusted to make the hour-hand point to 12; and the dial for minutes is adjusted so that the minute-hand indicates 60.

Johann Meyer's Automatic *alla turca* Watch

Doubtlessly, many people should have wished to buy *alla turca* watches and clocks that adjusted themselves automatically and clockmakers did probably investigate the possibility of designing such mechanisms.

By 1887, Johann Meyer succeeded in designing an automatic watch keeping both the European and Turkish time. According to European watch producers, the manufacture of an automatic self regulating watch keeping the *alla turca* time was impossible. Johann Meyer took up the challenge and created an appropriate mechanism, and after 8 years of work he successfully produced a watch which automatically regulated itself each day at sunset. He presented the watch to the Sultan Abdulhamid II, together with a technical report accounting for its mechanism and functioning.[6] The Sultan granted him a silver 'pride medal' in 1887 in appreciation of his work and design.[7]

J. Meyer did not start designing his automatic clock from scratch, but modified an ordinary European pocket watch mechanism. As is known, on the back of every clock mechanism there is a small scale marked with the letter *A* (*Advance*) on one side and the letter *R* (*Reverse*) on the other for adjusting the clock when it looses or gains time respectively. The main modification made by J. Meyer was the addition of a heart cam to turn this scale gradually each day so that it gains time during the winter months and looses time during the summer months. The distance between the center of the cam to its periphery is short for the period of January-June and long for the period July-December, with the result that between December 21st and June 21st the ratchet was pushed gradually toward *R*, causing the clock to slow daily by one or two minutes during this season and gain one or two minutes daily during the other six months of the year (Figs. 2a, 2b and 3).

---

[6] Document (MS.HHA-E.II/1501) kept at Dolmabahçe Palace Museum (Istanbul) archives, Istanbul. The document dated c. 1887, states that Johann Meyer had devoted much effort to the difficult task of designing an automatic clock showing prayer time.

[7] The inscription on the reverse reads "Sa'atci Meyer, sene 1305, fi 28 Rebi'ulevvel" (Watchmaker Meyer, 14 December 1887). The inscription on the obverse reads "El-mustenid bi-tevfîkat-ir rabbâniye -Melik-ül Devlet-ül Osmâniye- Abdülhamîd Hân" (On the divine right of the Sultan of the Ottoman Empire, Abdulhamid Han).

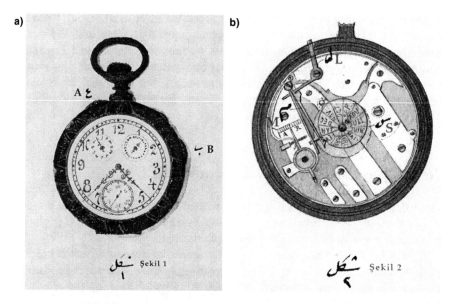

Fig. 2 (a) The front on the right, of J. Meyer's automatic watch keeping the *alla turca* time. **(b)** The back (with the continuous heart cam) of the same watch. (From J. Meyer's technical report dated 1887, Dolmabahçe Palace Museum Archives, MS.HHA-E.II/1501)

Fig. 3 J. Meyer modified the continuous heart cam he had designed in 1887 and later produced cams with 18 steps to overcome the frictional forces. Detail of such a heart cam from a Meyer *alla turca* watch (Ali Rıza Yılmaz Collection)

J. Meyer, in his petition to the Sultan, gave the drawings and below is a description of his automatic watch that he called *ezanî saat,* literally prayer-call watch:[8]

---

[8]This description is summarized from J. Meyer's petition kept in Dolmabahçe Palace Museum Archives, Istanbul, MS.HHA-E.II/1501.

"The upper part of the clock face bears two small dials. The one on the left shows the months and that on the right indicates the days. The small button **A** (*ayn*) which moves the hands of these two dials helps to adjust the hours according to the days. In case the clock stops functioning for a few days, one should adjust the month and the day when winding up the clock. The other button **B** (*be*) helps to adjust the hours and the minutes on the great dial. The cam **S** (*sin*) rotates once a year. Its movement is transmitted to the main arbor (*taksim çarkı*) which acts on the sliding bar **M** (*mim*) next to the spiral spring (*rakkas*) set on the cam **S**. The second sliding bar **L** (*lam*) designed to be used if necessary, helps the watch to function as a European watch. The sliding bar **M**, in its original position, helps to contact the main axis to the spiral spring and by the functioning of the cam indicating the months and the days, this sliding bar serves daily to decrease or increase the movement of the alteration pin on the cam. Therefore, since the cam **S** adjusts daily the movement of the watch, the sunset coincides with the 12 o'clock every day."

Astronomical Background of the *alla turca* Watch

Since the *alla turca* time is based on the apparent motion of the Sun, it can be designated as *Local Apparent Time* (*L.A.T.*). Local Apparent Time is local to a particular meridian and it takes the time of sunset as a reference point. It was the time used almost universally in daily life until about a century ago. Yet it suffers from disadvantages which have led most people to discard it in favor of another kind of theoretical fictive time system. Firstly, no two places have the same *L.A.T.* unless they lie on the same meridian. Secondly the length of the day varies during the year, partly because the Earth moves more swiftly in its orbit when close to the Sun, and partly because the Sun's apparent path lies on the ecliptic rather than on the equator.[9]

Instead of measuring time based on the irregular motion of the *apparent* or *real Sun*, it is more convenient to calculate it from an arbitrary *mean Sun* – a fictitious heavenly body which moves in the celestial equator at a constant speed which is just equal to the average speed with which the real Sun moves in the ecliptic. If the real Sun and the mean Sun start off together, the real Sun, moving irregularly, will sometimes run ahead of mean Sun and sometimes lag behind it, but by the end of the year, they will finish the course together.

Time measured by the hour angle of real or apparent Sun is called *apparent time*, whereas if we measure the hour angle of the mean Sun we find the *mean time*. The last step is taken by designing *standard meridians* at the boundary of these *standard regions* and defining a *standard time* (*S.T.*). In our present time system, there are on the world 24 well defined *standard time regions*.

---

[9]A. E. Waugh, *Sundials, Their Theory and Construction*. New York: Dover Publications, 1973, pp. 6–28.

Fig. 4 Diagram of the time equation

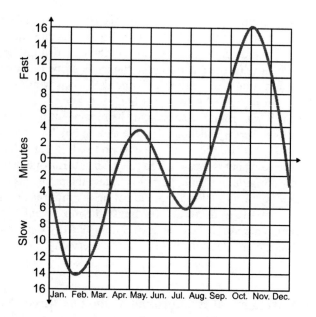

The Local Apparent Time (L.A.T.) can easily be calculated by the formula given below:

$$L.A.T. = S.T. + L + E$$

In this equation, the *Longitude correction (L)* is obtained by the difference of the longitude of Istanbul (29°) and the zone angle of Turkey (30°). Thus the longitude correction for Istanbul is a constant equal to $L = (29° - 30°) = -1° \equiv -4$ minutes. The time equation E is generally given by a diagram or table (see the below diagram) (Fig. 4).

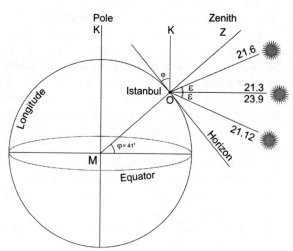

Fig. 5 Position and the variation of the Sun's elevation angle in Istanbul (above)

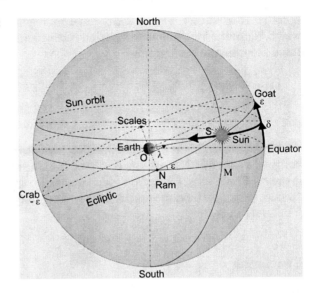

Fig. 6 The Sun's movement on celestial sphere (right)

The Sun rises in the East, moves across the sky from East to West and sets in the West in the evening. Istanbul is situated on a latitude $\varphi = 41°$ north. The Sun moves in one year on the ecliptic plane following the constellations circle. If one defines δ as the angle of declination, λ as longitude angle and ε as the angle of the ecliptic, one can write the relation $sin\ \delta\ 1/4 = sin\ \lambda \cdot sin\ \varepsilon$ (Figs. 5 and 6).

The duration of the day changes according to the angle of declination δ of the sun, the maximum is given by ε angle of the ecliptic ($0 \leq |\delta| \leq \varepsilon$). The half day remnant $\Delta t/2$ can be calculated from the spherical right triangle defined in the figure below as (Fig. 7)

$$\Delta t/2 = \text{arctg}\ [sin\ \delta \cdot \tan \varphi].$$

To obtain a watch indicating the *alla turca* or *ezânî* time, it is sufficient for a determined declination angle δ to fix the hour-hand and the second-hand to

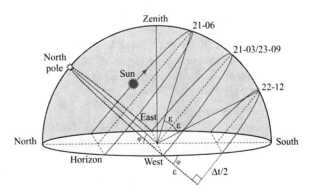

Fig. 7 Seasonal change of the half-day remnant $\Delta t/2$

point to 12 at sunset. This is easily done by regulating the heart cam of the watch. Since the longitudinal correction is constant, the heart cam is manufactured according to the formula:

(The change of time equation) + (The change of the day remnant Δt)

$$= \frac{dE}{d\delta} + \frac{d\Delta t}{d\delta} = \frac{dE}{d\delta} + \frac{2.\cos\delta.tg\varphi}{1 + (\sin\delta.\cos\varphi)^2}$$

Johann Meyer, in the technical drawing presented to the Sultan in 1887, had designed a continuous heart cam (Fig. 2b). But, the Meyer *alla turca* watch kept in the Ali Rıza Yılmaz Collection which we have examined, has a heart cam with 18 steps (Fig. 3). The cam with 18 steps was possibly designed to overcome the frictional forces.

Conclusion

The second half of the nineteenth century witnessed a tangible increase in Ottoman's political and economical relations with European countries. As a result, the 'European time', i.e. the mean solar time was introduced in the Empire and was first used for the timetables of railway and maritime companies. Composed mostly of Muslim subjects, the Ottoman society kept using the *alla turca* time (apparent solar time) in its daily life and religious practice. Therefore it was necessary for the Turks and the foreigners to consult and/or possess two separate watches, one showing the Turkish and the other keeping the European time. The invention of Johann Meyer, an automatic watch keeping both European and Turkish time in a single mechanism came just in time to overcome this need. As Ottoman government offices started to use the solar mean time from 1912 on, European time became widespread in the empire.[10] The last quarter of the nineteenth century saw the erection of clock towers in the capital Istanbul: they were mostly located in the grounds of imperial institutions and public buildings such as railway stations and hospitals. A decree issued in 1901 by Sultan Abdulhamid II, endorsed provincial cities to have clock towers showing both Turkish and European times.[11] The full adoption of the latter in the second quarter of the twentieth century hindered the popularization of Meyer's automatic *alla turca* watch. Thus its life on the market became confined to about 30 years. The lack of information regarding the number of such watches manufactured by Meyer, prevents us from arguing about its popularity and value.

---

[10]E. İhsanoğlu and F. Günergun, "Osmanlı Türkiyesinde 'Alaturka Saat'ten 'Alafranga Saat'e Geçiş," *X. Ulusal Astronomi Kongresi (Istanbul 2–6 Eylül 1996)*, Istanbul 1996, pp. 434–441.

[11]Ş. Gürbüz, "On the Presence and Absence of Clock Towers," in *Visible Faces of the Time – Timepieces*. Istanbul: Yapı Kredi Yayınları, 2009, pp. 133–147.

Appendix: The Clockmaker Family Meyer

As clockmakers, inventors and improvers of mechanical clocks and watches, the Meyer family over three generations has a distinguished place in the history of horology in Turkey. The company "Meyer Saat Tamirat ve Ticareti" (Meyer Clock Repair and Trade Company) was established in 1878 by Johann Meyer (1843–1920) and was active in Istanbul for a century until the death of his grandson Wolfgang Meyer in 1981. His shop was in Péra, Istanbul. Later on, the shop moved to Karaköy, the business district and pear of Istanbul. The workshop, located at Bakırköy-Bahçelievler district mostly produced large clocks for factories, schools and such institutions.[12]

Johann Meyer was sent to Istanbul by the German Emperor Wilhelm II (r.1888-1918) after his first visit to the Ottoman capital city in 1878, upon the request of Sultan Abdulhamid II (r.1876-1909). J. Meyer served as clockmaker in the Ottoman Palace until the end of his life. He invented the automatic watch keeping the *alla turca* (*ezânî*) time. This invention was recognized by Sultan Abdulhamid II by a silver medal and certificate on 14 December 1887. J. Meyer was also granted a *Mecidî* decoration of distinction with diamonds and a certificate on 21 May 1896 for his longstanding services (Fig. 8).[13]

Fig. 8 Three generations of family Meyer. Johann Meyer (1843–1920) (*left*), Emile Meyer (1883–1954) (*middle*), Wolfgang Meyer (1909–1981) (*right*)

---

[12]The first shop was located across the entrance of the *Tünel* (funicular) building in Karaköy. Following a project of public works in the era, the shop moved to a nearby address (Billur Sokağı, No.8/1) in 1958. Presently, the shop houses a hardware dealer. The workshop which was active in 1950s at Bakırköy-Bahçelievler district (Yayla Durağı Karşısı No.81) does not exist anymore. The company "Meyer Saatçilik" (Meyer Clockmaking) is presently run by Mr. Nahsen Bayındır at the "Mumhane Caddesi No.23, Karaköy."

[13]The Ottoman Archives at the State Archives of the Republic of Turkey, BOA, İ.TAL, 97/1313/Z-044, 08.Z.1313).

To Johann was born a son, Emile Meyer (1883–1954), in Istanbul. Emile was educated in the German School in Istanbul, and left for Germany to be trained as master clockmaker. He returned to Istanbul in 1908 to work with his father. At the end of the World War I, he left Turkey for Germany where he acted as the manager of the clock company Huber in Munich. He returned to Istanbul in 1922 and took over the business of his father who had died on 8 April 1920 in Istanbul. At the end of the World War II, the members of the family Meyer were interned as German nationals in Kırşehir, a town in central Anatolia, in 1944 when Turkey joined the allies. Back in Istanbul as the war ended, he took over the direction of the company that he run until his death on 7 March 1954.

Emile's son Wolfgang Meyer (1909–1981) was also educated in the German School, Istanbul and received his training as clockmaker from his father. He took over the direction of the company after his father died in 1954. He was interested in the theory of time keeping instruments, in repairing old clocks and watches. W. Meyer compiled and published the catalogue of the clocks and watches kept in the Topkapı Palace Museum (Istanbul) upon the suggestion of Hayrullah Örs, the director of the Museum.[14] Meyer also produced a number of articles on time keeping instruments, especially on sundials.[15]

After W. Meyer's death in 1981, the company's shares were left to his apprentices according to Meyer's will. One of them, Nahsen Bayındır, continued to run the company in a shop located in the vicinity of the Galata Port. Meyer's other former apprentice, the master watchmaker Recep Gürgen is actually the official clock repair expert of the Dolmabahçe and Topkapı Palace museums. At the grave of the Meyer family at the Feriköy Protestant Cemetery of Istanbul, are buried the Johann Meyer (18.11.1843–4.8.1920), Emile Meyer (3.9.1883–7.3.1954), E. Meyer's wife Martha Meyer (14.1.1886–14.8.1960), Wolfgang Meyer (23.5.1909–16.9.1981), Wolfgang Meyer's brother Ewald Meyer (Istanbul, 5.11.1910–Hamburg, 12.1.1998).[16]

Acknowledgements The authors are grateful to Mrs. Inge Traut (Hunger) Detmold, Wolfgang Meyer's spouse and to Mr. Nahsen Bayındır for providing documents and photographs related to the members of Meyer family and their horological activities.

---

[14] W. Meyer, *Topkapı Sarayı Müzesindeki Saatlerin Kataloğu / Catalogue of Clocks and Watches in the Topkapı Sarayı Museum*, İstanbul 1971.

[15] W. Meyer, *İstanbul'daki Güneş Saatleri* (Sundials of Istanbul), in H. Klautke, ed., Sandoz Kültür Yayınları, No.7, İstanbul 1985. See also; W. Meyer; "Sundials of the Osmanic era in Istanbul", *International Symposium on the Observatories in Islam, 19–23 September 1977*, Kandilli Observatory, İstanbul, 1980; W. Meyer, "Die Zeitbestimmung im Islam: historische Übersicht," in H. Wilschowitz, ed., *Deutschsprachige katholische Gemeinde in der Türkei 1954–1979: ein Überblick in Berichten, Aufsätzen und Geschichten*, [İstanbul, W. Blümel, 1979] S. 206–247; W. Meyer "Instrumente zur Bestimmung der Gebetszeiten im Islam", *Proceedings of the I. International Congress on the History of Turkish-Islamic Science and Technology, 14–18 September 1981*, Vol. I, pp. 9–32.

[16] Şinasi Acar, "Meyer'in ayar gerektirmeyen ezânî saati" (Meyer's alla turca watch that does not needs adjustments), *Yapı*, nr. 287 (Ekim 2005): 99–104.

The Adoption and Adaptation of Mechanical Clocks in Japan

Takehiko Hashimoto

This chapter examines the introduction and subsequent evolution of mechanical clocks in Japan. The mechanical clock was first brought by Jesuit missionaries as a gift to the rulers of Japan and China in the sixteenth century. After the prohibition of Christianity in Japan in the early years of the Tokugawa period, the clock evolved quite differently in the two countries. In China, increasingly elaborate Western clocks were either brought from the West or constructed in shops locally. In Japan, mechanical clocks were constructed exclusively by domestic craftsmen who modified them to indicate a seasonally variable system of hours. Although mechanical clocks were not owned and used widely, they were used by time-keepers at castles as well as at time-bell towers, to tell the time to city dwellers.

The mechanical clock was an important invention, sometimes even regarded as a prime contributing factor behind the modernization of Western society. It changed medieval society, which live in a seasonal time system, to a modern society based on the mechanical rhythm of the clock time system. The clock towers of city halls informed citizens of the precise hour, helping them to respond to demands more swiftly and punctually than before, and shaping their gradually accelerating civic lives.[1] The adoption of the clock time system in Western society prepared it for revolutionary change to come during the eighteenth and nineteenth centuries.[2]

The mechanical clock was first brought to China and Japan in mid-sixteenth century by Jesuit missionaries.[3] It was adopted and adapted in different ways.

T. Hashimoto (✉)
Department of History and Philosophy of Science, University of Tokyo, Tokyo, Japan
e-mail: takehiko.hashimoto@gmail.com

[1] G. Dohrn-van Rossum, *History of the Hour: Clocks and Modern Temporal Orders*. Trans. T. Dunlop. Chicago: University of Chicago Press, 1996.

[2] D.S. Landes, *Revolution in Time: Clocks and the Making of the Modern World*. Cambridge, Mass.: Harvard University Press, 1983.

[3] C.M. Cipolla, *Clocks and Culture*. London: Norton, 1967, pp. 83–114.

They were brought as gifts to the Chinese Emperor and to the Japanese feudal lords. As the Tokugawa government strengthened its rule over entire Japan, the Jesuits were expelled and the mechanical clocks had to be maintained by local craftsmen. These craftsmen eventually formed an expert group of clockmakers who began to make their own mechanical clocks, based on a unique engineering structure.

This paper will first examine the introduction of mechanical clocks to China and Japan, and show how its developments took different paths in the two countries, especially as a result of political decisions on Jesuits taken by Chinese and Japanese rulers, which helped create contrasting social and economic milieus. Secondly, it will focus on the special designs which enabled mechanical clocks to indicate the unique time-reckoning system in Edo Japan. Whereas the mechanical clock changed the time-reckoning system in Europe, it had to be adapted to the seasonal time system in Japan. One important use was in the system of time bells in Edo and other cities, some of which employed mechanical clocks to know the hours to gong. Lastly, the paper will introduce the notable craftsman Hisashige Tanaka in order to explain the social and technical aspects of time, clocks and clockmakers in late Tokugawa Japan.

The Introduction of Mechanical Clocks in China and Japan

In 1551, the Portuguese Jesuit missionary Francisco Xavier offered a mechanical clock to Yoshitaka Ōuchi, the feudal lord of Suō, the present-day Yamaguchi prefecture. He brought "a little striking clock" as a gift from the governor of Portuguese India which, for Japanese Buddhists, was a symbol of the distant sacred land.[4] Thanks to this tactical gift, Xavier received permission to engage in missionary activities in the territory.[5] Nobunaga Oda, the lord who successfully unified Japan, was also intrigued by the mechanical clock, and asked the Portuguese Jesuit Luís Frois to bring him a "self-sounding bell." Oda generously permitted Jesuits to organize a seminar to teach Western knowledge as well as the Christian religion, during which they also taught the construction of clocks. The mission sent by three lords from Kyushu to the Pope returned with mechanical clocks, which were offered to Hideyoshi Toyotomi. In 1606, the Portuguese interpreter João Rodrigues donated to Ieyasu Tokugawa, the third and final unifier of Japan, a mechanical clock with a solar and lunar model.[6] The next year, Rodrigues met Ieyasu's son Hidetada, accompanied by his assistant who was asked to remain in Edo to look after the clock sent from

---

[4] J. Crasset, *The History of the Church of Japan*. Vol. 1, trans. N.N., London: s.n. 1705, p. 76.
[5] Ryuji Yamaguchi, *Nihon no tokei* (Clocks of Japan). Tokyo: Nihon Hyoronsha, 1950, pp. 11–12; Taizaburo Tsukada, *Wadokei*. Tokyo: Toho Shoin, 1960, pp. 24–26.
[6] Crasset, op. cit., Vol. 2, 1707, p. 154.

Nagasaki.[7] The assistant, called Brother Paul, was probably a Japanese who had been trained at the Jesuit seminar in Nagasaki to construct and repair mechanical clocks. Rodrigues himself seems to have served as horologer to Ieyasu, to maintain his clocks.

The Jesuits' proselytizing project was brought to a close with the strict prohibition of Christianity by Hideyoshi, and then by the Tokugawa government. The harsh execution of Christian martyrs in 1622 and subsequent official ban of Christianity entirely terminated the activities of the Jesuits and expelled Portuguese missionaries and merchants.[8] The Tokugawa government permitted only Dutch merchants to work on a strictly limited scale on the small island off the port of Nagasaki.

In sharp contrast to the oppressed Jesuits expelled from Japan, those in China became successful in approaching the Emperor and establishing their social status at the Imperial Court. The gift of the mechanical clock was one way of pleasing the Emperor. According to the diary of the Italian Jesuit Matteo Ricci, the Emperor became especially interested in the chiming clock and accepted the petition of the European visitor to meet him.[9] After Ricci's successful gift-giving tactics, clocks and watches were considered indispensable gifts to gain access to the imperial court. Ricci's successor, Nicolas Trigault, had purchased numerous clocks in Europe, some of which being furnished with intricate automata mechanism. In the late seventeenth century, the Kangxi Emperor ordered the establishment of a workshop to make and repair clocks, where Jesuit craftsmen taught mechanical skills to Chinese craftsmen. In the eighteenth century, professional clock-makers were recruited and sent to work at the Clock-Making Office in Beijing.[10] Franz Stadlin, a Swiss clock-maker who worked in German cities as well as Vienna and Prague, joined the Society of Jesus and volunteered for the Chinese mission. From 1707 when he arrived in Beijing, until his death in 1740, he served as a master craftsman at the workshop, where the Kangxi Emperor himself often visited.[11]

On a much smaller scale, clock making skills must have spread from previous Jesuit schools in Japan. Craftsmen who received instruction at the Jesuit schools in Nagasaki and Owari are believed to have been engaged in making and maintaining the working of mechanical clocks even after the departure of the Jesuits. No sources remain on this Jesuit tradition being passed down to domestic craftsmen. We can only guess at the possibility of such a technological inheritance as a likely route.

---

[7] Crasset, op. cit., Vol. 2, 1707, p. 160.
[8] J. Elisonas, "Christianity in Daimyo" in *Cambridge History of Japan*. Vol. 4, Cambridge: Cambridge University Press, 1991, pp. 301–372.
[9] C. Pagani, *"Eastern Magnificence and European Ingenuity": Clocks of Late Imperial China*. Ann Arbor: University of Michigan Press, 2001, p. 2.
[10] C. Jami, "Clocks," in *Handbook of Christianity in China*. Vol. 1 (635–1800), N. Standaert, ed., London: Brill, 2001, pp. 844 f.
[11] Pagani, op. cit., pp. 50–51.

Other craftsmen became clockmakers independent of direct or indirect Jesuit instructions. Most notably, Sukezaemon Tsuda, originally an ironsmith in Kyoto, became a clockmaker after his successful restoration of Ieyasu's broken clock at the end of the sixteenth century. Tsuda became the official horologer to the Tokugawa family, then to the third son of Ieyasu, Tadayoshi, who subsequently dominated Owari, the region of the present Aichi prefecture. This successful restoration of the clock owned by the founder of the Tokugawa dynasty by a legendary domestic craftsman seems to be symbolic as the de-Christianization in Japan of the mechanical clock originating from the Jesuit gift. Official clockmakers like Tsuda and his descendants employed by the clan were called "*on tokeishi* (honorable clockmaker)." Not only the Owari clan but also numerous other clans employed such "honorable clockmakers" during the Edo period, which is considered the most important social factor enabling the subsequent development of clock making activities in post-Jesuit Japanese society.

The support of many feudal lords throughout the nation who were intrigued by the idea of a "self-sounding bell" was essential for developing the expertise of a generation of clockmakers who were specialized exclusively in making and repairing mechanical clocks. These feudal lords were not like the Chinese Emperor, capable of purchasing numerous exquisite and extravagant clocks and clockwork machines, but could afford to employ a special clockmaker together with his assistants. Most castles of the clans had a special room where one or more clocks made by official clockmakers were kept. These clocks told the time to the officers in charge of beating the time drums in the castle.

Adapting Clocks to the Seasonal Time System

After Ieyasu Tokugawa brought stability to Japan, the Tokugawa government formally established the seasonal time system. This system divides equally the length of daytime and nighttime, whereas the clock time system uniformly divides the length of an entire day into a certain number of units of time.[12] The length of hours therefore changed from season to season as well as from daytime to nighttime.

The seasonal time system of the Edo Period was defined as follows. Daytime from dawn to dusk was divided equally into six units of time called *koku*, and

---

[12]China adopted such a clock time system from ancient times and used it in cities. Medieval Japan introduced the Chinese system of counting time, and the activities inside the palace were conducted according to such a clock time system. However, it did not prevail in the provinces, and as the time descended, the seasonal time system became more common, and the time system itself was not rigorously determined. See Manpei Hashimoto, *Nihon no jikoku seido* (The Japanese System of Hours). Tokyo: Hanawa Shobo, 1966.

the night-time from dusk to dawn was divided equally into six *koku*s.[13] Beginning at midnight, which was numbered as 9, the Edo time system counted time from 9, 8, 7, to 6, which corresponded to the time of dawn; then descended to 5, 4, and back to 9, which was noon. The counting in the afternoon proceeded similarly, starting from 9 to 6, which was the time of dusk, and descending to 5, 4, and back to 9, the following midnight.[14] Although, precisely speaking, the time of dawn and dusk changed each day, the time of dawn and dusk on the clock only changed as 24 seasonal divisions called "*sekki*" changed. That is, the time of dawn and dusk only changed semi-monthly.[15]

The difficult task for Japanese craftsmen was to make a clock adjustable to such a seasonal time system. In other words, their task was to modify a uniformly moving clock into a device to designate a daily and monthly variable time system. Thus, they had to change the speed of the hand in the daytime and night-time and for different seasons, or had to change the display to designate time during the day and throughout the year. Whereas the clock served to reform the time counting system in the West, in Japan it had to be adapted to a seasonally variable time system.

All the clocks made during the 260 years of the Edo Period were thus designed to indicate time according to the seasonal time system.[16] There were several methods, both mechanical and non-mechanical, for this adjustment. The first and most representative method was to use a *tempu* or foliot on top of the clock (Fig. 1). It was a natural development from the earliest type of Western clock, which had an oscillating horizontal bar (foliot) whose motion

---

[13] It should be noted that the Japanese used the time of dawn and dusk, rather than sunrise and sunset, as the point to divide daytime and nighttime. They were defined as 36 minutes before the sunrise and 36 minutes after the sunset. The period, 36 minutes, came from 1/40 of the length of a day. They were later more precisely redefined as the time when the sun was 7 degrees 21 minutes below the horizon. See Kuniji Saito, *Nihon Chugoku Chosen kodai no jikoku seido* (The System of Hours in Ancient Japan, China, and Korea). Tokyo: Yuzankaku, 1995; Takehiko Hashimoto, "Kansei reki to wadokei: Yoake no teigi o megutte" (The Kansei Calendar and Japanese Clocks: On the Definition of Twilight," *Tenmon Geppo* (The Astronomical Herald), 98 (2005): 373–379.

[14] Manpei Hashimoto, op. cit., pp. 128–130. Hashimoto refers to the existence of the different system of counting time based on the 12 zodiac symbols and the confusion among the public on the use of this counting system. Ibid., p. 130.

[15] Manpei Hashimoto, op. cit., pp. 131–134.

[16] The exception was the clocks of the astronomers who used a different unit of time which uniformly divided the day. See M. P. Fernandez and P. C. Fernandez, "Precision Time-Keepers of Tokugawa Japan and the Evolution of the Japanese Domestic Clock," *Technology and Culture*, 37 (1996): 221–248. For the overview and explanation of various types of the Japanese clocks, see J. Drummond Robertson, *The Evolution of Clockwork with a Special Section on the Clock of Japan*. London: Cassell, 1931; N. H. N. Mody, *A Collection of Japanese Clocks*. London: Kegan Paul, 1932; Ryuji Yamaguchi, *Nihon no tokei* (Clocks of Japan). Tokyo: Nihon Hyoronsha, 1950; Taizaburo Tsukada, *Wadokei*, Tokyo: Toho Shoin, 1960; Sachiko Oda, ed., *Seiko Tokei Shiryokan zo wadokei zuroku* (A Catalogue of Japanese Clocks Preserved at the Seiko Clock Museum). Tokyo: Seiko Tokei Shiryokan, 1994.

Fig.1 A clock with one foliot (Oda, 1994, p. 28)

was regulated by the inner mechanism of verge-and-foliot engaged with a crown wheel. The left and right arms of the horizontal bar had teeth from which were hung two small weights symmetrically on both sides. The function of these movable weights was to control the speed of the foliot and thus the clock. If the clock was too fast, the position of the weights was changed from the original positions towards the outer ends of the bar to make it slower.

The Japanese craftsmen used this controlling function as a device to adjust the clock to the seasonal time system. They made a clock with just such a horizontal bar with teeth, on which a weight was hanged on each arm. In the daytime on the day of the summer solstice, the weights were hung on the positions farthest from the center of the foliot; at six in the evening of the same day, the weights were repositioned nearest to the center. After about two weeks, when the next seasonal division arrived, the weights were moved to the next notch, which slightly changed the oscillating speed. Repositioning the weights twice-a-day must have been cumbersome for the users of the clock. Sometime around the turn of the eighteenth century, a certain craftsman invented a clock with two separate foliots, each of which counting daytime and night-time (Fig. 2).[17] At daytime, one longer foliot was moving while the other shorter one was at rest; at night, the longer one was at rest and the shorter one in motion. At six, at dawn and dusk, an automatic mechanism inside the clock worked to stop the foliot in motion, and start the movement of the other foliot. Such a clock with two foliots with an automatic changing device was the representative Japanese clock, adjusting the mechanical rhythm to the natural rhythm of solar motion.

Another popular type was the *warikoma* clock. This clock had a face with movable number plates called *warikoma*. The face typically had twelve plates to designate the hours of the day. These plates were not uniformly distributed but

---

[17] Ryuji Yamaguchi, *Nihon no tokei* (Clocks of Japan). Tokyo: Nihon Hyoronsha, 1950, p. 104; Taizaburo Tsukada, *Wadokei*. Tokyo: Toho Shoin, 1960, pp. 130–131.

Fig. 2 A later clock with two foliots (Oda, 1994, p. 23)

positioned differently for day and night (Fig. 3). Like the repositioning of the weights on top of the *tempu* type of clock, these plates were moved twice a month to adjust to the seasonal time system. Some *warikoma* clocks were driven by a spring rather than by a weight. The weight-driven clock needed a fairly long vertical space to let the weight fall, but the spring-driven clock could be encased in a small contained space. This compact bracket clock was called in Japanese the *makura dokei* (pillow clock), because presumably it was often placed at the bedside of its user with an alarm mechanism to wake him or her. It could be also placed on a table, but the naming seems to suggest the ordinary usage of the clock.

The third adjusting method is found in the *shaku dokei* (a-foot-ruler clock or pillar clock).[18] It had only a uniformly descending weight and attached indicator, which designated on a plate the time during the day and night. Whereas its

Fig. 3 The face of *warikoma* clock (Photograph by T. Hashimoto)

---

[18] On the development of *shaku dokei*, see Ryuji Yamaguchi, *Nihon no tokei* (Clocks of Japan). Tokyo: Nihon Hyoronsha, 1950, pp. 183–196.

earlier type had movable *warikoma* on the indicator, its later type had several replaceable plates each of which showing the index of the corresponding seasonal division. In its modified version, a graph was drawn on the plate to enable the seasonal change of hours to be read on a single plate. This simple type of clock was less expensive, and economical versions were sold by merchants during the nineteenth century. Even so, clocks were still quite rare. Only wealthy merchants and high-class samurai could obtain and use these clocks either for practical or ornamental purposes. Ordinary people without clocks had a different means to know the time during the day and night.

The Time Bell and the Use of Clocks in Edo Japan

During the Edo period, people learnt the time by the beating of a drum inside the castle, and the striking of a bell outside the castle. The time bell (*toki no kane*) was set up in Edo in the early years of the Edo period, and later such bells were built in many towns and cities. Eventually villages also came to have such bells to tell the time to residents and travelers all over Japan.[19] Villages on the main roads connecting large cities especially needed such time keeping and information systems in order to coordinate the arrival and departure of transporters.[20] Some of those in charge of the time bell used traditional incense seals called *kōdokei*.[21]

In a large city like Edo, time bells were constructed in several places to cover an entire city area and to let the sound of gongs reach every resident. The well-known comedy tale, *tokisoba* (time noodle), tells us how well the system of time bells was established in Edo city. The story is about a seller of a bowl of noodle soup on a street and his two guests. A cunning guest first came and ate a bowl of *soba* at midnight. Having finished his bowl, he asked for the bill and was told the price was 16 *mon*s. He paid the seller coin by coin counting from one to eight; at eight, he asked the seller what time it was; hearing it was nine, he restarted counting from ten to sixteen and paid 15 *mon*s in total. Having successfully cheated one *mon*, he went away. The tale goes on to tell that the other client – less clever – wished to imitate the former guest, but failed because he asked to pay the bill at four *koku*, one *koku* before the midnight; when the

---

[19]Sakae Tsunoyama, *Tokei no shakaishi* (A social history of clocks). Tokyo: Chuo Koronsha, 1984, pp. 66–82.

[20]Toru Morishita, "Time in an early modern local community," in Shigehisa Kuriyama and Takehiko Hashimoto, eds., *The Birth of Tardiness, The Formation of Time Consciousness in Modern Japan*, Special issue of *Japan Review: Journal of the International Research Center for Japanese Studies*, 14 (2002): 65–78.

[21]S. A. Bedini, *The Trail of Time: Time Measurement with Incense in East Asia*. Cambridge: Cambridge University Press, 1994; Toru Morishita, op. cit., p. 69.

seller told him it was four, the customer continued to count from five. The tale ends there, leaving the audience only to imagine that the unlucky customer should have paid at the end four extra *mon*s in addition to the price.

How did this system of keeping and telling time develop and operate in Edo city? The historian Sachiko Urai provides a detailed description of the origin, management, and operation of time bells in Edo city.[22] When the new castle was built in Edo, a room was furnished with a clock. Based on the time told by the clock, a bell gonged at the time of dawn and dusk. Soon afterwards, a monk from Nara was permitted to set himself up to strike a drum at Hongokuchō to tell the time of dawn and dusk, and subsequently strike a bell twelve times a day, which marked the beginning of the time bell in Edo. As the city's population grew during the seventeenth century, other time bells were built elsewhere, including the three major temples of Kan'ei, Sensō, and Zōjō, thus covering the entire city.

Time bells were operated by an expert striker and his assistants, and their income was paid by the residents of the region. A document written by the time bell striker at Hongokuchō in 1725 explains that each householder throughout the region was expected to pay monthly either one *mon* of official Eirakusen or four *mon* of unofficial currency.[23] The striker of the same bell reported in 1736 on the expected costs for the next year, including maintenance for a mechanical as well as an incense clock.[24] Bell strikers relied on both kinds of time-keeping devices to know the time to gong the bell.

A document of the Kan'ei Temple is more specific on the use of clocks to measure time. It mentioned two types of mechanical clock, the *tempu* type and foot-ruler type. The latter was relied on as a standard time-keeper, and the former was stopped for a short while before dawn and dusk. The document also explains that length of time between the fifth hour in the late afternoon and the sixth hour at dusk was relatively longer than the length of time between the previous hours.[25] Whether this special custom of striking the bell in the important time-bell tower was shared by other time bell strikers is unknown. Another 1726 document, however, tells us that the time bell at the Kan'ei Temple in Ueno synchronized well with the three time bells located in Ichigaya, Akasaka, and Shiba.[26] The striker at Ichigaya just outside the northwest corner of the Edo castle had to wait to hear the first of the three preparatory gongs from Ueno, before he struck his own preparatory gongs. The gongs of Ichigaya were heard by the striker at Akasaka, southwest of the castle. The Akasaka bell's preparatory gong was then heard by the striker at Zōjōji in Shiba, to the south of the

---

[22]Sachiko Urai, *Edo no jikoku to toki no kane* (Time in Edo and the Time Bell). Tokyo: Iwata Shoin, 2002.
[23]Ibid., p. 29.
[24]Ibid., p. 108.
[25]Ibid., pp. 188–190.
[26]Ibid., pp. 173–174.

castle, who then started his own gongs. In this way these strikers formed a chain of gongs covering the areas surrounding the castle.[27]

In such a social circumstance, many citizens of Edo city may not have felt the need for time-keeping machines in their houses, or in public spaces. They heard the sounds of time bells, which might have been sufficient for regulating most of their daily activities. Nevertheless, as the intensity of social activities increased, newer types of clocks were developed and purchased. From the early nineteenth century, less and less expensive clocks were produced, and became more widely used than before. As mentioned above, some clocks, notably *makura dokei* placed by the bedside, were furnished with an alarm mechanism to help users be on time.

The Engineering Career of Hisashige Tanaka

This section will examine episodes from the career of Hisashige Tanaka, a well-known nineteenth-century inventor of various ingenious machines, to illustrate some historical and sociological aspects of Japanese clocks and their makers. Hisashige Tanaka, originally Giemon Tanaka, was born in 1799, the son of a craftsman who made objects from tortoiseshell. From childhood, he showed mechanical ingenuity into making curious and useful devices.[28] As will be explained, he later worked to repair clocks and constructed a magnificent clock by himself. With such ingenuity, independent craftsmen like Tanaka could become makers and repairers of mechanical clocks.

Tanaka then became an independent inventor and devised numerous entertaining automata – a teacup carrying doll, an arrow shooting doll, and so forth. He organized an exhibition to display the automata devices in the gardens of shrines and other places. The public display of entertaining automata was popular in various cities from the late eighteenth century. Tanaka's invention of intricate automata mechanisms and their public display can be contrasted to the ingenious automata clocks donated to the Chinese Emperor and kept in his private rooms. The Chinese Emperors welcomed and received ingenious and expensive mechanical machines like clockwork automata. Japanese feudal lords did not purchase such elaborate automata, or order the craftsmen whom they employed to devise and construct them. The automaton doll able to write Chinese characters was devised by a Swiss clock maker and sent to the Chinese Emperor. Tanaka also made a similar automaton, but it was made for public display at his entertaining show.

---

[27]The document did not specify how the bell strikers at other time bell towers determined the time stroke the bell, but it would be probable that the synchronized gongs of these four bells forming the inner circle around the castle were heard by the strikers of the other time bells which formed the outer circle around the castle and they followed the suit to gong the bell.

[28]Kenji Imazu, ed., *Tanaka Oumi Daijo*, unpublished revised edition, 1993, pp. 16–23.

Tanaka traveled widely around the western part of Japan, and settled down in Kyoto where he established a workshop to make various machines. Two time-related items brought him lucrative business. He found that he could repair the broken clocks that no other craftsmen could repair, and this business brought him a certain amount of profit. He also devised oil lamps with a pump mechanism and a glass shade, which provided brighter and more stable illumination than traditional candle lamps. This lamp called "*mujintō* (eternal lamp)" became extremely popular and was widely sold. Tanaka manufactured several different types of *mujintō* and succeeded in gaining enormous profits from this business. The popularity of the stable and bright light among city dwellers indicates how much they wanted to continue their activities during the time after sunset in the early nineteenth century.

With the profit gained by the *mujintō* business, Tanaka ventured to construct a magnificent clock system, "*Mannen Dokei* (Myriad Year Clock)".[29] This complex mechanical system, completed in 1851, had six faces, and a solar and lunar planetarium model on the top of the hexagonal body. The six faces on the main body included Japanese and Western clocks as well as a calendar and a lunar appearance display. Disassembling this *Mannen Dokei* revealed the details of its mechanism as well as technical characteristics of the design and construction of the clock. The clock has more than one thousand parts. Of these, one of the wheels remains to show the hundred angle-dividing lines which functioned to assist carving each cog of the wheel by a human hand. Without a machine tool, Hisashige or his assistants made most of the wheels by hand.

The Japanese clock controlled by the French watch is the *warikoma* type. Different from the ordinary *warikoma* type of clock, Tanaka's clock automatically moved the *warikoma*-plates throughout the year. Behind its face, the clock had an intricate mechanism consisting of tiny sets of cranks and oscillating wheels; these once-a-year oscillating cranks and wheels slowly slid the *warikoma* plates into position, to designate seasonal time. Besides this automatic mechanism of the *warikoma* clock, the Myriad Year Clock displayed the daily rotation of the sun and the moon above the earth represented by the map of Japan. The small white and red balls representing the two celestial bodies showed not only daily rotation, but also yearly change in their altitude. The astronomical model on the top also had a disk plate slightly below the plate of the map which corresponded with the twilight boundary time between daytime and night.

Tanaka intended to sell this clock, because he wrote an advertisement to explain how the system worked with the solar and lunar model. It was not purchased by any wealthy person, however. It is said that the lord at Matsue was quite intrigued by this clock and desired to purchase it. But his official advised against the purchase of such an enormously expensive machine. And

---

[29] Takehiko Hashimoto, "Mechanization of time and calendar: Tanaka Hisashige's myriad year clock and cosomological model," *UTCP Bulletin*, 6 (2006): 47–55, reproduced in idem, *Historical Essays in Japanese Technology*, UTCP Collection, vol. 6, 2009, p. 31.

the lord gave up his idea of purchase.[30] This episode indicates the limit of purchasing capacity of wealthy potential customers in Japan, which could be contrasted to that of the Emperor or to those who purchased luxurious good to donate them as a gift to the Emperor in China.

Conclusion

The mechanical clock, brought to Japan and China and adopted during the sixteenth and the seventeenth centuries, developed in strikingly different ways. In China, the Jesuits offered the chiming clock together with numerous other astronomical instruments to the Emperor and such gifts helped them to gain access to the Chinese Imperial Court. Indeed the gifts were so successful and crucial that all subsequent Jesuit missionaries and other visitors sought more and more elaborate mechanical clocks to offer to the ruler.

Jesuits who had arrived in Japan and who had brought clocks to local feudal lords in a warring state had to face an unfortunate fate after their half-century missionary activity. When their proselytizing activity in Japan ended, the maintenance and new construction of mechanical clocks had to be done by domestic craftsmen who, directly or indirectly, learned these skills from Jesuit craftsmen or acquired them by themselves. Though it is not certain whether the de-Christianization and indigenization of the mechanical clock were intentional or unintentional policy of the founders of the Tokugawa dynasty, they did bring a lasting influence on subsequent developments of the clock in Japan. All the clocks made and manufactured by Japanese craftsmen during the Edo period were simpler and more primitive than those imported and made in China. The spring-driven mechanism, hair-string control, and pendulum were not introduced until the late eighteenth century. However, unlike "self-sounding bells" displayed at the Emperor's private rooms, many clocks made by Japanese craftsmen were used for practical purposes.

The most important feature of the Japanese clocks was their adaptation to the seasonal time system, which was adopted from the beginning of the Tokugawa period and continued until its end. To adapt the clock to the seasonally variable time-reckoning system, the Japanese craftsmen first used weights hanging on a foliot as a variable control. Craftsmen later developed *warikoma*, the movable hour plates which were manually adjusted to the designated position at each seasonal division. The *shaku dokei* with a vertical scale and descending indicator later came to have seven scale plates or a single plate with a graph representing the seasonally varying hours. With these devices, the Japanese clock makers were able to adapt the mechanical clock to their seasonally variable time system. The mechanical clock had already become an important influence for change in Western society. However, in Japan, clock making had

---

[30] Kenji Imazu, ed., op. cit., 1993, p. 67.

to be adapted to the seasonal time system of pre-modern society. The clock time system was only achieved when the new Meiji government decided to adopt it as a part of the full fledged modernization of the nation.

Early customers for such mechanical clocks were feudal lords or high-ranking wealthy samurai bureaucrats, who could afford to purchase these expensive commodities. The relatively affluent clans could employ specialist clock making craftsmen and their assistants, who worked exclusively to make and maintain the official clocks possessed by the clan. The urban commoners living in the city furnished with time bell systems did not seem to feel the need for mechanical clocks as a time reckoning device, although, it should be noted, mechanical clocks were used by some officers in charge of striking time bells punctually. From the turn of the nineteenth century, mechanical clocks found more and more customers. The production of increased numbers of less decorative and more economical *shaku dokei* suggest the emergence of the practical need for a time reckoning and informing device among a certain class of wealthy citizens, who would probably become more time conscious towards the end of the Edo period. With the adoption of Western time and the calendar system in 1873, the history of Japanese clocks ended and a new history of the Japanese adaptation to modern social activities, controlled by the mechanical clock, began.[31]

---

[31] Shigehisa Kuriyama and Takehiko Hashimoto, eds., *The Birth of Tardiness: The Formation of Time-consciousness in Modern Japan*. A special issue of *Nichibunken Japan Review*, 14 (2002): 65–78; Takehiko Hashimoto, "The Japanese clocks and the history of punctuality in modern Japan," *East Asian Science, Technology, and Society: an International Journal*, 2 (2008): 123–133.

Adoption and Resistance: Zhang Yongjing and Ancient Chinese Calendrical Methods

Pingyi Chu

The fifteenth-century expansion of European science breathed life into hybrid scientific practices around the globe. The renowned historian of Chinese science Joseph Needham (1900–1995) memorably described an intellectual synthesis: "The older streams of science in different civilizations like rivers flowed into the ocean of modern science."[1] Needham's description sounds as if the Chinese should simply embrace Europe's "oecumenical" modern sciences and overlooks the new relations of power between the foreigners and their hosts. A broader perspective suggests that the landing of modern science on Chinese soil involved more than a few jolts.

Recent scholarship on colonial science has provided new ways of thinking about the export of European science.[2] Previous models, which often relied on a rudimentary dichotomy between the center and the periphery, failed to account for the resistance and contestation that accompanied scientific transactions. Sciences traveled through a number of different channels, maintained and modified by native users who contributed to the formation of colonial scientific practices. In addition, these users did not always view the European sciences as invaluable: they often attacked them vociferously. Instead of taking the universality of modern/European sciences for granted, these new findings have led

---

P. Chu (✉)
Institute of History and Philology, Academia Sinica, Taipei, Taiwan
e-mail: kaihsin.2007@gmail.com

[1] J. Needham, "The Roles of Europe and China in the Evolution of Oecumenical Science," in *Clerks and Craftsmen in China and the West: Lectures and Addresses on the History of Science and Technology*. Cambridge: Cambridge University Press, 1970, p. 397.

[2] Fa-ti Fan, *British Naturalists in Qing China: Science, Empire, and Cultural Encounter*. Cambridge: Harvard University Press, 2004; J. Canizares-Esguerra, "Iberian Colonial Science," *Isis*, 96, 1 (2005): 64–70; S. J. Harris, "Jesuit Scientific Activity in the Overseas Missions, 1540–1773," *Isis*, 96, 1 (2005): 71–79; M. Harrison, "Science and the British Empire," *Isis*, 96, 1 (2005): 56–63; M. A. Osborne, "Science and the French Empire," *Isis*, 96, 1 (2005): 80–87; L. Schiebinger, "Introduction," *Isis*, 96, 1 (2005): 52–55.

scholars to concentrate more on what the imported scientific practices were and who was involved when different traditions met.[3]

In the Chinese case, the Han Confucian literati had long recognized that religious and commercial interests mingled with the transmission of calendrical knowledge in the seventeenth and eighteenth centuries. They tended to possess an ambivalent attitude towards the European calendrical techniques employed by the Manchu conquest dynasty. They studied and discussed the new knowledge, and they came to appreciate the strengths of European astronomy and mathematics. However, they also sensed an essential incongruence in the blend of different European astronomical traditions presented to them as part of a coherent and continuous heritage. They also found parallels between their native traditions and European astronomy, leading some literati to conclude that the new sciences had Chinese origins. They discovered a lost Chinese tradition of astronomy and mathematics and integrated European astronomy and mathematics into their reconstructions of those traditions.[4]

Zhang Yongjing 張雍敬, the protagonist of the present inquiry, firmly rejected the notion that European methods had a place in traditional astronomy. He chose to reconstruct ancient methods without relying on Western learning or instruments, which he viewed as useless. Zhang was one of the limited number of Confucian astronomers who stood up to the new challenge from the West, determined to rejuvenate indigenous calendrical studies.

It is easy to dismiss Zhang as a stubborn conservative who did not even recognize that the earth was spherical. But *why* did he so persistently resist a system that had been endorsed by the emperor himself? What rationale underlay his discourse and what can this rationale tell us about science and culture in his time? By studying his research as well as his criticisms of European science, we will enrich our understanding of the complicated story of European science in China.

Zhang Yongjing and *Dingli Yuheng*

Zhang Yongjing is an obscure figure. Had not one of his manuscripts, *Dingli yuheng* 定曆玉衡 (Guidelines for Producing Calendars), recently been unearthed, we would probably know nothing about him. He came from Xiushui 秀水 county in Zhejiang 浙江 province, and he learned astronomy from his father. Zhang Yongjing was a versatile writer. Like most literati, he spent his early years preparing for the examinations and left some commentaries on Confucian classics. Zhang also wrote plays, though none of them has survived. In addition, he was a painter and poet, and his works received high praise from

---

[3]B. Elman, *On Their Own Terms: Science in China, 1550–1900*. Cambridge: Harvard University Press, 2005.
[4]Pingyi Chu, "Remembering our Grand Tradition: *Chourenzhuan* and the Scientific Exchanges between China and Europe, 1600–1800," *History of Science*, 41, 2 (2003): 193–215.

fellow scholars. Like most of his contemporaries, the "lost generation" immediately after Ming-Qing transition that has been depicted by Lynn Struve,[5] Zhang became a sojourning scholar. He once visited the capital, seeking patronage, but failed. He then traveled around the country and returned to his hometown. For the latter part of his life, he immersed himself in calendrical studies; his one surviving treatise was the most significant fruit of his endeavors.[6]

An early version of *Dingli yuheng* was probably finished around 1690. Zhang then visited the renowned astronomer Mei Wending 梅文鼎 (1633–1721), whom he trusted to offer sound criticism of his production. The extant manuscript of *Dingli yuheng*, probably finished around 1699, bears the traces of their discussion: Zhang cited Mei's reactions and made appropriate revisions. But he did not move any closer to accepting the European idea. The manuscript of *Dingli yuheng* preserves an original and a revised copy of the table of contents. In the former, Zhang indicated which sections he considered correct and satisfactory – most of these involve the reconstruction of ancient calendars. The revised table of contents included new sections related to European geographic and astronomical knowledge. But, in fact, few additions were made to the text: Zhang rearranged things, then added labels that might conciliate critics, but his arguments did not change at all. Like all previous scholars, he held that the earth was flat. Later in the Qing, scholars remembered Zhang principally for stubbornly refusing this idea which many Confucian astronomers had accepted as an essential assumption for astronomical calculations.[7]

Against Western Methods: The Problem of Vision

In his attacks on Western cosmology, Zhang Yongjing rejected the notion that the sun revolved at varying speeds, that the shadow of the earth caused lunar eclipses and that the earth was spherical.[8] If he could prove that the fundamentals of the foreign science were fallacious, he could fully justify a return to the ancient Chinese methods. Like many of his colleagues, Zhang believed that the real picture of the cosmos was attainable by recognizing the principle underlying

---

[5]L. Struve, "Ambivalence and Action: Some Frustrated Scholars of the Late K'ang-hsi Period," in S. Jonathon and J. E. Will Jr., eds., *From Ming to Ch'ing*. New Haven: Yale University Press, 1979, pp. 323–365.
[6]Shikai Zhu, 朱士楷, *Xincheng xianzhi* 新塍鎮志 (Gazetteer of Xincheng town). Shanghai: Shanghai shudian, 1992, pp. 817, 898–899, 989.)
[7]Pingyi Chu, "Trust, instruments, and cross-cultural scientific exchanges: Chinese debate over the shape of the earth, 1600–1800," *Science in Context*, 12, 3 (1999): 385–411; Yuan Ruan, 阮元, *Chourenzhuan* 疇人傳 (Biographies of Astronomers and Mathematicians). Taipei: Shijie shuju, 1982, p. 504.
[8]Zhang Yongjing, 張雍敬, *Dingli yuheng* 定曆玉衡 (Guidelines for Producing Calendars). In *Xuxiu siku quanshu* 續修四庫全書 (Addendum to the Complete Collection of the Four Treasuries), Vol. 1040, Shanghai: Shanghai guji chubanshe, 1997, pp. 485–494.

the celestial phenomena. He contended against Ptolemaic cosmology that if the planets simply revolved on concentric crystal shells, the solid planets would have fallen to the earth. And he harped on the inability of Ptolemaic models to explain comets and retrograde planetary motions. Although it is possible that Zhang learned about the improved model developed by Tyco Brahe (1546–1601) – court astronomers did use it during the Kangxi 康熙 period (1661–1722) – he referred only to the somewhat old-fashioned crystal-shell model, which set God in the outermost shell. Is it possible that this intertwining of astronomy and religion was Zhang's real target? By rejecting Ptolemaic cosmology, he also refused its religious implication: the existence of a 'European God'.

Curiously, many of Zhang Yongjing's arguments against the Western methods were somehow related to vision and visual representations. He confronted the numerical values given by European astronomers with their Chinese counterparts. For example, in a discussion of the size of the planets, Zhang argued that since European scientists had calculated the distance between the sun and the earth as roughly 16,000,000 *li* (about 8,000,000 kilometers), and the apparent size of the sun was roughly 160 times greater than the earth, the apparent size decreased 1 percent per 100,000 *li*. Applying this ratio to the size of the moon, which was, according to Europeans, 480,000 *li* from the earth, Zhang concluded that the moon ought to be about 4.8 times smaller than the earth. However, European cosmology declared the size of the moon was 38 times smaller than the earth. Moreover, according to the ratio he had worked out between the apparent size and the distance, when the sun came closer to the earth in the summer, it should seem far larger than it did. Zhang concluded that Western cosmology simply contradicted itself.[9]

To understand why ideas about the apparent size preoccupied Zhang Yongjing, we need to examine the works of Matteo Ricci and Manuel Dias. These two Jesuit missionaries wrote works in Chinese in which they discussed theories about human vision and employed visual representations to account for celestial phenomena. They also imported European optics to explain how certain graphs and instruments worked. The elaborate explications of refraction and, later, the telescope, were new to Zhang and his contemporaries, but rather than giving them confidence about phenomena previously inexplicable, they contributed to a mistrust of vision that permeated Chinese thought. Although Chinese scientists had grasped some elements of the behavior of light, the problem of vision, light, and shadow had never formed a coherent field of investigation. In other words, European optics did not exist in China.[10] This clash of visual cultures led some to criticize European visual representations, in particular the use of linear perspective.

---

[9]Zhang Yongjing, *Dingli yuheng*, p. 484.

[10]Daiwie Fu, "Problem Domain, Taxonomy, and Comparativity in Histories of Sciences: with a Case Study in the Comparative History of 'optics'." in Cheng-hung Lin and Daiwie Fu, eds., *Philosophy and Conceptual History of Science in Taiwan*, Boston Studies in the Philosophy of Science, Vol. 141, Dordrecht: Kluwer Academic Publishers, 1992, pp. 123–148.

The Picture of the Cosmos

As a substitute for Ptolemy's crystal-shell model, Zhang Yongjing proposed a revival of the ancient *xuanye* 宣夜 cosmology.[11] This obscure theory had been long lost even in the third century AD. From the few lines about this theory left by scholars in the third century, it appears to have asserted that the sky was composed of unbounded *qi*. The apparent retrograde motion of planets was possible exactly because the sky was not solid. No astronomical techniques based on this particular cosmology had ever been developed, however. The *xuanye* cosmology was mainly a source of fairly abstract speculations. Zhang Yongjing nonetheless managed to invest this long-lost cosmology with new vigor. He concocted a new version of *xuanye* cosmology in which *qi* acted as a creative force: it had created everything known to man, including a cosmos readily divided into nine parts (an inspiration from Western cosmology), with a flat earth at the center. The *qi* of the great harmony filled the space between the earth and the lowest heaven, and it was there that the myriad things proliferated.

Zhang Yongjing deliberately picked retrogression to demonstrate the explanatory power of his modified version of *xuanye* theory because he believed that the crystal-shell model faced profound challenges in accounting for it. According to Zhang, the planets moved counter-clockwise. Their speed of revolution decreased in proportion to their distance from the outermost layer of heaven, which rotated at the highest speed. The sun, the measure of the relative speed of other planets, took a year to complete a circle. Retrogression occurred when the earth, a planet, and the sun aligned. At such moments, Zhang alleged, the planet speeded up thanks to a boost of *yang qi* from the sun. At the other extreme, when the sun and the planet were in opposition, the planet moved fastest, so fast that its speed exceeded the outermost heaven, and it received the strongest *yang qi* from the sun.[12] An observer on earth thus found that the planet retrogressed vis-à-vis the outermost heaven at those two moments. Zhang dexterously mobilized resources from the Chinese astronomical tradition to solve an anomaly in Ptolemaic cosmology. If he could convince his readers that his solution was correct, he could legitimate a properly Chinese cosmology.

When he turned to ancient texts to determine the scale of the cosmos, Zhang soon found that the sizes of the earth and of other heavenly bodies were not congruent.[13] This led him to give up on finding precise measurements of the cosmos – precisely what European astronomy offered. A rough measurement with the aid of a gnomon was possible: that, for Zhang, was enough. The ancient sages simply never meant to create a method to survey the cosmos, and while some claimed that the methods recorded in the *Zhoubi suanjing* (the

---

[11]Zhang Yongjing, *Dingli yuheng*, pp. 451–456.
[12]Ibid., pp. 451–456.
[13]Ibid., pp. 458–460.

arithmetical classic of the gnomon)[14] provided a sound basis for mathematical astronomy, Zhang insisted that this text was too recent to qualify as part of the sagely tradition. Precise measurements, he said, were simply unnecessary.[15]

Zhang explained the physics of atmospheric distortion to deny the efficacy of mathematical techniques and European cosmology. Since our vision was limited and the universe was infinite, human eyes were incapable of grasping the cosmos. The key was refraction: all our eyes could make of light traveling to the earth from the stars was illusions (*xuxiang* 虛象), not reality. Westerners, Zhang insisted, considered the earth spherical because they had failed to take refraction into account. Moreover, Zhang argued that if the earth were spherical, people on the far side of the planet would be standing upside-down – such a thing was utterly fantastic.[16] Though Zhang cited his friend Wang Zhouzai 汪周載 as an expert on how refraction and visual distortions work, ironically, Wang's argument and example actually were drawn from Matteo Ricci's work.

Zhang then offered his own description of the cosmos. If the planets were accumulated *qi* rather than physical entities, the European measurement of distances between planets became meaningless; it was the relative positions between the planets, the angular measurements called *du* 度, that provided the key to understanding the cosmos. A *du* equalled to $\frac{360}{365.25}$ degree. The circumference of heaven, which was thought to be spherical, measured 365.25 *du*, with one *du* equal to the distance the sun moved in a day. The earth, a round disk with China at its center, was located, not on the central plane of the sphere of heaven, but beneath it and occupied one-third of the lower hemisphere of heaven. The intersection between heaven and earth was filled with water. Zhang claimed that the size of the earth was not much larger than China, about two thousand *li* in diameter. Since the boundaries of the earth were quite limited, the missionaries must have been lying when they said they came from a place ninety thousand *li* away. Based on this basic picture, Zhang then derived a whole series of figures for the cosmos.[17] After repudiating the quantitative fetish of Western scientists, he gave in and provided lots of numbers himself.

Zhang Yongjing did not fabricate his cosmography out of thin air – he relied heavily on various texts available to him.[18] He turned even to geomantic texts to justify his premise that China, a place where sages proliferated, was at the center of the earth. He firmly believed that ancient calendrical methods held the truth to understanding the stars and planets whirling overhead.

---

[14]Ibid., p. 461.
[15]Ibid., p. 457.
[16]Ibid., pp. 460–462.
[17]Ibid., pp. 470–476.
[18]Yuan Ruan, op. cit., pp. 341–345.

The Invention of Ancient Methods

After the dramatic arrival of European astronomy, Chinese scholars gradually came to believe in an ancient Chinese calendrical tradition. According to the classics such as *The Book of Documents*, calendars had been crafted by the sage kings. However, before the transmission of the Western astronomy, the so-called ancient calendrical methods were often considered lost or imprecise.[19] Later astronomers sometimes used *gufa* 古法 or "ancient methods" to refer to the calendrical methods used before their times. In the context of the Ming, *gufa* often referred to the methods bequeathed by Guo Shoujing 郭守敬 (1231–1316).[20] In any case, astronomers rarely mythologized the ancient methods or looked to them to provide comprehensive solutions to all calendrical problems. This situation changed, however, after the Western calendrical methods arose as competitors to the Chinese calendar.

The late Ming critics of European science had alluded to ancient methods, but they failed to specify the relationship between these practices and the sages. The association between the calendar and the Confucian tradition was solidified when Yang Guangxian 楊光先 (1597–1669) took on the German Jesuit Adam Schall von Bell (1591–1666) in 1664.[21] All worthy debates had to have a textual basis, Yang argued. In the case of the calendar, the "Yaodian" 堯典 (Canon of Yao) chapter in *The Book of Documents* had reliably documented an ancient sage's effort at calendar making. Yang warned that Schall's pitiless attacks on the Chinese calendrical tradition threatened the Qing's connection to its Confucian heritage.[22] He equated a viable Chinese calendar with a respectful attitude toward the ancient sages, the very source of Chinese culture.

Later Mei Wending expanded the role of the sages in making the calendar. He contented that calendrical techniques had come to be more precise with the passage of time, not because the moderns were cleverer than the ancients, but simply because celestial phenomena changed so subtly, so slowly, that it took a long time to notice. The ancient sages had recognized this profound subtlety, and therefore established a flexible method that could be modified over time. They had thus foreseen all later methods.[23] In other words, although the sages

---

[19] Zheng Wei, 魏徵 et al., eds., *Suishu* 隋書 (Sui Dynastic History). Taipei: Dingwen shuju, 1980, p. 523.

[20] Tingyu Zhang, 張廷玉 et al., eds., *Ming shi* 明史 (Ming Dynastic History). Taipei: Dingwen shuju, 1982, p. 518.

[21] Pingyi Chu, "Scientific dispute in the imperial court: the 1664 calendar case," *Chinese Science*, 14 (1997): 7–34.

[22] Guangxian Yang, 楊光先, *Budeyi* 不得已 (I cannot do otherwise), in W. Xiangxiang ed., *Tianzhujiao dongchuan wenxian xubian* 天主教東傳文獻續編 (Supplement to the Documents on Christianity Coming to the East), Vol. 3, Taipei: Xuesheng shuju, 1965, pp. 1157–1161.

[23] Wending Mei, 梅文鼎, Lixue yiwenbu. 曆學疑問補 (Supplement to the Questions on Calendrical Learning), in *Lisuan quanshu* (Complete Works on Mathematics and Astronomy), in *Siku quanshu* (Complete Collection of the Four Treasuries), Vol. 794, Taipei: Shangwu yinshuguan, 1983, pp. 6–7.

had not been able to produce a perfect, timeless calendrical technique, they had grasped the ultimate principle of calendar making. After the Kangxi emperor took an interest in Mei Wending, his views became very influential.

Zhang Yongjing employed an argument quite similar to Mei's. As though accepting a challenge, he traced the history of calendrical methods back much further than Yang Guangxian had. Zhang believed that the ability to make accurate calendars had functioned as a crucial governmental tool at the very beginning of human history. In addition, like most of his contemporaries, Zhang believed that all calendrical techniques could be traced to *Yijing* 易經 (The Classic of Changes). If the calendrical techniques had lost precision, this was because later generations had proven unable to uphold the principles bequeathed by the sages. Once human beings understood the universal calendrical principles, they would produce a calendar that would never need revision. Since there was no way to foretell whether any given technique would need revision in the future, the best way to verify the precision of a calendrical model was to check it against past records. Based on this argument, Zhang insisted that it was not difficult to make a calendar; the real challenge lay in detecting the shortcomings and strengths of various ancient calendrical techniques.[24]

Since Zhang believed that one had to learn from the ancients, he began his reconstruction of ancient calendrical methods by studying ancient astronomy. He listed thirty-three historical techniques, including measuring the position of the pole star, or *bu dou* 步斗; dividing the field, or *fenye* 分野; calculating the three orthodox calendars, or *santong* 三統; and so on.[25] Although these were common astronomical terms, Zhang Yongjing offered idiosyncratic definitions of each of them. Moreover, he often claimed that the common understanding of these terms was erroneous. It is difficult to understand what he really meant. While he boasted of having recovered the secrets of the ancients, Zhang never did explain how his methods could produce a contemporary calendar.[26] His achievement was the application of astronomical data in ancient texts to reconstruct the ancient methods; he left it to others to extrapolate a practical, modern calendar.

Metaphysical Foundations of the Calendar: *Li* and *Shu*

Although many Chinese mathematicians and astronomers insisted on the superiority of their own traditions, European mathematical and astronomical texts, techniques, and instruments had deeply penetrated their thinking. To justify their use of foreign techniques and evaluate the relations between Western and Eastern techniques, they attempted to build a firm metaphysical

---

[24]Zhang Yongjing, *Dingli yuheng*, pp. 436–440.
[25]Ibid., pp. 441–444, 512–513, 528–541, 638–642.
[26]Ibid., p. 645.

foundation for their studies. They meant reconstructing the relationship between *li* 理 (principle) and *shu* 數 (numbers).

When the missionaries came to China, they soon realized that Confucian scholars had a competing model to account for the phenomenal world: the idea that matter's constitutive element, *qi*, was governed by *li*. *Li* preceded *qi* and regulated its transformation; therefore *li* was above and beyond the phenomenal world constituted by *qi*. Consequently, one had to grasp *li* in order to understand the phenomenal world.

The missionaries persistently denied the validity of this neo-Confucian theory; since God was the only creator, it was impossible that *li* had "spawned the myriad things." The error, they explained, emerged as soon as one posited a dyad: *li* was inseparable from the things it purportedly animated.[27] The Western missionaries understood the concept of the *li* as something like the "causes" in Aristotelian philosophy. Therefore, *li* was used to account for how and why phenomena had occurred: it was bounded by the phenomena it sought to explain. Since the Aristotelian philosophical scheme had successfully served the missionaries' interpretation of phenomena, the missionaries tended to proudly proclaim that their understanding of *li* was superior to that of Confucian scholars.

Did *li* produce *qi*? Did *li* inhabit *qi*? Chinese astronomers chose one theory or the other, depending on their relation to Western science. Regardless of their explicit ideological attitude towards Western learning, those who applied European mathematics and astronomy to their research often held the position that *li* existed in *qi*. Such a stance enabled them to recognize the significance of instruments and highly quantified methods. On the other hand, those Confucians who emphasized the priority of *li* tended to look down upon the technical aspects of calendar making and demeaned instruments and observations. This metaphysical foundation even led to Yang Guangxian's mocking claim that he understood the *li* of calendar making without having the slightest idea of how to produce a calendar.[28]

Like Yang, Zhang Yongjing emphasized the priority of *li* over numbers. Still, he argued, along with many of his Chinese contemporaries, that to make proper calendars one needed an understanding of both. Although the European missionaries were good at manipulating numbers, he explained, their minds were clouded when it came to discussing *li*. They might astound their audience with facile displays, but their grasp of the root cause of these phenomena fell far behind Chinese astronomers.

For Zhang, *li* alone rendered the phenomenal world logical and explainable. And, like Yang Guangxian, he scoffed at the idea that all had to be quantified. Such material aspects of astronomy as telescopes and mathematical and

---

[27]Yongtang Zhang, 張永堂 *Mingmo Qingchu lixue yu kexue guanxi zailun* 明末清初理學與科學關係再論 (Reinvestigating the Relationship between the Study of Principle and Science), Taipei: Taiwan xuesheng shuju, 1994.
[28]Guangxian Yang, op. cit., pp. 1263–1264.

astronomical tables interested him not at all. A single principle governed such devices and calculations, and if one only indulged in the instruments and numbers without understanding the principle, one was doomed to be lost in trivia and incapable of producing a good calendar.[29]

In sum, Zhang's quite conventional view of the relationship between *li* and *shu* enabled him to defend the ancient concepts he had unearthed but also led him to discard the significance of tools, of calculations based on repeated observations, and of the manipulations that instruments could accomplish. While many of Zhang's contemporaries had adopted the astronomical techniques unveiled by Ricci and other Jesuit missionaries, Zhang's complex textual operations were driven by the longing to save neglected traditions and to undermine the very foundations of Western science.

Conclusion

After the arrival of Christianity in China, European sciences percolated into mainstream Chinese astronomical discourse along a variety of routes.[30] No picture of this dramatic series of events would be complete without taking into account resistance. Zhang Yongjing's story, like that of Yang Guangxian, shows that Chinese astronomical traditions were not simply washed away by an allegedly superior rival in the seventeenth and eighteenth centuries. Moreover, Confucian astronomers who resisted European calendrical methods worried about other things besides efficacy. Reactions to a paradigmatic crisis varied based on such personal considerations as relations with the Manchu court, commitment to Confucian values, and a general sense of the essential nature of the cosmos. The migration of European sciences was thus embedded in a complicated politico-cultural context, not mere efficacy and rationality.

After the imperial adoption of European calendrical methods, Confucian astronomers suddenly realized that the Chinese astronomical tradition faced a crisis. They reacted to the crisis with a combination of ideological struggles and technical appropriations. Many chose to adopt techniques from Western astronomy while insisting that the new methods had a Chinese origin. Compared to that, Zhang Yongjing's efforts to reconstruct the ancient methods appear as an interesting anomaly.

He stuck to the Confucian belief that *li* ultimately governed *shu*: if an astronomer had a duty, it was to understand *li*. Zhang cautiously avoided a direct ideological competition with the European missionaries, attacking the flank of European astronomy by reconstructing ancient indigenous methods

---

[29]Zhang Yongjing, *Dingli yuheng*, p. 644.
[30]Pingyi Chu, "Archiving knowledge: a life history of the *Chongzhen lishu* (Calendrical Treatises of the Chongzhen Reign)," *Extrême-Orient, Extrême-Occident* (2007): 159–184.

Adoption and Resistance

from various textual resources. This search for the past inevitably became a reinvention of tradition.

Zhang grafted European cosmology to traditional *xuanye* cosmology, yielding a new picture of the cosmos. Arguing that the stars were accumulated *qi* without physical form, Zhang denied the effectiveness of the visual theory the missionaries relied on. He then accused them of mastering the mechanics of turning out an accurate calendar without appreciating the underlying *li*. Interestingly, the missionaries blamed Chinese astronomers on the same ground. To recover what was missing, Zhang studied the ancient calendars and drew from them the concepts that he considered significant. He believed that the best way to verify the effectiveness of a calendrical technique was to test its applicability to ancient data and cosmological events. Since the *li* at work governed both past and future events, the ability of a calendrical technique to generate accurate readings could, as an endorsement, hardly be bettered. Zhang claimed that the astronomical constants that he had obtained using his methods surpassed all preceding estimates.[31] Moreover, Zhang's emphasis on the priority of *li* also led him to trust the words of the ancients more than the results of modern technical practices and findings. His failure to appreciate the strong material aspects of the new science undermined his efforts to save the Chinese astronomical tradition.

Zhang was probably the last person to reconstruct ancient methods without relying on Western astronomy. But the intellectual and social environments in the years around 1700 did not favor his ideas and his book does not appear to have ever been published; it certainly did nothing to drive Chinese astronomy away from the foreign methods.

In the years that followed, Zhang's research program evaporated without a trace. Like Zhang, many Confucian astronomers turned to ancient texts for help when reconstructing ancient calendrical methods, but if Zhang had studied ancient calendars in order to discover the timeless methods of the sages, others wished only to prove that China had given birth to what eventually became Western astronomy. The Confucian astronomers of the eighteenth century were aware, it seems, of the limitations of the Chinese astronomical tradition; the only way to salvage it was to link it to the alien concepts and instruments they had gradually come to accept. Zhang Yongjing's quixotic call for a return to ancient astronomy hung in the air for a time, provoking some raised eyebrows and murmured comments, but very soon he and his intemperate diatribe were forgotten.

---

[31] Ibid., pp. 437–438.

Part III
On Localizing, Appropriating and Translating New Knowledge

Travelling Both Ways: The Adaptation of Disciplines, Scientific Textbooks and Institutions

Dhruv Raina

The historiography of sciences, we have been informed, has been parasitic upon the sciences, and as a genre has been the most conservative of the genres of historical writing.[1] However, the externalities and internalities that the history of sciences is bent on identifying, shape the historiography of science itself; and the so called parasitic determinants may well belong to the domain of internalities. Returning to the theme of the present paper, I look at the system of colonial education in British India as a site for the 'expansion of European science' in non-Western contexts. Thus while post-colonial theory has since dispensed with and extensively critiqued the notion of 'European science',[2] the question of the localization of so-called Western science has been addressed from the perspective of the practitioners of scientific disciplines in research environments. This leaves out the schools and institutions of higher learning, which are veritably the sites for the localization or encounter of the new knowledge form and existing ways of knowing and acting upon the surroundings.

In moving towards the context of localization the focus of discussion gradually slides away from the epistemological contest between different knowledge forms to one on processes of cultural redefinition, translation, the invention of

---

D. Raina (✉)
School of Social Sciences, Jawaharlal Nehru University, New Delhi, India
e-mail: d_raina@yahoo.com

[1] S. Shapin, *A Social History of Truth: Civility and Science in Seventeenth Century England*, Chicago and London: University of Chicago Press, 1994; S. Shapin and S. Schaffer, *Leviathan and the Air-Pump: Hobbes, Boyle and the Experimental Life*, Princeton: Princeton University Press, 1985.

[2] D. Chakrabarty, "Postcoloniality and the Artifice of History: Who Speaks for "Indian" Pasts", in R. Guha, ed., *The Subaltern Studies Reader: 1986–1995*. Delhi: Oxford University Press, 1998, pp. 261–293; S. Harding, *Is Science Multicultural? Postcolonialisms, Feminisms, and Epistemologies*, Indiana: Indiana University Press, 1998; G. Prakash, *Another Reason: Science and the Imagination in Modern India*, Princeton: Princeton University Press, 1999; D. Raina, "Multicultural and Postcolonial Theories of Science and Society", *Sandhan*, 5, (2005):1–32; D. J. Hess, *Science & Technology in a Multicultural World: The Cultural Politics of Facts & Artifacts*, New York: Columbia University Press, 1995.

new meanings and modes of communication.[3] This creation of a new form of socialization is both practical and cognitive, requiring an inventiveness on the parts of its proponents that is often seen in the expanding phase of religious movements. The idea that cognitive movements exhibit a similar logic of propagation is not a new one and is encountered in an important way in the *The Structure of Scientific Revolutions* of Thomas Kuhn.[4] Scientists committed to a paradigm resist changing their research orientations in the face of mounting anomalies; and Kuhn likened the 'inevitable shift' of paradigm to an act of religious conversion.[5]

While recently reading a programmatic statement for the study of the career of an Eastern religion in a Western context, I was surprised to see that mechanisms similar to the spread of Western science in non-Western contexts were at work.[6] Much of what I say in the next three paragraphs attempts to adapt Bruinssen's argument for religion to science. Thus wherever modern science was introduced and institutionalized it had gone on to develop local forms. These local forms were not to be analyzed as the persistence or signature of pre-modern knowledge forms and practices that had not yet been expunged or purged by the new forms and practices. Consequently, regional variations in these forms of knowledge and practices correspond to the diversity of roles they come to serve in a variety of contexts. For example, pedagogy and evidential cultures could be organized differently; in similar fashion the contract between state and science could be enacted and enforced differently. When viewed under a microscope the purely normative account of science may then exhibit distinct regional adaptations. Furthermore, recipient cultures and nations could proffer varying constraints and possibilities for the development of scientific institutions, thoughts and practices. In this manner local forms of science grounded in locally acquired knowledge of science develop.

The sources of this knowledge, read modern science, may have been originally located in Europe and transnational European networks that claimed this knowledge to be of European origin, further claimed that it was a purer and more universal knowledge. The gradual shift in non-European countries from vernaculars to English as the language of scientific discourse was reflected in the changing patterns of authority and changes in discourses themselves. Over the centuries non-Western nations have produced

---

[3]S. L. Montgomery, *Science in Translation - Movements of Knowledge through Cultures and Time*, Chicago: University of Chicago Press, 2000; B. Metcalf, "Hakim Ajmal Khan: Rais of Delhi and Muslim leader", in R. E. Frykenberg, ed., *Delhi through the Ages*, Oxford: Oxford University Press, 1986.

[4]T. Kuhn, *The Structure of Scientific Revolutions*, Chicago: University of Chicago Press, 1962.

[5]S. Fuller, *Thomas Kuhn: A Philosophical History of our Times*, Chicago: University of Chicago Press, 2000, p. 2.

[6]M. van Bruinssen, "The production of Islamic knowledge in Western Europe", *ISIM Newsletter*, 8 September 2001, p. 3.

considerable scientific information revealing a multiplicity of voices, arising from the diversity of the countries of origin of this information as well as variations in socio-cultural contexts and patterns of social interaction. So while each region displays a degree of autonomy that defines its identity, there emerge a number of scientists whose authority transcends regional and national boundaries.

The production of local knowledge commences when modern science is planted outside its 'original' heartland. Modern science emerged in a specific historical context of Western Europe. On expanding into other cultures it undergoes a dual process of universalization and localization. The producers and brokers of scientific knowledge selected elements that could be discarded and those considered essential and non-negotiable – this is the process of universalization or the separation of what was considered universal in science from the contingent. Localization required the adaptation of universalized sciences to local practices and needs.[7] The process of production of local scientific knowledge continues as local traditions of scientific knowledge develop. Scientific knowledge in non-Western regions is produced on the basis of universalized versions of local knowledge of the home countries and other prestigious centers. There is a process of abstraction and another of adaptation of discourse and practices to local conditions.

This paper looks at some cases of adaptation of modern science within the contexts of education in colonial India. Historians of science, with some exceptions,[8] have paid little attention to the educational dimensions to the expansion of Western science, assuming all along that once the scientific community and the literati class had accepted modern scientific theories, it was but natural in the course of events that these theories would become part of educational curricula; or on the other hand they overemphasize the hegemony of the colonial administrators in enforcing their writ on science education. Both these theoretical assumptions need clarification; for it may well be that the truth resides not just somewhere in between but somewhere else.

Several models for the transmission of scientific knowledge from West to non-West have been in circulation for sometime now.[9] These have been situated in historiographic frames of either the transmission of scientific theories or technologies; or their uptake explained in terms of a diffusionist model. Postcolonial theory of science is seen as offering a historiographic departure in the

---

[7]D. Raina and S.I. Habib, *Domesticating Modern Science: A Social History of Science and Culture in Colonial India*, Delhi: Tulika Books, 2004.

[8]S.N. Sen, "The Character of the Introduction of Western Science in India during the Eighteenth and Nineteenth Centuries," *Indian Journal of History of Science*, 1 and 2 (1988): 112–122; D. Kumar, *Science and the Raj*, Delhi: Oxford University Press, 1995; D. Raina and S.I. Habib, *Domesticating Modern Science: A Social History of Science and Culture in Colonial India*, Delhi: Tulika Books, 2004.

[9]D. Raina, *Images and Contexts: Studies in the Historiography of Science in India*, Delhi: Oxford University Press, 2003.

transmission of scientific knowledge.[10] There is nevertheless a rich literature on the subject examining the transmission of scientific knowledge from the West that seeks to locate knowledge and power in British India drawing insights from the work of Foucault and Said. However, there are important distinctions in the understanding of how colonialism transformed the coordinates of understanding Indian knowledge systems. Despite the differences between these Foucauldian approaches, Orientalism appears as a shorthand for the 'imperial sociology of knowledge.'[11] These differences arise from differences in the understanding of the establishment of colonial rule in India. Thus Chris Bayly[12] addresses the importance of the knowledge received in the making of empire; and how a new information order that is created subsequently structures commerce and politics.[13] Whereas Cohn, Inden[14] and others elaborate how the creation of knowledge about India contributed to the furtherance of the imperial project. Colonialism as post-colonial scholarship has detailed for us, produced 'complicated' forms of knowledge that 'Indians had constructed but were codified and transmitted by Europeans.'[15]

The task of offering a taxonomy or a clustering of the different narrative families is germane. This paper draws upon some recent studies more so in the history of education or the history of the setting up of disciplines in different cultural contexts to reopen discussion of localization, to suggest that localization within the realm of higher education was more or less a precondition for the expansion of science. Thus the attempt here is to look at specific scientific activities that are then mapped onto a specific narrative frame of transmission. The multiplicity of narrative forms informs us what localization means for each of these activities; whether it involves the production of new curricula, the creation of new organizational structures, or even the production of new kinds of knowledge.

Institutionalization of Psychoanalysis

Studies on the reception of Freudian psychoanalysis in India in the late nineteenth century reveal important features of the localization of a discipline. The field of study was first opened up through the studies of Christiane Hartnack,

---

[10]D. Raina, "Multicultural and Postcolonial Theories of Science and Society", *Sandhan*, 5, 1 (2005):1–32.

[11]W. R. Pinch, "Same Difference in India and Europe," *History and Theory*, 38, 3 (1999): 389–407 (pp. 390–393).

[12]C.A. Bayly, *Empire and Information: Intelligence Gathering and Social Communication in India, 1780–1870*. Cambridge: Cambridge University Press, 2000.

[13]Pinch, "Same Difference in India and Europe," p. 395.

[14]R. Inden, *Imagining India*. Oxford, Cambridge: Blackwell, 1990.

[15]B.S. Cohn, *Colonialism and its Forms of Knowledge: The British in India*, New Delhi: Oxford University Press, 1997, p. 16.

followed by that of Ashis Nandy.[16] Since then, the floodgates have opened. Some of these studies have focused upon the persona and efforts of Girindrasekhar Bose, Freud's first Indian disciple, in institutionalizing the discipline in India, and the subsequent evolution of the discipline following upon the institutional legacy left behind by Bose. Equally important have been the larger cultural issues and concerns that structured the reception and cultural reappropriation of the discipline in India.

On investigating the emergence of psychoanalysis, a striking fault line becomes evident, wherein psychoanalysis is seen to offer a critique of bourgeois society. And yet its enlightenment vision rendered it incapable of providing a critique of colonial practices, rendering it a tool in the hands of the colonizers as a technology of colonial domination. For the first generation of British psychoanalysts in India, the Freudian science offered a tool for comprehending the dark side of the worshippers of Kali, a discipline to control and colonize the subject population. On the other hand, Indian scientists within the frame of the nationalist struggle re-appropriated psychoanalysis as a technology of emancipation and possibly as a weapon in anti-colonial struggle. However, the modality of cultural appropriation of psychoanalysis within this context is important for it is contended that the process of appropriation produced a set of psychoanalytical practices that reproduced the traditional authoritarian hierarchies of Indian society. The cultural appropriation of psychoanalysis entailed redefining the goal of psychoanalysis as oriented towards the higher brahmanical goals of self-realization. Here as was done with several other scientific disciplines, counter-colonial discourse, worked out of a frame of Western exteriority and an Indian interiority.[17] Psychoanalysis was thus homologous with the theories of Samkhya and Yoga. The need of the Indian nationalists to reassert a modern identity stimulated the challenge to the universality claims of Freudian psychoanalysis.[18] This triggered processes of cultural appropriation that exposed the traditional hierarchies within Indian society, while at the same time embedding it within the moral economy and discourse of that very social order.[19]

Furthermore, several practices were redefined as the goals of psychoanalytic practice were reset. The relationship between the analyst and the patient was

---

[16]C. Hartnack, "British Psychoanalysis in Colonial India" in M. G. Ash and W. Woodward, eds., *Psychology in Twentieth Century Thought and Society*, Cambridge: Cambridge University Press, 1987; A. Nandy, "The Savage Freud: The First Non-Western Psychoanalyst and the Politics of Secret Selves in Colonial India", in *The Savage Freud and Other Essays on Possible and Retrievable Selves*, Delhi: Oxford University Press, 1995.

[17]See paper entitled "The Moral Legitimation of Science" in D. Raina and S.I. Habib, *Domesticating Modern Science: A Social History of Science and Culture in Colonial India*, Delhi: Tulika Books, 2004; P. Chatterjee, *Nationalist Thought and the Colonial World: a Derivative Discourse*. Minneapolis: University of Minnesota Press, 1993.

[18]Hartnack, op. cit.

[19]P. Kharbanda, The Cultural Reception of Psychoanalysis in India: A Study of the Institutionalization of a discipline (1900–1950), M.Phil. dissertation, Jawaharlal Nehru University, 2004.

reworked in terms of the *guru-sisya* relationship; the training protocol by the early decades of the twentieth century in India moved away from institutes of psychoanalysis to a university department. These moves framed the institutionalization of the discipline within a moral universe.[20] And here paradoxically we see an overlap between the colonial and anti-colonial agendas. The colonizer's often evoked the universality claims of most sciences to press for the agenda of modernization. The anti-colonial agenda re-situated this very moral discourse within the ideology of burgeoning nationalism. In which case, psychoanalysis could be mobilized as a technology in re-fashioning the self, and a multitude of projects of social engineering. Having pointed out that the career of psychoanalysis is almost of the same vintage as it was in Europe; and yet psychoanalysis in the post-independence era is a marginal discipline and practice on the Indian sub-continent. Scholars have offered a variety of reasons to explain 'the decline of psychoanalysis in India', some even going so far to suggest that the discipline makes no sense in a society where the modern-self is very loosely defined.[21]

Geology Textbooks in Tamil

The nineteenth century witnessed the expansion of the European 'empire', not in terms of Christianity but in terms of an imagined civilization. The term civilizing mission came to connote the imperial ideology and official doctrine of imperialism in the nineteenth century.[22] As ideology it meant different things to different actors. To colonial administrators it meant good governance and political order, for Christian missionaries it translated into the projects for converting heathen and the promotion of moral education. The general assumption that swept all these meanings into a unitary frame was that of the intrinsic superiority of the colonizers and the perfectibility of mankind. The related corollary was that it denigrated the colonial subjects by considering them specimens for moral up liftment and civilization through education, which included science education as well. By the second half of the nineteenth century a scientific and technological assessment of human capacity and social development were central to the ideology of the civilizing mission. This new ideological variant was premised on rationality, empiricism, systematic inquiry, industriousness and adaptability.[23]

---

[20] Ibid.

[21] Sudhir Kakkar, *The Indian Psyche*, Delhi: Oxford University Press, 1996.

[22] H. Fischer-Tiné and M. Mann, eds., *Colonialism as Civilizing Mission: Cultural ideology in British India*, London: Wimbledon Publishing Company, 2004, p. 4.

[23] M. Adas, "Contested Hegemony: The Great War and the Afro-Asian Assault on the Civilizing Mission Ideology," *Journal of World History*, 15 (2004): 31–64 (p. 32).

Challenging the historiography of the pure hegemonic diffusion of European science, a number of scholars sought to examine the reordering of indigenous knowledge (the methodological imperative) within the European canon.[24] Chris Bayly examined the debates in the Indian public sphere in the 1840s and rejected the portrait constructed by European and Indian reformers that the growth and communication of knowledge was stunted by hierarchy, and indicates that Indians had begun adapting their practices to the modernist idiom and literary technologies, and further were reflecting upon the status of Indian and Western learning.[25] The relationship as argued by Bayly during these early decades between science and colonial rule was fairly complex and driven from the metropolis by a variety of concerns. On the one hand, within the sphere of the modern nation state, as much as in the colonial state the colonial rulers sought political legitimacy through the patronage of Indian learning. On the other hand, Western science was expected to promote Christian values.[26] Finally, as far as European officials in India were concerned, competition among amateurs stimulated scientific research. This complex of arrangements stimulated by the sheer asymmetry inherent in the colonial experience prompted a positive evaluation of Indian scientific traditions by Indians themselves.[27] Studies focusing upon the first century of British colonial rule revealed that the construction of modern science involved greater complexity and reciprocity 'even in the asymmetrical colonial situation'. This resulted in the development of heterogeneous networks in forging research and teaching traditions even in the early nineteenth century. Local knowledge was thereby enrolled into a global science. However, the calibration of scientific instruments and the standardization of scientific practices played a crucial role in universalizing material and cultural practices.[28]

The liminal status of the so called historical sciences[29] among the natural sciences is reflected in the modes of staging these sciences both in Western and non-Western contexts. In the history of education in colonial India, this is starkly revealed in the processes leading up to the naturalization of the revolu-

---

[24] D. Raina and S.I. Habib, *Domesticating Modern Science: A Social History of Science and Culture in Colonial India*, Delhi: Tulika Books, 2004.

[25] Bayly, *Empire and Information*, p. 247.

[26] D. Gosling, *Science and Religion in India*. Madras: The Christian Literary Society, 1976; see chapter 1 of Raina and Habib, *Domesticating Modern Science: A Social History of Science and Culture in Colonial India*, Delhi: Tulika Books, 2004; T.V. Venkateswaran, "The Topography of a Changing World: Geological Knowledge during the Late Nineteenth Century Colonial Madras Presidency," *Indian Journal of History of Science*, 37, 1 (2002): 57–83.

[27] Bayly, *Empire and Information*, p. 253.

[28] K. Raj, "Colonial Encounters and the Forging of New Knowledge and National Identities: Great Britain and India, 1760–1850," *Osiris*, 15 (2001): 119–134.

[29] S. J. Gould, *Time's Arrow, Time's Cycle*, Cambridge MA: Harvard University Press, 1987.

tionary geological knowledge of the nineteenth century in popular writing in Tamil. These popular science books are not to be conceived through the optic of haute vulgarization. Rather the knowledge form itself both in West and non-West was nested within theories of cosmology, origin myths, ideas of the beginning of time and Christian chronology. Consequently, in this translation from the scientific realm into the vernacular, it was facile to resort to tropes designed to civilize the native population. In his study of popular books on geology translated into Tamil, Venkateswaran points out that the translators cleverly subverted the civilizing intentions of the missionaries by substituting the East-West dichotomy with that of the ancient and the modern.[30] This facilitated the naturalization of the cultural import of modern science. Science in vernacular translation frequently subverted the tropes of cultural imperialism. By reading Greek, Roman and Hindu origin myths and cosmologies through the optic of modern science, the Tamil interlocutor's symmetrical approach to the past of West and East eliminated the cultural denigration immanent in the missionary textbooks introduced into the region. This historicization of the Indian and European pasts[31] was faithful to the spirit of the historical sciences.

Thus strategies of localization and naturalization varied across scientific disciplines and domains of written scholarship. A point in contrast would be the introduction of English literature into India. A monetary commitment was made under the 1813 Charter Act enacted by the British Parliament delegating the East India Company to undertake the promotion of the sciences in India. This responsibility to the education of the native subjects had not been made for its own citizens.[32] However, education during the first half of the nineteenth century was taken to mean predominantly literary education. As a discipline introduced in colleges and universities, the career of English literature commenced in the British colonies, and it was institutionalized far more rapidly as a discipline at these outposts of empire than it was back home in England. The discipline had acquired a place in Indian curricula by the 1820s.[33] The pressure to reshape Indian education, forced the colonial government to intervene and promulgate a policy of non-interference in religious matters of the local population. This gave cause for protests among the missionaries who visualized the broader goals of their mission as being advanced through the education of the local populations. If the missionaries were to be restrained from performing their activities, how would the colonial state continue to proceed with its civilizing mission? The productive resolution lay in the introduction of English literature as the bearer of the

---

[30]Venkateswaran, "The Topography of a Changing World," pp. 57–83.

[31]Ibid, pp. 76–7.

[32]G. Viswanathan, *Masks of Conquest. Literary Study and British Rule in India*. London: Faber and Faber, 1989, p. 23.

[33]Ibid., p. 23.

cultural values and superiority of the colonizing culture. Two of the most visible objectives that underpinned the introduction of English literature was the need to impart a knowledge of the mechanics of the English language, while on the other through an appropriate selection of literary texts new exemplars for the inculcation of industriousness, trustworthiness and compliance among the native subjects were set into circulation.[34] In the missionary publications English literature had been depicted as an exalted from of intellectual or cultural production. These were contrasted with the 'scriptural' Oriental literature with its focus on divine authority. Access to Western literature was enabled through a new hermeneutic of reading that required the exercise of reason as opposed to faith. English literature as a discipline constantly portrayed the distance separating colonizer from the colonized and thereby de-naturalised itself from the source of its origin.

Technical Education in Nineteenth-Century India

The growth and development of technical education in India cannot be studied in isolation from colonial policies relating to technical and general education. These policies were conditioned by the politics of colonial expansion and consolidation of the British rule in India. Ambirajan has argued that the system of colonial education has been viewed either as a 'valuable legacy' or as an outcome of the metropolitan's deliberate project of exploitation.[35] As valuable legacy it was 'an unintended outcome' of the efforts of a variety of agencies that were involved in the establishment of the system. The idea of the system being an exploitative instrument is also problematic, as has been argued by Krishna Kumar, for it does not distinguish between the practical ends of the educational system and the motivations that prompted its establishment.[36] Other scholars have argued that in the growth of technical education in India both these projects were interfoliated.[37] In either case, the inadequacy of what are classically considered 'opposing schools of thought – the imperialist school and the anti-imperialist or the nationalist school', has been outlined by both Ambirajan and Kumar. However, the two camps had very divergent perspectives and appreciations of the barriers to economic development and theories of

---

[34]Ibid., p. 23.

[35]S. Ambirajan, "Science and Technology Education in South India, in R. Macleod and D. Kumar, eds., *Technology and the Raj. Western Technology and Technical Transfers to India.1700–1947.* New Delhi: Sage Publications, 1995, p. 112.

[36]K. Kumar, *Political Agenda of Education: A Study of Colonialist and Nationalist Ideas.* New Delhi: Sage Publications, 1991, p. 23.

[37]S. Bangaru, Debates in Technical Education; A Prelude to the Foundation of the Indian Institutes of Technology (1930–1950), M.Phil. dissertation, Jawaharlal Nehru University, 2002.

economic development.[38] Within the frame of economic history this divergence can be schematized from the work of Bangaru[39]:

	Imperialist	Anti-imperialist/ nationalist
Economic agenda	Economic colonialism	Economic self-reliance
Educational agenda	Colonial administration	Emancipatory

Back home in England close to the end of the nineteenth century there were no formal institutions for imparting a technical education or producing engineers.[40] Inspired by a college in Glasgow, a Civil Engineering College was established at Roorkee, India in 1847, while the Imperial College, London was founded in 1879. The shortfall of textbooks and teaching materials for engineering schools in England was initially met by the periodically revised lecture – notes, examples and drawings, and college manuals circulated amongst students at Roorkee.[41] – these books codified Indian engineering practices as well and were 'hailed as the most complete and satisfactory work on the subject in the English language.'[42] While there was an immanent connection between instruction on science and technology in India and the emergence of the late colonial capitalist state, this required that the colonial state be innovative in the founding of formal technical institutions. Dionne and Macleod had established that these colleges founded in India served as models for replication in England in the late nineteenth century and the colonial encounter contributed to the development of technical education in England and the state supported model of science.[43] Thus while the trajectory of technical education was structured by the imperatives of colonial governance, the model of efficient governance required that the state play a proactive role in the construction of society – this is evident in the relation between the Public works Departments and the first teaching universities of Calcutta, Bombay and Madras in the 1850s. The requirements of the PWD placed a constant demand on technical education and indirectly structured engineering courses throughout

---

[38]B. Chandra, *Nationalism and Colonialism in Modern India*. New Delhi: Orient Longman Ltd., 1979, p. 183.

[39]Bangaru, op. cit.

[40]Z. Baber, *The Science of Empire: Scientific Knowledge, Civilization and Colonial Rule in India*, New York: State University Press, 1996, p. 207.

[41]K.V. Mital, *History of the Thomason College of Engineering (1847–1949)*, Roorkee: The University of Roorkee, 1986, p. 17.

[42]Ibid., p. 98.

[43]R. Macleod and R. Dionne, "Science and Policy in British India, 1858–1914: Perspectives on a Persisting Belief", *Proceedings of the Sixth European Conference of Modern South Asian Studies, Colloques Internationaux du CNRS, Asie du Sud: Traditions et Changements*. Paris: CNRS, 1979, pp. 55–68.

the nineteenth century.[44] This relationship was never a purely speculative one, for the Roorkee College, while awarding its own degrees came under the jurisdiction of the PWD, while the universities were under the jurisdiction of the Department of Education.[45]

Conclusion

Disciplinary boundaries often pose impediments to the understanding of deeper social processes, since the connections between the different components of the system are snapped between disciplines. This paper has suggested that the history of science has insufficiently addressed the context of localization of science since it is disconnected from the world of education. On the other hand, the history of education in colonial India has recently begun to engage with the possibility that the encounter between English educational practices and the South Asian ones may have resulted in innovations that altered educational practices back in England. This insight was available to historians of science working within a variety of research programmes, the Science and Empires being one of them. This lack of engagement with the two way flows of knowledge and practices probably arose from the preoccupation of historians of education with the colonial system as a system of enforcing hegemonic control and to ensure the efficient governance of empire. While pointing this out, it is important that there have always been exceptions to this trend and now we see the burgeoning of new trends that break away from this preconception.[46]

There is then a pressing need to reckon with the fundamental asymmetry that separates the discourse on localization and globalization. The source of this asymmetry is historical practice wherein historians are tuned to see localization or cultural reception as a process initiated within recipient cultures by the march of a triumphalist modern science. The possibility that this triumphalist science could itself be transformed by the encounter is something that has

---

[44]A. Kumar, 'Colonial Requirements and Engineering Education: The Public Works Department, 1847–1947' in R. Macleod and D. Kumar, eds., *Technology and the Raj. Western Technology and Technical Transfers to India, 1700–1947*. New Delhi: Sage, 1995, pp. 216–234.

[45]Mital, op. cit., pp. 18–19; For a more detailed discussion on the subject see the second chapter in S. Bangaru, op. cit.

[46]Dharampal, *The Beautiful Tree. Indigenous Indian Education in the Eighteenth Century*, Mudra: Biblio Impex, 1983; R. Macleod, "On Visiting the Moving Metropolis: Reflections on the Architecture of Imperial Science," *Historical Records of Australian Science (Australian Academy of Science)*, 5, 3(1982): 1–16; J. Tschurenev, "Diffusing Useful Knowledge: The Monitorial System of Education in Madras, London and Bengal, 1789 – 1840," *Paedagogica Historica*, XLIV, 3 (2008): 245–264.

been reckoned within the historiography of the south Asian region over the last two decades.[47]

This brings us to this other concern with the study of strategies of localization. Globalised science and the globalization of science is itself premised on the enlisting of local sciences into the process called 'globalisation' and in order to do so it must adopt its own strategies of localizing knowledge that will be retrospectively considered global. A second asymmetry arises from the modernist historical practice of framing localization in terms of accommodating modern scientific knowledge to the pre-modern or of substituting the former with the latter. This is done at the cost of valuable insights into the dynamics of the flow of knowledge and its adaptation to different institutional contexts. One of the features of post-Kuhnian studies of science has been its primary focus upon research communities and environments and the deconstruction of disciplinary discourses while at the same time creating a space for disciplinary pluralism.[48] Consequently, studies of science outside the aforementioned sphere are delegated to the spheres either of science popularization or science education. These studies veritably reveal the context of the globalization of scientific knowledge, as opposed to studies focusing on collaborations that go on between small communities of scientists. It must be remembered that these communities were much smaller or were of a radically different nature in the nineteenth century from what they are today. Or to put the point in a more contemporary context, a study of the emergence of India as one of the new giants of information technology cannot be limited to the establishment of elite centres of research in theoretical computer science and information technology. Historians will one day have to study the explosion of fly-by-night teaching shops calling themselves institutes of information technology and now spreading their tentacles into China, played a central role in taking India into cyberscape. Another kind of cultural history then, and another kind of history of science perhaps.

---

[47]A. Nandy, *Alternative Sciences*, Delhi: Allied Publishers, 1980; K. Raj, "Colonial Encounters and the Forging of New Knowledge and National Identities: Great Britain and India, 1760–1850," *Osiris*, 15 (2001): 119–134; D. Raina, *Images and Contexts: Studies in the Historiography of Science in India*, Delhi: Oxford University Press, 2003; D. Raina and S.I. Habib, *Domesticating Modern Science: A Social History of Science and Culture in Colonial India*, Delhi: Tulika Books, 2004; G. Prakash, *Another Reason: Science and the Imagination in Modern India*, Princeton: Princeton University Press, 1999.

[48]Fuller, *Thomas Kuhn*, p. 2.

Between Translation and Adaptation: Turkish Editions of Ganot's *Traité*

Meltem Akbaş

In a petition sent to the Ministry of Education in 1870, Tahsin Bey, the Rector of Darulfunun (University) listed the instruments needed for physics lectures. The list included 130 instruments of 108 different types which were to be bought in France. At the end of the list, he added an unprecedented note concerning a specific instrument, the *resonance pipe*. According to this note, this instrument "should be the same as the one shown in figures 178, 180–184 of the 13th edition of Ganot's book published in 1868".[1] Imagine an official of the Ottoman government walking through the streets of Paris with Ganot's book in hand and visiting the shops of instrument-makers. Does this not suggest to us the influence of Adolphe Ganot's (1804–1887) physics book, in the nineteenth century? That book of course was the *Traité élémentaire de physique expérimentale et appliqué*.[2]

The book went into several editions, and the sales reached 20,000 copies in the 18th edition of the *Traité* (1880).[3] Undoubtedly, it was one of the most widely known physics books of the nineteenth century with its many French editions and translations spread over many countries. As far as is known, it had been translated into 12 languages.[4]

M. Akbaş (✉)
Department of the History of Science, Faculty of Letters, Istanbul University,
Istanbul, Turkey
e-mail: akbas.meltem@gmail.com

[1] E. İhsanoğlu, "Darülfünun tarihçesine giriş: İlk iki teşebbüs", *Belleten*, LIV, 210 (August 1990): 725; Ottoman Archives of the Turkish Prime Ministry (Başbakanlık Osmanlı Arşivi, BOA), İ.DH, 618/43038.

[2] The first edition of the book: A. Ganot, *Traité Élémentaire de Physique Expérimentale et Appliquée*, chez l'auteur, éditeur, Paris 1851.

[3] J. Simon, "The Franco-British communication and appropriation of Ganot's *Physique* (1851–1881)", in J. Simon, N. Herrán et al., eds., *Beyond Borders: Fresh Perspectives in History of Science*. Newcastle upon Tyne: Cambridge Scholar Publishing, 2008, p. 141.

[4] J. Simon and P. Llovera, "Between teaching and research: Adolphe Ganot and the definition of electrostatics (1851–1881)", *Journal of Electrostatics*, 67 (2009): 536.

The *Traité* became also popular in the Ottoman Empire. Even today, Turkish libraries together have at least 11 different editions of the *Traité*: the earliest edition going back to 1860 (9th edition) and the latest to 1913 (25th edition).[5] The diversity of these editions points to the degree and continuity of interest in the book. In the last quarter of the nineteenth century, the book circulated among Ottoman intellectuals associated with science as testifies the correspondence of Beşir Fuad (1852–1887) and Fazlı Necip (1863–1932) between 1885 and 1887.[6]

In one of his letters, Fazlı Necib who trusted Beşir Fuad's knowledge in scientific issues asked him for advice about books to read. The latter recommended various books on topics ranging from astronomy to psychology.[7] The physics book he suggested was Ganot's *Traité*. He even offered brief explanations as to how the book was to be read properly. Evidently, *Traité* was the first book that crossed his mind as far as physics was concerned.

The immediate connection between physics and the *Traité*, however, was cause for misinterpretations in some cases. A number of writers[8] alleged that one of the books on physics written by the Turkish mathematician Salih Zeki (1864–1921), namely *Muhtasar Hikmet-i Tabiiye* (1894/95), was a translation of Ganot's work.[9] Indeed, it was not a translation, but a compilation from at least three different sources including the *Traité*.[10] This perception arose from the presence of the illustrations used in Salih Zeki's book: a considerable number of them originated from Ganot's work. Writers who knew the *Traité* well but were unfamiliar with other physics books misinterpreted the situation. There is one more fact which they were not aware or forgot: the *Traité* had already been translated into Turkish by Dr. Antranik Gırcikyan (1819–1894) and published under the title *İlm-i Hikmet-i Tabiiye* in Istanbul, in 1876.

---

[5] The relevant data was compiled from the catalogues of the libraries of Istanbul University and Istanbul Technical University as well as the National Library of Turkey (Ankara), Beyazıt State Library and Atatürk Library in Istanbul.

[6] *İlk Türk Materyalisti Beşir Fuad'ın Mektupları*, C. Parkan Özturan, ed., İstanbul, pp. 28, 29, 36, 80.

[7] Among his recommendations were Camille Flammarion's (1842–1925) *Astronomie Populaire*, Louis Figuier's (1819–1894) *Les Merveilles de la Science*. He also advised Faustino Malaguti (1802–1878) for chemistry, Auguste Axenfeld (1825–1876), Fréderic Paulhan (1856–1931) and Théodore Ribot (1839–1916) for medicine, physiology and psychology. See *İlk Türk Materyalisti...*, and O. Okyay, *Beşir Fuad: İlk Türk Pozitivist ve Natüralisti*, 2nd ed., İstanbul: Dergah pub., 2008, pp. 43–47.

[8] These writers were Mehmed İzzet, İsmayıl Hakkı Baltacıoğlu and Tevfik Sağlam. See Mehmet İzzet, "Konferans 1- Salih Zeki", *Talebe Mecmuası*, nr. 41 (March 1935): 6; C. Saraç, *Salih Zeki Bey Hayatı ve Eserleri*, Y. Işıl Ülman, ed., İstanbul, 2001, p. 172; Tevfik Sağlam, *Nasıl Okudum*, 3rd ed., H. Hatemi, A. Kazancıgil, eds., Istanbul: Nehir pub., 1991, p. 84.

[9] Salih Zeki, *Muhtasar Hikmet-i Tabiiye*, İstanbul: Karabet Matb., 1312 (1894–1897), Vol. I: 426 pp. and Vol. II: 470 pp.

[10] These books are *Hikmet-i Tabiiye* (trans. Hafız Mehmed, İstanbul, 1878/1879) and *Dürus-ı Hikmet-i Tabiiye* (trans. Ahmed Tevfik, İstanbul 1884/1885). They were translated from the textbooks of Prytanée National Militaire de la Flèche for Darüşşafaka high school in İstanbul.

Antranik Gırcikyan: Translator of the *Traité* and Physics Teacher at the Imperial School of Medicine

Antranik Gırcikyan, an Ottoman Armenian, completed his primary education at the community school in Beşiktaş, Istanbul.[11] Then, he went to Paris for further education and studied natural philosophy or physics. He returned to Istanbul in 1844 and started to work at the Imperial School of Medicine (*Mekteb-i Tıbbiye-i Şahane*, a military medical faculty) as an assistant to the physics teacher in 1851.

On the other hand he was registered at the same school, as a student.[12] Following his graduation with a medical diploma, he was sent to Europe by the State to advance his education: He traveled to England, France and Italy. From 1866 on, he continued his perennial career in teaching as the assistant of Dimitraki Balasides, the physics teacher of the Imperial School of Medicine. In addition to his duty, he was appointed to the Civil School of Medicine (*Mekteb-i Tıbbiye-i Mülkiye*) as physics teacher in 1872/1873. After 1879, he became a full-fledged physics teacher at the Imperial School of Medicine. He successfully worked at these two schools until his death in 1894.

While working as an assistant in early 1850s, his assignment was to introduce physics instruments to medical students and account for their use and functioning. Furthermore, with knowledge of taxidermy acquired in Paris, he stuffed and mounted animals for the Natural History Museum of the Imperial School of Medicine. He was very adept and capable of handling instruments, in addition to which he was talented in engraving and painting. During his working life, he personally fixed and repaired physics instruments, although an

---

[11] The biography of Antranik Gırcikyan was compiled from following sources: Ռ. Չաքար, Sopp; "Sopp. Անդրանիկ Փաշա Հրճիկեան", Հանդես Մշակոյթի, Թիւ 1, Սեպտեմբեր 1948; Ա.Ն. Մեգայիրեան, Sopp; "Հայ Եի Ծագիմով Հայ Բժշկներ", Իսթ. 1950; Y. Çark, *Türk Devleti Hizmetinde Ermeniler 1453–1953*, İstanbul, 1953, pp. 105–106; Rıza Tahsin, *Tıp Fakültesi Tarihçesi (Mir'ât-ı Mekteb-i Tıbbiye)*, İstanbul, 1991, Vol. I, p. 94; Y. Öztuna Şirin, "Osmanlı Sâlnâmelerinde 1908 tarihine kadar tıp eğitimi", *Yeni Tıp Tarihi Araştırmaları*, nr. 5 (1999): 208–323; Arsen Yarman, *Osmanlı Sağlık Hizmetlerinde Ermeniler ve Surp Pırgiç Ermeni Hastanesi Tarihi, Surp Pırgiç Hastanesi Vakfı*, İstanbul 2001, p. 782; Meltem Akbaş, *Osmanlı Türkiyesi'nde Modern Fizik (19. Yüzyıl)*, unpublished PhD dissertation, İstanbul University, İstanbul 2008.

[12] It is intriguing why Antranik Gırcikyan studied medicine after the age of 30 and became a military doctor although he was an assistant physics teacher. In the Ottoman Empire military school teachers were generally recruited among the graduates of the school they were employed or other military schools. The ranks and salaries of those military officer-teachers increased in time as those of other military officers and they gained prestigious positions in terms of career and social status. Nevertheless, Gırcikyan did not graduate from a military school and started his career as a civilian assistant teacher. Perhaps for the advantage of being both an officer and a doctor, or due to the fact that physics was not yet recognized as a professional field at that period, he opted for obtaining a diploma from the Military School of Medicine, too.

apparatus-keeper was employed specifically for this purpose.[13] He cherished working with scientific instruments, an uncommon characteristic amongst Ottoman physics teachers of the time.[14]

When Gırcikyan published *İlm-i Hikmet-i Tabiiye*, he was "assistant teacher" at the Imperial School of Medicine and "teacher" at the Civil School of Medicine. The second edition,[15] on the other hand, was published in his late career. Since both editions are different in content, they worth to be studied separately.

Turkish Editions of Ganot's *Traité*'s: An Analysis of the Preface and Textual Content

As a first step in this article, we analyse the prefaces that Gırcikyan wrote for the first and second editions of *İlm-i Hikmet-i Tabiiye*, the translations he made from Ganot's *Traité*. The preface of a book is one of the paratextual elements like the author's name, title, notes, illustrations etc.[16] According to Gerard Genette, paratexts "surround and extend the text precisely in order to present it, in the usual sense of this verb but also in the strongest sense: to make present, to ensure the text's presence in the world, its reception and consumption in the form (nowadays, at least) of a book".[17] Recently, paratexts in general, and

---

[13] The apparatus-keeper (*muhafız-ı âlât*) of the Imperial School of Medicine was responsible for the cleaning and the protection of physics and chemistry instruments. Carrying them to the classroom and setting them up for demonstration were among his duties. He was expected to have skills in fixing instruments. The duties of apparatus-keepers are defined in the 1857 Regulation of the Imperial School of Medicine. See F. Günergun and N. Yıldırım, "Cemiyet-i Tıbbiye-i Şahane'nin Mekteb-i Tıbbiye-i Şahane'ye getirdiği eleştiriler (1857–1867)", *Osmanlı Bilimi Araştırmaları*, III, (2001): 53–54. There have been similar positions in the other higher educational institutions such as Military Academy (*Mekteb-i Harbiye*). See Mehmed Esad, *Mirat-ı Mekteb-i Harbiye*, İstanbul 1310 (1892–1894), pp. 187, 205, 206.

[14] In his autobiography, Dr. Tevfik Sağlam (1882–1963) provided a humorous reminiscence about a classroom experiment in physics held at the Imperial School of Medicine in late 19th century. This memory is illustrative of the roles played by the teacher, the apparatus-keeper and the student in the experiment: the apparatus-keeper brings the instruments to the classroom, sets them up for the experiment, operates them according to the directions of the teacher and does all the handiwork. The teacher gives directions to the apparatus-keeper and explains the physical phenomena observed while students act as spectators. The teacher mentioned above was İsmail Ali (1866–1913), the successor of Antranik Gırcikyan. See Tevfik Sağlam, *Nasıl Okudum*, pp. 85, 86.

[15] A. Ganot, *İlm-i Hikmet-i Tabiiye* (trans. Antranik Gırcikyan), 2nd ed., Mahmud Bey Matb., Vol. I, Istanbul, 1886 (1303), 714 pages; Vol. II, Istanbul 1891 (1308), 959 + 32 pages.

[16] A. Ganot, *İlm-i Hikmet-i Tabiiye*, trans. Antranik Gırcikyan, 3 vols., İstanbul: Mekteb-i Tıbbiye-i Şahane Matb., 1293 (1876), 44+668 pp., 765 pp., and 52 double pages.

[17] G. Genette, *Paratexts: Thresholds of Interpretation*, trans. J. Lewin, Cambridge: Cambridge University Press, 1997, p. 1.

prefaces in particular, are the subject of research on their own in various disciplines.[18] In the history of science studies, however, paratexts are mostly regarded as complementary elements providing valuable information about the main text. For its novel character, the exploration of prefaces in history of science studies may yield fruitful results.

In case the text to be studied is a translation, how can a preface written by the translator be interpreted? From the point of view of translation studies, translators as composers of prefaces "can be considered as mediators that become visible in paratexts. Although they do not become visible physically in the target text in the process of mediation, their presence becomes discernible and visible in the thresholds of the texts, especially in the prefaces. Translators mediate between source and target texts, between source and target cultures and between target texts and readers".[19] For this reason, prefaces are valuable tools to concieve the role of translator as a mediator in the transmission of science between two cultures. In this article, we understand the translator's preface as a space where she/he could construct her/his original discourse which enabled her/him to take a position as a translator, writer, scientist, teacher, transmitter of science or whoever she/he wanted to be identified as.

In the following parts of the article, we analyse some extracts from the prefaces written by Gırcikyan. The criteria for selecting the extracts depend on their relatedness to the the target text, the source text and connected issues. In other words, we chose sentences suitable for answering the following questions: Why was the translation made? Who were the intended readers of the target text? With whom and how did the target text communicate? How was the translator's relation with source text and source author?

As a second step, we comparatively analyse the content of the translated texts with the source text. Our purpose is to investigate the interventions made by the translator in the text. This is done not only for the verification of translator's discourse about his own work but also to understand the pattern of this active intervention and the scientific motivations that underpin them.

---

[18] For a study in Canadian literature see S. T. de Zepetnek, in A. Barsch, ed., *The Social Dimensions of Fiction: On the Rhetoric and Function of Prefacing Novels in the Nineteenth Century Canadas*. Wiesbaden: Vieweg 1993.

[19] S. Bozkurt, Tracing Discourse in Prefaces to Turkish Translations of Fiction by *Remzi Publishing House* in the 1930s and 1940s. Unpublished MA dissertation, Boğaziçi University, Istanbul 2007, p. 3.

The First Edition of *İlm-i Hikmet-i Tabiiye* (IHT[1], 1876)

The preface (*mukaddime*[20]) written by Antranik Gırcikyan to the 1st edition of *İlm-i Hikmet-i Tabiiye* is two pages long and begins with a *besmele* and *hamdele*.[21] It bears some of the stylistic features observed in the Ottoman literary tradition, such as phrases glorifying the reigning Sultan. On the other hand, it has expressions peculiar to its own genre like offering a definition of physics. Prefaces provide an opportunity for translator to introduce himself. Like many other translators/authors of science books, Gırcikyan took this opportunity to enumerate his official designation and duties,[22] signs that establish his credibility in the readers' eyes.

From the preface discourses oriented to the target text, the extracted portions are crucial for our research since they reflect the genesis of the text or the reasons that motivated the translation.

> ...although this high science [physics], together with many other sciences, was in fact conveyed from Asians who were the illuminated source of knowledge, and had attained its present status in Europe, a rather detailed physics book in Turkish doesn't exist; and because the medical sciences are being taught in the Ottoman language which is sweet in expression...

In this extract, Gırcikyan sets forth two justifications for the translation. The first one is his claim about the absence of "rather detailed" books on physics in Turkish. He implicitly admits the existence of other books on physics in Turkish, but does not dwell on them, in other words he avoids communicating with them. His discourse creates an impression that he does not appreciate other books in circulation; because they are not comprehensive enough. Thus, by translating the *Traité*, he fills a gap in the Turkish physics book market.

Another statement displaying the motivations leading to the translation comes from the remark that "medical sciences are being taught in the Ottoman language". This remark needs to be unpacked here. By filling in the blanks, we arrive at the second justification for the translation: "medical sciences are being taught in the Ottoman language," [physics courses are given in the medical

---

[20] Several terms are employed to denote preface in Ottoman Turkish: *mukaddime* (Arabic, a preliminary discourse, a preface, an introduction), *dibace* (Persian, a preface to a book, a prologue) and *ifade-i meram* (the exposition of wish, thought or intention). James W. Redhouse, *A Turkish and English Lexicon*, Constantinople 1890 (Facsimile 1992), pp. 153, 932, 1942.

[21] *Besmele* is the Bismillah phrase meaning "in the name of God, the most gracious, the most merciful". *Hamdala* is the abridged version of "elhamdülillâh" meaning "thank God". It's also the name of the part in a literary text where the author presents his/her praises and gives glory to the God.

[22] In the preface, he introduces himself in the following way; "...the present servant who has little knowledge, namely the assistant physics teacher at the Military School of Medicine and the physics teacher at the Civil School of Medicine and a permanent member of the Ottoman Medical Society, Lieutenant Colonel, Doctor Antranik Gırcikyan".

education], [a physics book written in Ottoman does not exist for medical students], [the reason why I, Gırcikyan, translated this book].

Who were the intended readers of the target text? Only a tiny clause of the preface talks about that:

> It [the book] was approved to be taught in the School [of Medicine].

This indicates that the potential readers were the medical students. Apart from that, one cannot find an apparent clue on the reader's profile. But some implicit references embedded in the preface denote that Gırcikyan communicated with other people too. For instance, as noticed by a careful reader, a compliment was paid to the Ottoman language in the previous quotation. This compliment refers to a controversy that erupted in medical circles, in the 1870s. Again, the first part of the sentence (starts with "although this high science...") refers to the same controversy. To analyze the same we require a historical perspective of the controversy.

Modern medical education was introduced in the Ottoman Empire in 1827 and aimed to train Muslim physicians for the army. Teaching was supported by language courses in French and Italian, so that students could read the European medical texts. Education was both in Turkish and French. Medical education was reformed in 1839 with the establishment of the Imperial School of Medicine under the supervision of the Austrian physician C. A. Bernard (1808–1844). The staff was composed of European, Ottoman Muslim and non-Muslim professors and there were Ottoman Muslims and non-Muslims of different ethnic origin among the students, too. The language of instruction in this military institution was French.[23] Muslim students and teachers began displaying signs of discomfort towards the end of 1850s.[24] They opposed the teaching of medicine in French and voices of displeasure became increasingly louder. The controversy extended beyond the borders of the School and led to the establishment of the Ottoman Medical Society (1866).[25] The objective behind the establishment of the Society was to create the necessary conditions for launching medical education in Turkish and enriching the Turkish medical literature by translations from European books.

As a consequence of these efforts the Civil School of Medicine was founded in 1867 to impart medical instruction in Turkish[26] while teaching continued in

---

[23] A. Altıntaş, "Osmanlılarda modern anlamda tıp eğitiminin başlaması- Tıphâne-i Âmire", *Osmanlı*, Vol. VIII, Ankara 1999, p. 528; E. İhsanoğlu and F. Günergun, "Tıp eğitiminin Türkçeleşmesi meselesinde bazı tespitler", *Acta Turcica Historiae Medicinae* I, A. Terzioğlu, ed., İstanbul 1994, p. 127.

[24] Rıza Tahsin, *Tıp Fakültesi Tarihçesi*, pp. 34–35.

[25] For the history of Society see N. Sarı, "Cemiyet-i Tıbbiye-i Osmaniyye ve tıp dilinin Türkçeleşmesi akımı", *Osmanlı İlmi ve Mesleki Cemiyetleri*, E. İhsanoğlu, ed., İstanbul, 1987, pp. 121–142.

[26] E. K. Unat and M. Samastı, *Mekteb-i Tıbbiye-i Mülkiye (Sivil Tıp Mektebi) 1867–1909*, İstanbul, 1990, p. 8.

French in the military medical school. An imperial decree dated 1870 established Turkish as teaching language in both schools.[27]

On the other side, a group of professors composed of Europeans and Ottoman non-Muslims advocated the idea that a qualified medical education could only be in French. Most of them were members of Société Impériale de Médecine de Constantinople[28] and published their opinions in the *Gazette Médicale d'Orient*,[29] the journal of the Society. One of the main arguments proposed by the defenders of French was that Turkish was not a language of science and it was not sophisticated enough for medical education. On the contrary, defenders of Turkish education claimed that, scientific terminology in Turkish could be improved by resorting to native words in Turkish, and to medical texts in Arabic and European languages when necessary. Opponents of medical education in Turkish rejected the possibility of benefiting from Islamic texts claiming that there was no Islamic medicine, and the so-called Arabic science was in fact Greek science. In their opinion, it was absurd to claim that Islamic physicians contributed to the development of medicine in the Middle-Ages.[30]

Against this backdrop Antranik Gırcikyan started to translate Ganot's *Traité* to Turkish. Discussions on the language of instruction, about coining scientific terms in local languages and controversies about Islamic science were still raging. Gırcikyan was a physics teacher lecturing in Turkish at the civilian school of medicine and he was a member of the Ottoman Society for Medicine founded by Ottoman physicians. Besides, his book had the privilege of being the first Turkish physics text book prepared for medical education. Evidently, he too was among the supporters of medical education in Turkish.

The rhetoric of the preface makes sense, if we consider altogether the above historical context surrounding the language of instruction and Gırcikyan's position. His praise for the Ottoman language can be interpreted as an objecting to the unfavorable judgment that Ottoman could not serve as a language of science. Especially, his emphasis on Asians' role in the development of sciences – an irrelevant rhetoric to the context – is noteworthy. His response to the assertion that "there wasn't an Islamic medicine" reaffirming the non-European contribution to medicine and redefining it from Islamic to Asian, might be an attempt to integrate his own identity as a non-Muslim Asian. As a result, the translation

---

[27] N. Sarı, "Cemiyet-i Tıbbiye-i Osmaniyye ve tıp dilinin Türkçeleşmesi akımı", p. 125.

[28] Société Impériale de Médecine (Cemiyet-i Tıbbiye-i Şahane) was established by the English and French physicians who came to Istanbul during the Crimean War. For further information see H. Hüsrev Hatemi, A. Kazancıgil, "Türk Tıp Cemiyeti (Derneği) Cemiyet-i Tıbbiye-i Şahane ve tıbbın gelişmesine katkıları", *Osmanlı İlmi ve Mesleki Cemiyetleri*, E. İhsanoğlu, ed., İstanbul, 1987, pp. 111–119; F. Günergun and N. Yıldırım, "Cemiyet-i Tıbbiye-i Şahane'nin Mekteb-i Tıbbiye-yi Şahane'ye getirdiği eleştiriler (1857–1867)", pp. 20–63.

[29] The journal was published between 1857 and 1925. H. Hatemi and A. Kazancıgil, "Gazette Médicale d'Orient'ın ilk sayıları", *Tıp Tarihi Araştırmaları*, nr. 15 (2007): 33–39.

[30] E. İhsanoğlu and F. Günergun, "Tıp eğitiminin Türkçeleşmesi meselesinde bazı tespitler", pp. 127–133.

implicitly communicates with a circle of physicians-intellectuals that constitute the different sides of the language controversy.

Source text and source author were not notable parts of the preface discourse. The only place indicating that they were of concern is the following extract:

> approved to be taught in the School [of Medicine], [the book] is translated from the 14th but also from the 13th, 16th and other editions of the physics book of a person named Ganot who is among the famous authors of France...

In this statement, it is not obvious which book of Ganot was meant; in other words, the book title (*Traité*) was not indicated. It is highly possible that the readers would have known which book was meant, especially when taking into account the popularity of *Traité* among Ottoman readers. But would it not be more practical to give the title of the book instead of mentioning it indirectly? Why did Gırcikyan prefer this way of expression? Another missing statement oriented to the source text is the selection of *Traité* for the translation. What makes it superior to other physics books? Why did Gırcikyan translate *Traité* and not another book? The preface does not help us to acquire an implicit or explicit answer because Gırcikyan did not "present" the source text.

In the preface, Gırcikyan paid only a few words on his own interventions to the source text: he added 32 more figures to the 776 figures borrowed from the source text. Moreover, he asserted that he used at least three editions of Ganot's book in his translation.

How did Gırcikyan use the 13th, 14th and the 16th editions of *Traité* for his translation? If we compare the source texts with IHT[1], we see that, he basically translated the 14th edition and occasionally resorted to the other editions. An example can be found in the section on *électricité animale*. Ganot removed this section from his book after the 13th edition, but Gırcikyan preferred to translate some parts of it.[31] Moreover, he used the other editions to enrich the illustrations. The 32 figures mentioned in the preface are from the 13th and 16th editions of the *Traité*.

Figures of IHT[1] were not included in the text but gathered in a separate volume (Vol. III). In all likelihood, the printing technology of the Imperial School of Medicine's printing press could not integrate them into the text. Or, their insertion was regarded as a time consuming and troublesome work. In the first pages of this volume, Gırcikyan provided an inventory of physics instruments, experiments and illustrations and gave their Turkish names or brief descriptions. They were sorted according to their figure numbers. The reader interested in the figure of the instrument had to refer to the third volume. Although running from one volume to the other complicated the reading of the book, the presence of this list helped the reader to learn the Turkish names of the instruments. Such a list does not exist in the source text. This points to Gircikyan's meticulous work and may be related with the movement of conducting the medical teaching in Turkish.

---

[31] He translated only the two sections: "Courant propre des animaux" (pp. 846–847) and "Poissons électriques" (pp. 847–848).

Gırcikyan did not apply his enriching strategy to the numerical tables as he did for the illustrations. When compared with the 14th edition, some tables of 16th edition are more substantial such as the table of "melting points of various substances".[32] Gırcikyan, however, preferred to borrow this table from the 14th edition.

The translator made a few cultural interventions into the text as well. He added a list of the metric equivalents of the Ottoman units of length. The list was inserted within the section on the measuring instrument "Vernier Caliper".[33] Thus, Gırcikyan aimed to disseminate the use of metric system among students, a similar effort can be appreciated in other textbooks of the time.[34] Another intervention appears in the chapter on electrodynamics: Gırcikyan added a historical note about the reception of the first telegraphic message in Istanbul on 29 August 1855 from Şumnu, announcing the fall of Sevastopol.[35] It seems that, these historical-cultural insertions were necessary to augment the "familiarity" of the text for the reader and make it more attractive.

The only part of the *Traité* (14th edition) that Gırcikyan did not translate was the final section of the book concerning physics problems and their solutions. This may have nothing to do with the disapproval for a mathematically intensive text in a physics book. In all likelihood, it was a naive act of a teacher unwilling to share the exam questions with his students or to discourage "other" potential readers by overloading the text with exercises.

Despite the presence of such minor interventions by the translator, we can assert that IHT[1] is not an adaptation but almost a full translation of the *Traité*'s 14th edition. Gırcikyan even kept the *Traité*'s system of numbering chapters, sections and subsections. This can be interpreted as a sign of his willingness to stay within the limits of the source text.

The Second Edition of *İlm-i Hikmet-i Tabiiye* (IHT^2, 1886–1891)

Ten years after the publication of the first edition (IHT^1), the first volume of the second edition was released.[36] Between these two Turkish editions, new French editions of the *Traité* had been published.[37] Thus Gırcikyan was able to revise his

---

[32] The table of "températures de fusion de diverses substances" included 31 items in the 14th edition. In the 16th edition two substances (sperma ceti ve thallium) were removed and three new substances (protoxyde d'azote, acide carbonique, brome) were insterted (for comparison see p. 307 and p. 298 respectively).

[33] IHT^1, p. 9.

[34] For more information see F. Günergun, "Introduction of the Metric System to the Ottoman State", *Transfer of Modern Science & Technology to the Muslim World*, E. İhsanoğlu, ed., İstanbul: IRCICA, 1992, pp. 297–316.

[35] IHT^1, Vol. II, p. 588.

[36] A. Ganot, *İlm-i Hikmet-i Tabiiye*, trans. Antranik Gırcikyan, 2nd ed., Mahmud Bey Matb., Vol. I, Istanbul, 1886 (1303), 714 pp.; Vol. II, Istanbul 1891 (1308), 959 + 32 pp.

[37] These were the 17th (1876), 19th (1884) and 20th (1887) editions.

translation before sending IHT² to press. He composed a new preface for the second Turkish edition, proper to its new content. It begins with statements emphasizing the importance of physics and contains the following extract:

> Upon those eagerness-giving benefactions that we have witnessed with joy, resorting to the goodness of God and admitting my impotence, I, for the purpose of serving humbly to the sacred homeland, have translated the physics book of Ganot which has been published nineteen times. I previously prepared a book and it drew more interest than I expected. Hopefully, this book which is my second publication will be fortunate to receive the attention of the public as the previous one did. I would like to express that as medical and industrial subjects are very few in the Ganot's physics book – many selected issues about these subjects have been added [to the Turkish translation] from the books of famous authors like Gavarret, Graham, Jamin, Gariel and Brodie, and 845 figures – over 1200 with their subfigures – were inserted. And because most of the names of the chemical substances and instruments have recently been translated into Turkish [and] considering that it is impossible to be fault-free, a short glossary including the French terms was appended.

Gırcikyan presented the second Turkish edition with a new discourse. According to his statement, the reason that motivated the publication of the second edition was "to serve the sacred homeland". Gırcikyan's discourse has been altered with the passing of years. This might reflect the change in his intended readership. According to the extract above, the readership of IHT¹ extended beyond Gırcikyan's intended audience. This situation might have led him to rethink his initial readership in translating the book. Preparing a book on physics for a broader public thus meant "to serve the sacred homeland". The form of presenting the scientific material, however, was not altered. Moreover IHT² was more comprehensive than IHT¹. It could be said that, his expectation of the attention of the public didn't lead him to transform the translation into a popular physics book.

There was only one sentence in the preface, concerning the source author and the source text. This related to the insertion of medical and industrial subjects into the translation, relying on the books of Jules Gavarret (1809–1890), Thomas Graham (1805–1869), Jules Célestin Jamin (1818–1886), Charles Marie Gariel (1841–1924), and Benjamin Brodie (1817–1880). Gırcikyan expressed his opinions about the "weak" aspects of the source text, but never revealed its positive sides. He did not present the source text as he had avoided presenting it in the preface of IHT¹. This time he presented the text by stressing his own efforts to complete the text with medical and industrial issues from other sources and by introducing his glossary.

The Turkish-French glossary he compiled includes 269 physics terms and names of the instruments. It constitutes one of the most significant parts of IHT². Gırcikyan was not the first Ottoman author who prepared a glossary for a textbook – the preparation of such glossaries was not conventional either.[38] In

---

[38] Hayrullah Efendi (1818–1866) appended a small glossary to his physics book which he prepared for the students of *rüşdiye*s (middle school). However it is different in approach than that of Gırcikyan. Hayrullah Efendi's glossary was organized to account for the meanings of the scientific terms in Turkish. Apparently he prepared it for students who were to meet physics for the first time. Hayrullah Efendi, *Mesail-i Hikmet*, İstanbul: Darü't-tıbbaatü'l-amire, 1265 (1849).

the glossary of Gırcikyan, the main entries were listed in alphabetical order in Turkish, whereas their equivalents were given in French. He presented his terminology to the readers knowledgeable in French, especially to his colleagues, noting the very recent beginnings of Turkish scientific terminology and stressing the non-existence of a fault-free human being. This is a repeated rhetoric of translator's of scientific texts in the nineteenth-century Ottoman Empire.[39] In our opinion, it indicates the existence of a scholarly community endeavoring to improve scientific terminology in Turkish. Furthermore, it suggests that the members of this community were ready to assert their expertise by criticizing each other's competence in translation.

If Gırcikyan's statements are accepted, IHT^2 cannot be considered as a full translation but an adaptation of *Traité*, as it resorted to other physics books. How was the translation of *Traité* intercalated with medical and industrial issues from different sources, in Gırcikyan's hands? It is significant for us to figure out his methods of adaptation as well as to understand the reasons behind it. Dealing with this issue, we made a comparative content analysis in the following pages. We picked a chapter to analyze: the one on "Heat", one of the most dynamic fields of nineteenth-century physics. Comparing IHT^2 with the source text is rather challenging, because the "source text" is blurred by Gırcikyan's use of multiple editions of *Traité* as well as of "unknown" sources. Therefore, we considered the four editions of *Traité* (13th, 14th, 16th and 19th editions) as a single source text.

Based on our study of the chapter on "Heat", we can assume that Gırcikyan, when preparing IHT^2 for publication, revised the IHT^1 from the beginning to the end. The changes he made in IHT^2 show that he critically examined the four editions of the *Traité*, instead of purely translating the latest edition available to him (19th edition). He preferred to use earlier editions in some cases; in others he exploited various editions simultaneously. For instance, in the 13th and 14th editions, Ganot accounts for the Ferdinand Carré's (1824–1894) device which freezes water by the "distillation of ammonia". This section was removed from the 16th edition. In the 19th edition Ganot gives a different model of Carré's device which freezes water by the "evaporation of sulphuric acid". What Gırcikyan did in IHT^2 was to account for the two models.

Gırcikyan made other interventions remaining "inside" the *Traité*. The limits of the source text, however, were not inviolable for him. Occasionally he omitted several sections although they appeared in all the four editions of the *Traité*. On the other hand, he sometimes inserted subjects from the

---

[39] A teacher at the Ottoman Military Academy, Ali Rıza Efendi (d. 1937), while modestly underlying in the preface of his mechanics book his incompetency, pointed out that Turkish scientific terminology had not been properly formed. He asked his knowledgeable colleagues to correct his possible mistakes that might be found in his terminology. For this purpose, he included French equivalents of some Turkish terms within parentheses into the text. See Ali Rıza, *Fenn-i Mihanik-i Riyâzî ve Makineler*, İstanbul: Karabet Matb., 1306 (1889).

"outside" of the *Traité*. We classified these two groups of interventions in the below table:

1st group of interventions: *Subjects existing in* IHT^2 *but not in the four editions of Traité*	2nd group of interventions: *Subjects existing in all four editions of Traité but not in* IHT^2
Maximum and minimum thermometer of Six and Bellani	Thermomètre à maxima de Negretti et Zambra
Maximum thermometer of Constantin Paul	Formules et problèmes sur la dilatation des gaz
Anesthesia instrument of Richardson	Expériences de Boutigny
Thermal conductivity of textiles	
Animal heat	

Gırcikyan preferred to leave out the "maximum thermometer of Negretti and Zambra" and inserted information on the "maximum and minimum thermometer of Six and Bellani" instead.[40] Although Angelo Ludovico Negretti (1818–1879) and Joseph Warren Zambra (1822–1887) successfully modified Rutherford's thermometer to eliminate its handling difficulties, their thermometer was not precise enough. According to Gırcikyan, the thermometer of James Six (1731–1793) and Angelo Bellani (1776–1852) was more favored by scientists.[41] Seemingly, he adapted this section from M. J. Jamin's *Cours de Physique de l'École Polytechnique*.[42] The two other sections of the 2nd group at the table, "Formules et problèmes sur la dilatation des gaz" and "Expériences de Boutigny", were among the small-fonted parts of *Traité* which Ganot had written for advanced readers. Although Gırcikyan made such sections part of his book, he did not translate the two above-mentioned ones.

The section about the maximum thermometer of Constantin Paul (1833–1896) in the 1st group, deals with the surface thermometer measuring the temperature of the human body.[43] Another intervention made by Gırcikyan

---

[40] IHT^2, Vol. I, pp. 388–389.

[41] The instrument was designed by the English scientist James Six at the end of the 18th century and was later improved by Angelo Bellani. It was also known as "thermometrograph". The major advantage of this thermometer was that it supplied various measurements without requiring constant presence of an observer: http://www.uniurb.it/PhysLab/strumenti/Heat.html

[42] M. J. Jamin, *Cours de Physique de l'École Polytechnique*, Paris 1868, Vol. II, pp. 97–98.

[43] This thermometer was constructed by Henri Galante for Dr. Constantin Paul who showed it first at the meeting of the Société de Médecine (Paris). A drawing of the instrument was published in the bulletin of French Academy of Medicine in 1884. *The British Medical Journal*, Sept. 20, 1884, p. 559 and also see; *Bulletin de L'Académie de Médecine*, 12 février 1884, pp. 260–261. For other models of the instrument see H. Galante, Fils, *Catalogue Illustré des Instruments et Appareils de Chirurgie, Appareils de prothèse, Orthopédie, Bandages, etc.*, Paris 1885, pp. 142–143.

concerned an instrument for administering anesthesia. This was an ether spray used for local anesthesia and was invented by Benjamin Ward Richardson (1828–1896) in 1866. The thermal conductivity of textiles was another section he inserted from "outside". He added a table showing the thermal conductivity of basic materials used in human clothing such as cotton, wool and angora. The last item in the 1st group was a chapter on animal heat which dealt with subjects like the temperatures of animals and the distribution of animal heat on their bodies. Gırcikyan went beyond the scope of the *Traité* not only for borrowing the texts but also for adapting illustrations and tables. Contrary to the IHT[1], Gırcikyan preferred to use extensive numeric tables in IHT[2] such as the table for "the coefficients of linear expansion of the body most used in the arts". Richer in content and presented in a different way, this table did not appear in the *Traité* but was borrowed from an unknown book.[44]

The last intervention we would like to mention refers to one of the sections of the *Traité* omitted by Gırcikyan. It deals with the concept of energy and the principle of energy conservation. Among the newest theoretical issues treated in *Traité*'s 19th edition, this concept and principle do not appear in the previous editions (13th, 14th and 16th).[45] In the section titled "Heat as a Form of Energy", energy was defined as follows[46]:

> *Energy* is the total quantity of *force vive* a materialistic system *possesses* or *can develop at a given moment*. It is measured by the amount of mechanical work that the system is able to produce at this moment. Energy is either *kinetic* or *actual*, or *potential*.

In his translation, Gırcikyan neither gave a definition for the energy nor explained the principle of energy conservation. The glossary he prepared does not include the term "energy".

The content analysis we have made shows that Gırcikyan's efforts focused on making his book more comprehensive in physiology and in medical applications of physics. His endeavour matches with his preface discourse. He does not approach the subject in a logic of "the newest is the best", especially when theoretical issues are considered. He neither introduces a discourse on novelty: he does not present his book as the transmitter of the newest developments in physics.

Conclusion

Studies conducted so far points to the absence of research and scientific publications in physics in the nineteenth-century Ottoman Turkey. There were no research centers, laboratories, institutes or science academies where physics

---

[44] For comparison see IHT[2], Vol. I, p. 402 and *Traité*, 16th ed., p. 268.

[45] We don't claim that the "energy" concept was first employed in the 19th edition of *Traité*, since we have not had access to the 17th and 18th editions.

[46] *Traité*, 19th ed., pp. 542–543.

research could conducted. No historical figure famous with his knowledge in physics flourished in this century. On the other hand there existed several educational institutions where physics was taught and there were physics teachers who gave those lectures. For someone interested in physics, the only career opportunity was to be a teacher and the highest accomplishment in his career was to write a textbook. For this reason, textbooks dominated nineteenth-century Ottoman physics literature. These textbooks were compiled or translated from western languages, particularly from French. Their authors or translators were physics teachers who prepared these books to be used for teaching purposes in the schools where they lectured. Modern physics was transferred from Europe and its seeds were sown in the educational institutions. As a discipline, physics tried to find a place for itself in the schools. Being a physics teacher in this century meant to create a field of specialization for himself as well as for the future generations. Professional life of Antranik Gırcikyan and his translation of A. Ganot's *Traité Elementaire* into Turkish were both shaped in this general landscape of the nineteenth-century Ottoman history of physics.

Things were difficult for Gırcikyan as they were for other scholars of this century. Everyone who translated a scientific work into Turkish had to confront new terms and concepts flowing from foreign languages and sciences, and they had to coin new terms in Turkish. We can say that, in the acquisition of scientific knowledge by means of translation, the actors of this process created a mechanism of observation and inspection of each other's efforts and Gırcikyan was a part of this scholarly community.

In the eyes of others, a person who performed a competent scientific translation or compilation was considered as knowledgable about the subject and having good skills in foreign languages, especially in French. This profile made him/her an appreciated scholar in the Ottoman Empire. For Gırcikyan, to translate Ganot's *Traité* into Turkish might have been a mean of gaining recognition outside the School of Medicine. The first edition of *İlm-i Hikmet-i Tabiiye* had a better reception than Gırcikyan expected; this indicates that he made a place for himself in the field of physics within the Ottoman society. Gırcikyan who enriched the second edition using other source texts tried to expose his knowledge of physics with the adaptations he made. As the distance between his Turkish text and *Traité* increased, his preface discourse became bolder and stronger. Thus his status moved from the "translator" to that of a "man knowledgable man in physics".

Acknowledgements I would like to thank Feza Günergun for her constructive suggestions as well as her assistance in translating the texts in French. I am also thankful to Şeref Etker for lending me his personal copy of *İlm-i Hikmet-i Tabiiye* and for sharing publications; to Arsen Yarman for providing biographical information on Antranik Gırcikyan; to Tomas Terziyan for translating the texts in Armenian into Turkish; and to Darina Martykánová for sending copies from *Traité*'s 16th edition and for her comments on this chapter.

Eclecticism and Appropriation of the New Scientific Methods by the Greek-Speaking Scholars in the Ottoman Empire

Manolis Patiniotis

Implicitly or explicitly, a great deal of recent historiography of science takes the distinction between scientific centers and scientific peripheries as granted. Historians who inquire into the emergence of modern science primarily focus on areas and events that gave birth to what we now consider "original" science, and confine the rest of the story to a more or less straightforward process of distribution of the sciences to areas which did not participate in the formation of the "original" theories and practices. Due to the lack of local innovation, those areas are described as importers of "new products, new technologies, new ideas" which emanated from the centers and were transferred to the periphery by means of migration.[1]

Usually, the story goes as follows. Peripheries were conquered by the sciences thanks to the self-evident explanatory power of their methods and the obvious social usefulness of their findings. The scholars of the periphery, actually unable to fully assimilate the new methodological developments, contented themselves with simply copying and mechanically reproducing the original research. Indeed, many of them were at pains to compromise their religious and scholastic convictions with the spirit of modernity that started emerging in their local social contexts. As a result, they picked up and combined ideas and practices they considered important for upgrading their intellectual profile, but their adherence to the contemplative dimension of the philosophical discourse and the rejection of experimental and mathematical methods in natural philosophy bear witness to their inability to unequivocally embrace the dynamics of modern science.

The standpoint taken in this paper is different. First and foremost, it aims at problematizing the notions of scientific center and scientific periphery. Instead of

---

M. Patiniotis (✉)
Faculty of Philosophy and History of Science, Athens University, Athens, Greece
e-mail: mpatin@phs.uoa.gr

[1] P. Selwyn, "Some Thoughts on Cores and Peripheries," in D. Seers, B. Schaffer and M. L. Kiljumen, eds., *Underdeveloped Europe: Studies in Core-Periphery Relations*. Hassocks: Harvester Press, 1979, pp. 37–39.

examining the spread of the sciences from their birthplaces to a more or less receptive periphery, one might examine how the very notions of scientific center and scientific periphery resulted from the fact that a certain way of philosophizing about Nature dominated over a variety of other such ways.[2] In order to investigate this process one should examine in strictly historical terms the social and intellectual status of these other ways of philosophizing and the means by which they were displaced by the sciences. To what extent did pre-scientific perceptions of Nature correspond to the needs of the various local societies? Which were the factors that stimulated the change of perspective and what kind of social or intellectual demands did they serve? Did all places and all scholars react in the same way to the advent of modern scientific discourse? Did they unconditionally subscribe to its truth and to its social usefulness? Beyond the superficial level of acceptance or rejection, which were the attitudes of the various groups of local scholars towards the attainments of the "moderns"? How did they appreciate the new ways of philosophizing about Nature? And, most importantly, which were the characteristics of the philosophical discourses they articulated in order to accommodate the new ideas and practices in their intellectual universe?

The case study presented in this paper addresses, of course, only some aspects of the above questions. In what follows I will be dealing with eighteenth-century Greek intellectual life and I will try to show how a change of perspective may help us reconsider the relation of a local culture with the emergent scientific spirit of modern times.

Historiographical Remarks

In the eighteenth century, the greatest part of the Greek Orthodox populations lived in the territories of the Ottoman Empire. Due to the geopolitical fragmentation of the Empire, however, these populations lacked the continuity that would allow for a coherent organization of their social activities. What gradually came to be modern Greek society at that time basically consisted of a network of communities where Greek Orthodox populations developed various social, economic and political activities.[3] Besides the Balkans, the Greek

---

[2] K. Gavroglu, M. Patiniotis, F. Papanelopoulou, A. Simões, A. Carneiro, M. P. Diogo, J. R. Bertomeu Sánchez, A. García Belmar, A. Nieto-Galan, "Science and Technology in the European Periphery: Some Historiographical Reflections," *History of Science*, xlvi (2008): 153–175.

[3] Γ. Τόλιας, "Η Συγκρότηση του Ελληνικού Χώρου 1770–1821 [The Construction of the Greek Space, 1770–1821]," in Β. Παναγιωτόπουλος, ed., *Ιστορία του Νέου Ελληνισμού, 1770–2000 [History of Modern Hellenism, 1770–2000]*, vol. 1. Athens: Ελληνικά Γράμματα 2003, pp. 59–74; Γ. Τόλιας, "Ιερός, Κοσμικός και Εθνικός χώρος στην Ελληνική Γεωγραφική Φιλοσοφία κατά τον 18° αιώνα [Divine, Secular and National Space in Greek Geographical Philosophy of the 18th century]," in *Η Επιστημονική Σκέψη στον Ελληνικό Χώρο, 18ος-19ος αι [The Scientific Thought in the Greek Space, 18th–19th Centuries]*. Athens: Τροχαλία, 1998, pp. 147–172; M. Mazower, *The Balkans*. London: Weidenfeld & Nicolson, 2000.

communities were dispersed along the main commercial routes of Eastern Europe, and within the most important cities of the Northern Italian peninsula, Habsburg Empire, and the German states.[4] As indicated by the term "Greek Orthodox", the elements that played the most prominent role in unifying these populations were fundamentally cultural and ideological. Christian Orthodox faith and Greek-speaking education managed to provide a quasi common cultural identity to such different social groups as the Phanariots of Istanbul, the Vlach merchants of Epirus, the Greek fraternity of Venice, and the administrative elite of the semi-autonomous Danubian regions. Both Christian faith and Greek-speaking education were under the jurisdiction of the same authority, the Ecumenical Patriarchate of Constantinople; but in the light of the eighteenth-century developments both were also heavily colored by the particularities of the various local communities. This was especially the case with education because, due to the lack of other (state) institutions, Greek-speaking education became the ground upon which all negotiations and collective pursuits concerning the emergent society's identity converged. At the same time, however, education was the context wherein the local intellectual traditions came in contact with eighteenth-century natural philosophy, and this interweaving is of special importance to our story.

The second half of the eighteenth and the first two decades of the nineteenth century witnessed the publication of many scientific and philosophical books aiming to cross-fertilize Greek intellectual life with the achievements of the European Enlightenment. The protagonists of this initiative were almost exclusively teachers. But at that time the profile of teacher was gradually redefined. The image of the teacher-priest whose work was a religious mission gave way to another kind of scholar: The great majority of these teachers were also priests, but their educational agenda became more secular and their actual work tended to be more *in tandem* with their contemporary philosophy. The scholastic teaching of the patristic literature and of ancient Greek philosophy, gave way to a curriculum determined through negotiations with the communities which had established and catered for the schools. The fact that teaching began to serve the social, political and ideological agendas of the local communities strengthened the relative autonomy of the scholars from the Patriarchate and reinforced their role as independent thinkers.

---

[4] Ο. Κατσιαρδή-Hering, "Η Ελληνική Διασπορά [Greek Diaspora]," in Β. Παναγιωτόπουλος ed., *Ιστορία του Νέου Ελληνισμού, 1770–2000 [History of Modern Hellenism, 1770–2000]*. vol. 1. Athens: Ελληνικά Γράμματα, 2003, pp. 87–112; Ι. Κ. Χασιώτης, *Επισκόπηση της Ιστορίας της Νεοελληνικής Διασποράς* [A Survey of the History of the Modern Greek Diaspora]. Thessaloniki: Βάνιας, 1993; Απ. Βακαλόπουλος, "Ο Ελληνισμός της Διασποράς" [Hellenism of Diaspora] in *Ιστορία του Ελληνικού Έθνους [History of Greek Nation]*, vol. 11, *Ο Ελληνισμός υπό ξένη κυριαρχία (περίοδος 1669–1821) Τουρκοκρατία - Λατινοκρατία [Hellenism under foreign dominance (1669–1821) The Turkish and Latin rule]*. Athens: Εκδοτική Αθηνών, 1975, pp. 231–243; T. Stoianovich, "The Conquering Balkan Orthodox Merchant," *Journal of Economic History*, XX (1960): 234–313.

Thus, from the outset of the eighteenth century, Greek-speaking scholars started moving all over Europe, and Italy ceased to be the almost exclusive place they would go for studies. They were acquainted with a multitude of intellectual traditions and schools, related mainly to the recent developments of the European Enlightenment. When these people returned to their homelands and were appointed teachers in the local schools, they looked forward to gaining social recognition corresponding to their intellectual qualifications. The production of a new philosophical and scientific discourse played a significant role in the legitimization of their upgraded authority. In many cases this program was carried out through the translation of philosophical and scientific books, or through the compilation of original works wherein the attainments of European thought were creatively combined with elements of the local cultural and philosophical traditions. Such undertakings did not, of course, serve a homogeneous educational agenda neither did they gain the general consent of the local authorities; they were, however, in tune with the aspirations of the most dynamic social groups who counted upon these scholars to shape their distinctive cultural and political profile. But the constituents of this profile were still under negotiation. As a result, the Greek-speaking scholars of the time found themselves at the intersection of multiple cultural traditions and social interests. The textbooks they wrote and the philosophical discourses they elaborated (primarily for the use of their students) reflected exactly this ambiguous situation.[5]

Starting with the fact that, throughout this long and intricate period, philosophy and the sciences were exclusively practiced in the context of education, many historians studying the introduction of the new scientific and philosophical ideas into the Greek intellectual life arrive at a rather trite conclusion: The scientific and philosophical attainments of the European thought were inserted into the Greek context exclusively for educational purposes, and *thus* they represented only a simplified version of the European science and philosophy. Neither hard-core science and philosophy nor original intellectual production did occur in the particular context. As a matter of fact, most of the Greek works of the time were either translations or "multilayer compilations", where pieces of knowledge and methodological declarations representing Enlightenment science mingled with the traditional philosophical discourse and moral instructions for achieving "true felicity".[6] The scholars themselves overtly admitted this attitude in the titles of their works: Expressions implying the selection from the whole range of the available wisdom – "ancient and modern" – occur very

---

[5] M. Patiniotis, "Scientific Travels of the Greek Scholars in the 18th Century," in A. Simões, A. Carneiro, M. P. Diogo, eds., *Travels of Learning. A Geography of Science in Europe*. Dordrecht: Kluwer Academic Publishers, 2003, pp. 49–77; M. Patiniotis, "Textbooks at the Crossroads: Scientific and Philosophical Textbooks in 18th century Greek Education," *Science and Education*, 15 (2006): 801–822.

[6] Π. Κονδύλης, *Ο Νεοελληνικός Διαφωτισμός: Οι φιλοσοφικές ιδέες [Neohellenic Enlightenment: The Philosophical Ideas]*. Athens: Θεμέλιο, 1988, p. 10.

often in the titles of the Greek scientific and philosophical textbooks. In this respect, the major intellectual task of the authors seemed not so much to be the acquisition of new knowledge through the impartial empirical study of Nature, as the use of already acquired knowledge to upgrade the philosophical traditions of their local cultural context.

Indeed, one issue that often puzzles historians about the intellectual disposition of the eighteenth-century Greek-speaking scholars is the latter's attitude towards experimental philosophy as it was practiced by their contemporary European natural philosophers. Greek philosophical and scientific textbooks contain a great deal of references either to specific experiments or to the value of experimental study of Nature in general. Beyond the verbal level, however, we have no evidence that Greek-speaking scholars conducted actual experiments. They mentioned experiments made by others, they commented on remarkable observations taken in European laboratories and observatories, they argued for the acquisition of experimental devices for the use of their pupils, and they declared their adherence to the new empirical method of investigation as opposed to the infertile scholastic explanations; but, as far as we know, they had never conducted actual experiments. At most (and according to scarce evidence) they organized some experimental demonstrations for the elucidation of their students or maybe of a wider learned public. The heuristic role of experiment and its instrumental use in the quantitative investigation of an external natural world was outside their scope.[7]

In fact, the ambiguous relationship of Greek-speaking scholars with experimental philosophy forms part of a broader discussion concerning the kind of philosophical discourse about Nature developed by these scholars. According to most historians, eighteenth-century Greek science altogether lacked originality and creativity. It was a vague reflection of the developments that took place in the centers of the Enlightenment, used in the Greek context almost exclusively for educational and ideological purposes. However, due to the Ottoman rule over the Greek-speaking populations of the Balkans, even the mere attempt to bring Greek education in contact with the Enlightened Europe is considered a heroic endeavor. Thus, historians elaborate an argument, according to which the assumingly low level of the philosophical and scientific production of the time reflects the real conditions of the specific society, and thus the question of originality is literally and metaphorically untimely.[8] The fact, however, that specific scholars assimilated and spread the new ideas in the

---

[7] Χ. Ξενάκης, "Το Πείραμα ως Επιστημονική και Φιλοσοφική Μέθοδος Γνώσης [Experiment as Scientific and Philosophical Method of Knowledge]," in Γ. Καράς ed., Ιστορία και Φιλοσοφία των Επιστημών στον Ελληνικό Χώρο (17^{ος}–19^{ος} αι.) [History and Philosophy of the Sciences in the Greek Space (17th–19th Centuries)]. Athens: Μεταίχμιο, 2003, pp. 514–555 (see pp. 518–520, 535–536 (see. n. 713), 552–555).

[8] Ν. Ψημμένος, ed., Η Ελληνική Φιλοσοφία από το 1453 ως το 1821 [Greek Philosophy from 1453 to 1821], vol. 1, Η κυριαρχία του Αριστοτελισμού [The Dominance of Aristotelianism]. Athens: Γνώση, 1988, p. 31.

Greek intellectual space, countering popular ignorance on the one hand, and the established authorities on the other, was not only important for the revival of the Greek intellectual life, but also determined the subsequent political and ideological developments until the Greek war of independence.[9]

The tacit premise behind such considerations is that Greek scholars were, at best, enlightened teachers. Due to particular historical circumstances, their intellectual activity was confined to the limits of education, and this confinement decisively marked the character of their scientific and philosophical production. For reasons that did not depend on their will or their capabilities, Greek scholars were unable to share the creativity of modern European thought, but one should properly appreciate the pedagogical and ideological consequences of their work. In this respect, a most characteristic aspect of the historiography holding such views is that it persistently links the introduction of the sciences with the enlightenment of the "nation" in anticipation of national emancipation.[10]

Apparently, this historiographical approach takes the distinction between scientific centers and scientific peripheries – production and distribution of science – as granted, without examining the particular historical circumstances of its establishment. Our intention in what follows is to reconsider this approach. But before doing so we should briefly comment on an important issue. The mere fact that the Greek scholars directed their scientific and philosophical considerations towards educational purposes does not necessarily bear witness to the low level of their intellectual production neither does it prove that the only role they assumed for themselves was that of the popularizer or propagandist of science. One should very seriously take into account that for the greatest part of the eighteenth century, education and knowledge production were still inseparable. Notwithstanding the establishment of scientific societies and the spread of experimental philosophy, the practice of modern science remained mostly in the private sphere and whenever it was practiced in public it was either for popularization purposes or for a strictly limited audience of experts. Numerically speaking, the overwhelming number of professional eighteenth-century philosophers were university professors who taught philosophy according to the inherited scholastic models.[11] And these models demanded that the acquisition of knowledge should be pursued by means of literary devices contrived and applied in front of one's students or written and

---

[9] G. P. Henderson, *The Revival of Greek Thought 1620–1830*. Albany, NY: State University of New York Press, 1970, introduction.

[10] Γ. Καράς, "Η Επιστημονική σκέψη κατά την Περίοδο της Νεοελληνικής Αναγέννησης [Scientific Thought during the Neohellenic Revival]," in Γ. Καράς ed., *Ιστορία και Φιλοσοφία των Επιστημών στον Ελληνικό Χώρο (17ος–19ος αι.) [History and Philosophy of the Sciences in the Greek Space (17th–19th Centuries)]*. Athens: Μεταίχμιο, 2003, pp. 45–101 (see pp. 48 (esp. n. 9), 49–50, 74).

[11] C. Lüthy, "What to do with Seventeenth-Century Natural Philosophy? A Taxonomic Problem," *Perspectives on Science*, 8 (2000): 164–195 (see pp. 171–172).

diligently analyzed for the sake of one's students. The pursuit of knowledge, in other words, was part and parcel of the teaching process and vice versa. Therefore, the only conclusion one could draw from the educational orientation of the Greek-speaking scholars is that they conformed to the general disposition of the time.

L'Éclectism

Let us, now, make a short detour that will take us to a seemingly irrelevant issue. In the fifth volume of the *Encyclopédie*, Denis Diderot (1713–1784) published one of his many influential essays. It is a long account on the history of *Éclectism*, a philosophical trend initiated in the ancient era by some of the most prominent philosophers.[12] Diderot follows the subsequent generations of philosophers who represented *l'Éclectism* throughout centuries, discusses the development of various sets of principles and expresses his ambiguous sentiments about the achievements of particular thinkers. What is important to our discussion, however, is not so much the historical account itself, as the programmatic ideas Diderot articulated as a general context for his narrative.

The definition he gives for *l'Éclectism* is not historical, as one might expect. It refers to contemporary philosophy and stresses the fact that eclecticism is a philosophical attitude rather than a specific belief or system of doctrines. It is characterized by impartiality and insistence on selecting from other philosophical systems only those ideas that are in agreement with reason and experience. The purpose of this selection is not the building of a new system or the rescuing of an old one, which falls apart; in fact this is exactly what syncretism tries to achieve by means of loans from every available source, leading to grotesque multicolor constructions. Quite the contrary, eclecticism is an active intellectual stance aiming at philosophical self-realization. The people who practice it borrow from the various existing systems because they believe that everybody should first get acquainted with the existing wisdom and then try to enrich it with new principles and findings. Thus, they do justice to whatever other people or systems of knowledge have contributed to philosophy, but they are aware that all systems with no exception have, in the course of time, fallen apart. In search of a new land where philosophy will be practiced beyond the limitations and the distortions of specific sects they set reason and experience as the ultimate criteria either for the selection and compilation of existing philosophical doctrines or for the suggestion of new

---

[12] D. Diderot and J. Le Rond d'Alembert, eds., *Encyclopédie ou Dictionnaire raisonné des sciences, des arts et des métiers*, vol. 5, *Discussion-Esquinancie. Par une société de gens de lettres; mis en ordre et publié par M. Diderot,... et quant à la partie mathématique, par M. d'Alembert*. Paris: Briasson, David, Le Breton, Faulche, 1755, pp. 270–293.

ones. In either case, philosophizing in one's own means is what best expresses the eclectic way.

The greatest philosophers in history had always been eclectic. Their followers, however, failing to share the originality and the intellectual independence of their teachers confined themselves to sectarian systems which do not serve anymore the progress of philosophy. In modern times, people like Giordano Bruno (1548–1600), Girolamo Cardano (1501–1576), Francis Bacon (1561–1626), Thommaso Campanella (1568–1639), Thomas Hobbes (1588–1679), René Descartes (1596–1650), Gottfried Wilhelm Leibnitz (1646–1716), Christian Thomasius (1655–1728), Andreas Rudigerus (1673–1731), Johann Jacob Syrbius (1674–1738), Jean Leclerc (1657–1736), Nicolas Malebranche (1638–1715) gave new impetus to eclecticism. To take up Diderot's metaphor, systematic eclecticism tries to build the new edifice of philosophy using stones spread on the ground from the collapse of the old philosophical systems. Soon, though, modern philosophers realized that many stones were unfit for their purposes and even more were missing. Thus, they started looking for new material to fulfill their mission. They searched into the depths of the earth, in the waters and in the atmosphere. This quest (along with its methodological developments) initiated *l'éclectism expérimentale*, which aims at accumulating as much new material as possible for the future building of new philosophy.[13] It is in this sense that Diderot considers Francis Bacon, actually, "le fondateur de l'*Eclectisme* moderne".[14] There are two kinds of eclecticism, though. Experimental eclecticism is occupied with natural investigation, without, for the time being, venturing into major theoretical syntheses. The other kind is *l'éclectism systématique*, which places emphasis on the selection and the combination of truths, either those recently unearthed or those originating in the philosophical systems of the past. Reasoning and cognitive manipulation are crucial for this kind of eclecticism, as it spends most of its time and effort in examining all the possible combinations among the available materials. It is a time-consuming and, actually, inconclusive process but it results from the persuasion that it is already possible to start erecting at least some parts of philosophy's edifice, even though this may overstretch the resources. Thus, concluding his programmatic contemplations, Diderot states:

> On voit qu'il y a deux sortes d'*Eclectisme*; l'un expérimental, qui consiste à rassembler les vérités connues & les faits donnés, & à en augmenter le nombre par l'étude de la nature; l'autre systématique, qui s'occupe à comparer entr'elles les vérités connues & à combiner les faits donnés, pour en tirer ou l'explication d'un phénomene, ou l'idée d'une expérience. L'*Eclectisme* expérimental est le partage des hommes laborieux, l'*Eclectisme* systématique est celui des hommes de génie; celui qui les réunira, verra son nom placé entre les noms de Démocrite, d'Aristote & de Bacon.[15]

---

[13] Ibid., pp. 283–284.
[14] Ibid., p. 271.
[15] Ibid., p. 284.

On Defining Science

Let us now turn back to Greek intellectual life and try to contemplate on some key-features of eighteenth-century natural philosophy. The first of these features relates to the fact that almost all major eighteenth-century Greek-speaking scholars, who had dealt with natural philosophy, had also published at least one book on logic and or metaphysics.[16] Most European scholars who placed themselves in the context of the new natural philosophy adopted more and more a discourse built around experimental philosophy and mathematics, and one might argue that the adherence of Greek-speaking scholars to the traditional philosophical discourse depicts their inability to assimilate the new methodological and philosophical developments of the Enlightenment. The point, however, is that the content of their works on logic and metaphysics was *in tandem* with these developments: Overt support to empirical research of Nature, denunciation of the fruitless scholastic methods, re-definition of the principles of logic on the basis of recent philosophical discussions, incorporation of the scientific findings into the metaphysical discourse, re-arrangement of the traditional fields of metaphysical contemplation according to the emerging disciplines of modern science. Therefore, the conclusion one could draw from the adherence of Greek-speaking scholars to the traditional form of philosophical discourse is not about their support or rejection of the new natural philosophy, but about the way they perceived their role in the new context. In this respect, if one examines in detail their compilations and translations it becomes quite clear that their main concern was to secure the unity of philosophy. Being aware of the new methods of natural investigation, of important findings and theories that subverted the received world-image, and of the limits of traditional philosophy's speculative character they were willing to embrace all these developments, but only as long as they would not sacrifice the universal character of philosophy. Consequently, the reason why these people considered it important to "frame" their scientific contribution with metaphysical and logical works was their deep conviction that even this new and revolutionary kind of knowledge was apt to comprise part of an integrated

---

[16] Βενιαμίν Λέσβιος, Στοιχεία της Μεταφυσικής [*Elements of Metaphysics*]. Vienna, 1820; Ε. Βούλγαρης, Η Λογική εκ παλαιών τε και νεωτέρων [*Logic, from ancient and modern sources*]. Leipzig, 1766; Ε. Βούλγαρης, Στοιχεία της Μεταφυσικής [*Elements of Metaphysics*], 3 vols. Venice, 1805a; Κ. Κούμας, Σύνοψις της ιστορίας της φιλοσοφίας [*An Abridged History of Philosophy*]. Vienna, 1818a; Κ. Κούμας, Σύνταγμα φιλοσοφίας [*Constitution of Philosophy*], 4 vols. Vienna, 1818b; Γ. Κωνσταντάς, Στοιχεία της Λογικής, Μεταφυσικής και Ηθικής [*Elements of Logic, Metaphysics and Moral Philosophy*], 4 vols. Venice, 1804; Ι. Μοισιόδαξ, Ηθική Φιλοσοφία [*Moral Philosophy*], 2 vols. Venice, 1761–1762; Χ. Παμπλέκης, Περί Φιλοσόφου, Φιλοσοφίας... [*On Philosophers, Philosophy...*] Vienna, 1786; Δ. Φιλιππίδης, Η Λογική παρά του Κονδιλλιάκ [*Logic by Condillac*]. Vienna 1801; Α. Ψαλίδας, Αληθής Ευδαιμονία [*True Felicity*]. Vienna, 1791.

synthesis, which would maintain the qualitative features of traditional philosophical discourse.[17]

The second feature of Greek-speaking scholars' scientific activity refers to the *subject* of this activity. Matter is the unshakable background upon which they tried to organize all available knowledge. Historians of the Scientific Revolution have long ago agreed that the metaphysical category of matter was one of the first notions to be expelled from the realm of Newtonian natural philosophy.[18] As the mechanization of the world-picture proceeded, the attention of natural philosophers shifted to quantitative relationships between pieces of matter devoid of any causal, intentional or qualitative content. This wasn't quite the case with Greek-speaking scholars, however. Being the heirs of the seventeenth-century neo-Aristotelianism they insisted on perceiving matter as the *substratum* of all natural phenomena. And although, according to this tradition, it was motion that gave life to Nature, even motion was a "passion" of matter. As a result, matter in both senses, either as prime matter, conceived as the bearer of all natural transformations or as material body, substantiating particular physical qualities and inclinations, was the necessary ontological basis of every natural discourse. This philosophical preconception retained its central position even when Greek-speaking scholars came to deal with the new natural philosophy.

In fact, there was no actual reason to break with this notion, since it is in principle possible to attribute all the observed phenomena to the inherent qualities of natural bodies. Indeed, and contrary to a widely held assumption, in the mid-eighteenth century some of the most prominent supporters of Newtonian physics had almost a similar perception about the ontological importance of matter. Quite a few of them opened their works with an account of "those things which we find to be in all bodies", that is the essential qualities inherent to natural body's material substratum: "extension, solidity, inactivity, mobility, a capacity of being at rest or having a figure, [...] gravity and the power of attraction".[19] In this respect, it is not only legitimate but also in accordance with, at least, some aspects of their contemporary natural philosophy for Greek-speaking scholars to retain the ontological predominance of matter in their natural discourse. It is

---

[17] Κ. Γαβρόγλου, "Οι επιστήμες στον Νεοελληνικό Διαφωτισμό και προβλήματα ερμηνείας τους [The sciences during the Neohellenic Enlightenment and Problems in their Interpretation]," Νεύσις, 3 (1995): 75–86.

[18] F. Cajori, "An Historical and Explanatory Appendix," in A. Motte, trans., F.Cajori, ed., *Sir Isaac Newton, Principia*, vol. 2. Berkeley, Los Angeles, London: University of California Press 1962 (First edition 1934), pp. 627–680 (esp. pp. 633, 638); I. B. Cohen, "A Guide to Newton's *Principia*," in *Isaac Newton, The Principia. Mathematical Principles of Natural Philosophy. A New Translation*. trans. I. B. Cohen and A. Whitman. Berkeley, Los Angeles, London: University of California Press, 1999, pp. 1–370 (see pp. 89–94 and elsewhere).

[19] P. van Musschenbroek, *The Elements of Natural Philosophy. Chiefly intended for the Use of Students in Universities, by Peter van Musschenbroek, M.D., Professor of Mathematicks and Philosophy in the University of Leyden. Translated from the Latin by John Colson, M.A. and F.R.S., Lucasian Professor of Mathematicks in the University of Cambridge.* 2 vols. London: J. Nourse, 1744 (First Latin edition: 1734), p. 10.

important to stress this point, because the notion of matter comprises the "material" counterpart of the philosophical attitude described above: The unity of the philosophical discourse about Nature pursued by the Greek-speaking scholars could only be achieved on the basis of such an all-encompassing and deep-rooted concept as matter. The reduction of all natural phenomena to the essential qualities of material bodies might help organize the laws and findings of modern natural philosophy upon a concrete and homogeneous ground.[20]

The third feature that pertains to the intellectual activity of the Greek-speaking scholars has to do with the *aim* of natural philosophy. What was the purpose of natural investigation, and what were the prospective results of such an undertaking? In the early nineteenth century, Theophilos Kairis (1784–1853), one of the most erudite scholars of the time, ventured to give the definition of scientific knowledge.

> Knowledge is the perspicuous understanding of beings. *Partial* or *individual* knowledge results from individual observations or experiments; *empirical* knowledge results from many such experiments and observations; *scientific* knowledge, finally, is the knowledge which [on top of these] also includes the *reason* of the being and can be combined with other such pieces of scientific knowledge.[21]

Although this is one of the clearest statements of its kind, one can find a great deal of similar theoretical declarations in the philosophical works of the time. No doubt those Greek scholars honored the new experimental philosophy and were eager to represent its findings and its cognitive dynamics in their work. But how did they appreciate this particular cognitive enterprise? What value did they attach to, and to what extent did they perceive themselves as part of it? The picture one draws from their various statements is that, beyond the manifest praise of the moderns, they perceived themselves as seekers after a deeper kind of natural truth, which would transcend the level of mere appearances and would guide them to the heart of Nature's secrets. The word Nature (φύσις) itself retained in the philosophical language of the time a great deal of its original Aristotelian sense denoting the deep *causal structure* either of individual beings or of all beings as an efficiently organized whole.[22] According to the above definition, thus, the goal of moderns was, mostly, *empirical* knowledge, while the goal of the Greek-speaking scholars, who pursued understanding through the *principles* of beings, was real *scientific* knowledge.

Neither this attitude, however, was entirely untypical to the highly diversified eighteenth-century natural philosophy. It echoed a speculative tradition that dated back to the sixteenth-century Paduan Aristotelianism and retained its currency until at least Newton's time. Speaking of Zabarella's methodological investigations, H. J. Randall cites his reflections about the distinction between "resolutive" and "demonstrative" method. Both the phrasing and the conceptual content of Zabarella's contemplations are tellingly reminiscent of Kairis' definition.

---

[20] Καράς, op. cit., pp. 72–73.
[21] Cited in Καράς, op. cit., p. 77; translation and emphasis are mine.
[22] Cf. G. E. R. Lloyd, "The Invention of Nature," in idem, *Methods and Problems in Greek Science*. Cambridge: Cambridge University Press, 1991, pp. 417–434.

Since because of the weakness of our mind and powers the principles from which demonstration is to be made are unknown to us, and since we cannot set out from the unknown, we are of necessity forced to resort to a kind of secondary procedure, which is the resolutive method that leads to the discovery of principles, so that once they are found we can demonstrate the natural effects from them. Hence the resolutive method is a subordinate procedure, and the servant of the demonstrative... The end of the demonstrative method is perfect science, which is knowledge of things through their causes; but the end of the resolutive method is *discovery* rather than science. [...] It is certain that if in coming to any science we were already in possession of a knowledge of all its principles, resolution would there be superfluous.[23]

Greek Eclecticism?

Let us now recapitulate by bringing all pieces together. When in 1993 Andrew Cunningham and Perry Williams published their programmatic paper about the reorientation of Scientific Revolution studies, they aptly noted:

It is necessary to identify the particular and specific 'projects of enquiry' in which people in the past were engaged in their investigations of nature. [...] When we read texts from the past, we need to ask ourselves, "to what *question* – both what immediate question, and what project of enquiry – in the life and world of the person who wrote it, was this text the *answer* for its author?". For without knowing the project that a particular historical actor was engaged on, the results arrived at by that historical actor are meaningless to us; the answer is meaningless without the question. [...] [This principle] suggests that we should direct our attention, not simply to statements about the natural world in past texts, but to the precise enterprise of which these thoughts and statements were part and which gave them their identity and meaning.[24]

The extended reference to Diderot's account on eclecticism in the previous section did not, of course, aim at serving an assessment of whether or to what extent eighteenth-century Greek-speaking scholars implemented his intellectual program. On the contrary, it aimed at offering a contemporary testimony which, in the sense of Cunningham and Williams, would allow us to better understand the intellectual atmosphere wherein Greek-speaking scholars articulated their cognitive enterprise.

Two major conclusions which can be drawn from Diderot's account are the following: Firstly, that the predominance of experimental philosophy was questionable even in the late eighteenth century. Although it was clear that the impartial experimental investigation of Nature could free philosophy from its Sisyphean destiny, the weight of the past was huge and people could not easily ignore the legacy of ancient wisdom. Thus, not all natural philosophers in Europe had yielded to the irresistible appeal of experimental philosophy, although this

---

[23] Cited in H. J. Randall, Jr., *The School of Padua and the Emergence of Modern Science*. Padova: Editrice Antenore, 1961, p. 52.

[24] A. Cunningham and P. Williams, "De-centring the 'big picture': *The Origins of Modern Science* and the modern origins of science," *British Journal for the History of Science*, 26 (1993): 407–432 (see p. 420).

does not mean that those who departed from it were necessarily incapable or unwilling to appreciate the new developments. Secondly, and most importantly, that there were two kinds of cognitive enterprises to be pursued by the scholars of the time. One aimed at understanding Nature's workings either through the collection or through the combination of particular pieces of knowledge. The other aimed at producing an overarching synthesis, which would unite the contemplative with the empirical dimensions of natural philosophy leading to a sounder way of knowing. In Kairis' phrasing, the former comprised "individual" and empirical "knowledge" while the latter pursued real "scientific" knowledge. Those who would achieve the latter goal, Diderot noted, would see their names glow among the names of humanity's greatest philosophers.

The cultural dispositions, of course, were decisive in the distribution of these roles among the various communities of philosophers. Originating in an antischolastic Aristotelian tradition, which emphasized the inquiry into natural causes and even placed physics above metaphysics, Greek-speaking scholars were apt to participate in the discussions about the new ways of natural investigation.[25] The close ties of neo-Aristotelian philosophy with Renaissance naturalism, and the empirical background of many treatises resulting from this relationship helped these scholars get familiar with the inductive method of modern science and appreciate its findings. At the same time, though, it is a fact that Greek-speaking scholars never ventured into the empirical investigation themselves. What they valued as knowledge proper was a well organized and hierarchical set of principles that would allow them to reduce all observed phenomena to the deeper layer of natural causes.[26] They did not reject the notion of natural law that could be reached through experimental investigation, but they were convinced that such formulations sufficed only for the description of the appearances. To use the familiar Aristotelian terminology of the time, real physics was only "*justified* physiology" as opposed to "*historical* physiology" — causal accounts as opposed to mere qualitative or quantitative correlations.[27]

Eighteenth-century Greek-speaking scholars belonged to a community who shared with many other readers of *l'Encyclopédie* the desire to set up an intellectual enterprise that would meet the challenges of the time. In this respect, the decision they needed to make was not about the acceptance or rejection of a certain method of inquiry, but about the way they would get involved in the consolidation of their contemporary philosophy. And this decision was, to a great extent, dictated by their cultural dispositions: Many European natural

---

[25] Cl. Tsourkas, *Les débuts de l'enseignement philosophique et la libre pensée dans les Balkans. La vie et l'oeuvre de Théophile Corydalée (1570–1646)*. 2nd revised edition, Thessaloniki: Institute for Balkan Studies, 1967; Κ. Θ. Πέτσιος, *Η περί φύσεως συζήτηση στη νεοελληνική σκέψη. Όψεις της φιλοσοφικής διερεύνησης από τον 15ο ως τον 19ο αιώνα [The Discussion about Nature in Neohellenic Thought. Aspects of the Philosophical Investigation from the 15th to the 19th centuries]*. Ioannina, 2002, pp. 137–176.

[26] Καράς, op. cit., pp. 63–66.

[27] Ε. Βούλγαρης, *Τα Αρέσκοντα τοις Φιλοσόφοις [Philosophers' Favorites]*. Vienna, 1805, p. 4.

philosophers confined themselves to systematically unearthing new stones for philosophy's future edifice; others laboriously combined old and new stones trying to retrieve the missing pieces of "les plans perdus de [l']univers".[28] Both enterprises were rather distant from Greek-speaking scholars' style of philosophizing. This style emphasized the investigation into the *principles of beings* as a means for achieving integrated and, thus, real scientific knowledge. The resulting intellectual endeavors might take advantage of the above enterprises but, at the same time, transcended their limited and inconclusive character. Thus, it was quite natural for Greek-speaking scholars to assume a place amongst the philosophers who, in line with Diderot's suggestion, would undertake the other major task of the time: Restoring the unity of the cognitive enterprises, simultaneously liberating philosophy from sectarianism and dogmatism. Having this in mind, they attempted to articulate *their own* syntheses, accommodating the most precious pieces of ancient and modern knowledge as parts of a perpetual philosophical inquiry. This was the way Diderot perceived modern eclecticism[29] and this was, actually, the way most Greek-speaking scholars seemed to perceive their role in the context of their contemporary philosophy. Their contemplative naturalism and the good command of ancient sources coupled with the knowledge of the new scientific attainments rendered them suitable for the specific intellectual task.

Taking this perspective may significantly change the idea we have about the intellectual attitude of the eighteenth-century Greek-speaking scholars towards modern science. As we saw, many historians believe that although Greek-speaking scholars did not really embrace the new scientific method, they did their best to propagandize it and, under the specific historical circumstances (Ottoman rule, poor material conditions, lack of proper institutions), this suffices to offer them a kind of historical vindication. In light of the above discussion, however, it becomes clear that it was not their difficulty, inability or unwillingness to follow the new developments that kept them at the periphery of modern scientific discourse. Quite the contrary, they even assumed a *patronizing* role for themselves, and *it was this role* which, in the course of time, resulted in their marginalization. To paraphrase Diderot's metaphor, they were the "genius (es)" who aimed at supervising the work of the "industrious" ones. Why and how these roles came to be reversed, why and how the quest for a systematic organization of natural philosophy yielded to the formal organization of empirical knowledge is, apparently, a matter of further historical investigation. It is worth stressing, though, that eighteenth-century Greek-speaking scholars considered themselves fit to embark on an ambitious intellectual project built around the question of the unity of philosophy. Understanding how this question came to be outmoded in the specific historical context will contribute to better understanding how modern science came to dominate over other forms of knowledge and other cognitive priorities.[30]

---

[28] Diderot & d'Alembert, op. cit., p. 283.
[29] Diderot & d'Alembert, op. cit., p. 271.
[30] Cunningham and Williams, op. cit., pp. 429–431.

Part IV
On Medicine and Medical Practices

Conveying Chinese Medicine to Seventeenth-Century Europe

Harold J. Cook

As is now well-known, Europeans of the early modern period were eager to learn about medical practices used elsewhere in the world. The so-called voyages of exploration were famously undertaken not only in the hope of finding more Christian allies to help in the struggle for the Mediterranean but in the hope of finding new sources of wealth, particularly gold and spices. By the sixteenth century, the quest for spices had led the Portuguese around Africa and into trading networks of the Indian Ocean, while on their way to East Asia the Spaniards had bumped into a New World. Well-known spices arrived in Europe in larger quantities and at cheaper prices than ever before, while many new comestibles were introduced. Along with the spice trade went the trade in medicinals. The European pharmacopoeia was, therefore, enriched with many medicines from overseas, both those more easily available and those previously unknown. Several historians have recently drawn our attention to the importance of the search for useful medicines by Portuguese, Spanish, English, and Dutch investigators. Less well-known, however, are European discussions about medical ideas and practices they encountered elsewhere.

The more materialistic a medical practice, the more easily it crossed cultural boundaries. For instance, when employees of the Dutch East India Company (or VOC) such as Jacobus Bontius enthusiastically sought out the medicinally useful plants in the Dutch Indies, they eliminated whatever they also gathered about the medical rituals and ideas of the indigenous practitioners who used such remedies, calling them "superstitious." Or again, in the cases of Herman Busschoff, who introduced Europeans to the practice of moxibustion, and Willem ten Rhijne, who

---

H.J. Cook (✉)
John F. Nickoll Professor, History Department, Brown University,
Providence, RI, USA
e-mail: halcook7@mac.com

I would like to thank Sharon Messenger and Caroline Overy for tracking down some of the references for me, and the Institute for Advanced Study in Princeton and the Descartes Centre in Utrecht for the time away from other duties that allowed me to undertake much of the research for this paper.

did the same for the practice of acupuncture, descriptions of the employment of the moxa and needle were straightforward, whereas the explanations offered for why these methods worked were put in terms of the latest Dutch understanding of pathological processes, with only a few garbled glimpses of East Asian theory.[1] For the most part, then, Europeans were excited about useful new remedies, but only curious about, or even averse to, the medical views of other peoples.

There are nevertheless a few examples of European interest in medical concepts developed elsewhere. For the most part, these come from missionaries, who were the people most engaged with the beliefs held by other people. For example, the best historical information remaining about the medical ideas of the indigenous people of Central America stems from Franciscan missionaries and their converts. Francisco Hernández, the physician sent by Philip II to investigate the medicine and natural history of New Spain, came back with mountains of information about medicinal substances and their uses, but not with much record of the understanding of the indigenous peoples' views of the reasons for their practices.[2]

China was a somewhat different case, however, since Europeans understandably took it to be a very great and powerful empire that possessed great learning. Members of the Society of Jesus, who famously managed to launch a missionary effort among the Chinese, depicted it as an ideal state ruled by philosopher-kings, lacking only Christianity to make it the most glorious place on earth, while even merchants of the Dutch East India Company (VOC) – who tried unsuccessfully to gain entry to it – found it a most curious place. Large numbers of Chinese merchants and other residents, mainly from south-eastern China, also inhabited the VOC's headquarters in Asia, Batavia (now Jakarta), where they lived in a community with their own leaders and continued to live as much as they could according to their own customs. Some of the medical employees of the VOC, like some of the Jesuits, therefore also took an interest in finding out more about the resources of the Chinese, whether they be in the form of medicinal remedies, medical practices, or medical ideas.

The first European publications of and commentaries on Chinese medical texts appeared in France and Germany, in the 1670s and 1680s: *Les Secrets de la Médecine des Chinois, Consistant en la Parfaite Connoissance du Pouls* (The Secrets of Chinese Medicine, Which Consists in a Perfect Understanding of the Pulse), first published at Grenoble in 1671; the *Specimen Medicinæ*

---

[1] For these episodes, see H. J. Cook, *Matters of Exchange: Commerce, Medicine and Science in the Dutch Golden Age*. New Haven: Yale University Press, 2007.

[2] B. R. Ortiz de Montellano, *Aztec Medicine, Health, and Nutrition*. New Brunswick: Rutgers University Press, 1990; S. Varey, R. Chabrán, and D. B. Weiner, eds., *Searching for the Secrets of Nature: The Life and Works of Dr. Francisco Hernández*. Stanford: Stanford University Press, 2000; A. Barrera-Osorio, *Experiencing Nature: The Spanish American Empire and the Early Scientific Revolution*. Austin: University of Texas Press, 2006; and esp. D. Bleichmar, P. De Vos, K. Huffine, and K. Sheehan, eds., *Science in the Spanish and Portugese Empires, 1500–1800*. Stanford: Stanford University Press, 2009.

Sinicæ, sive Opuscula Medica ad Mentem Sinensium (An Outline of Chinese Medicine, or a Short Work on Medicine According to Chinese Thought), published at Frankfurt in 1682; and the *Clavis Medica ad Chinarum Doctrinam de Pulsibus* (Medical key to the Chinese doctrine on the pulse) published in 1686 as an issue of the Miscellanæ curiosa (or Ephemerides), a publication series of the Academia Naturae Curiosorum, a German medico-scientific society (later the "Leopoldina").

These texts have long been noticed as curiosities to historians of medicine in both Europe and China, and for almost a century several well-known historians have attempted to illuminate the process by which they arrived in Europe. Most of the evidence has come from Jesuits sources, with some from the German side as well. But new evidence from the Dutch part of the European network allows us to correct and revise portions of the accepted story. The projects that resulted in the publications of the 1680s were in effect a sometimes fractious but collaborative effort between the Jesuits in China and the Dutch and German medical employees of the VOC in Asia, and their local informants, over a period of at least two decades. A close study of the background to these two books can therefore help to illuminate the reasons for, and limits of, European interest in the medical concepts and practices of East Asia.

From the early eighteenth century, there has been a debate about the authorship of the three works. The first is anonymous, although the copy preserved in the library of the Faculté de Médecine de Paris has the name of "Louis Augustin Allemand" written on the title page, probably meaning Louis Augustin Alemand (c. 1653–1728) of Grenoble, an advocate and lawyer in the Parliament of Paris interested in church history who later became a physician.[3] He may therefore have been the person who saw it through the press. The title page of the second of them, the *Specimen*, listed the author as Andreas Cleyer, indicating that he was a medical licentiate from the German city of Kassel (in Hesse), and chief physician, head pharmacist, and overseer of the surgeons for the VOC in Batavia. The subsequent *Clavis* named Michael Boym – a Polish Jesuit who had been a member of the China mission – as its author, with Cleyer credited as the compiler of the works from dispersed manuscripts over a period of 20 years, and with editing performed by Philippe Couplet, the chief advocate ("Procurator") for the Jesuit China mission in Europe. It would have been more accurate if the second work had also specified that Cleyer had been its chief editor rather than "author." For reasons to be examined further below, as early as 1730, the antiquarian and philologist Gottlieb Siegfried Bayer, commenting from St. Petersburg on Chinese studies, thought that the authorship of both works belonged to Boym, so that by putting his name on the first one Cleyer had become a "plagiarist." It was a charge that was picked up by the French Sinologist Jean-Pierre Abel-Rémusat in 1829, and it has continued to echo

---

[3] W. Frazer, "Histoire Monastique d'Irelande, by Louis Augustin Alemand," *Notes and Queries*, 5th S., III, June 5, 1875, p. 456.

down to the present.[4] In 1933, Robert Chabrié argued that Boym had translated an edition of a Chinese medical work into Latin, but that because in 1658 it was sent to his confreres in Europe via Dutch Batavia, the Dutch confiscated it and later published it under Cleyer's name.[5] The accusation has had its most recent airing in the late twentieth century by two patriotic defenders of Boym, Boleslaw Szczesniak and Edward Kajdański, who believed that both the *Specimen* and the *Clavis* were posthumous works by Boym. Writing in 1949, Szczesniak considered them to be two separate editions of a single compilation of medical manuscripts by Boym.[6] Kajdański similarly concluded that "both editions – that published in 1682 and that of 1686 – comprise two parts of the same medical work: Michael Boym's Medicus Sinicus."[7] Another defender of Boym suggested that *Les Secrets* had also been composed by him.[8] Three other learned commentators of the twentieth century, however, have given good reasons for questioning whether Boym had been the author of all and for doubting the accusation of plagiarism.[9]

As I will show below, Cleyer was indeed not the only person to have been involved with the difficult work that went into the publication of the *Specimen*, although the *Clavis* was, as indicated on the title page, mainly composed of translations once in the hand of Boym. It is easy to suspect Cleyer of bad practice, since he is well known for at least occasional shady dealing and for taking credit for the work of colleagues and employees.[10] But there is no good evidence to think that all of the books are simply different editions of a work by Boym or that the second of them, the *Specimen*, simply plagiarized Boym's work. In fact, despite his personal reputation, Cleyer deserves credit for helping with the effort to bring some knowledge of Chinese medicine to Europe. He was helped by others in the East Indies, too, including a Dutch physician who visited Japan, Willem ten Rhijne. More importantly, given the many people involved with the collection, translation, editing, and publishing of the works that

---

[4] On Bayer and Rémusat, see M. D. Grmek, "Les Reflets de la Sphygmologie Chinoise dans la Médecine Occidentale," *Biologie Médicale, Numéro hors série*, 51 (1962), pp. 1–129, quotation from pp. 66–68; E. Kraft, "Christian Mentzel, Philippe Couplet, Andreas Cleyer und die chinesische Medizin: Notizen aus Handschriften des 17. jahrhunderts," in *Fernöstliche Kultur: Wolf Haenisch zugeeignet von seinem Marburger Studienkreis*. Marburg: N.G. Elwert, 1975, pp. 159–161, 187.

[5] Robert Chabrié, *Michel Boym: Jésuit Polonias et la Fin des Ming en Chine (1646–1662)*. Paris: Pierre Bossuet, 1933, p. 236.

[6] B. Szczesniak, "The Writings of Michael Boym," *Monumenta Serica: Journal of Oriental Studies*, 14 (1949): 481–538.

[7] E. Kajdański, "Michael Boym's *Medicus Sinicus*," *T'oung Pao* 73 (1987): 161–189, quotation from p. 169.

[8] R. Chabrié, *Michel Boym*, p. 237.

[9] P. Pelliot, "Michel Boym," *T'oung Pao*, 31 (1935): 95–151; M. D. Grmek, "Les Reflets de la Sphygmologie Chinoise," pp. 1–120; E. Kraft, "Christian Mentzel et al.," pp. 158–196.

[10] The best study of Cleyer remains F. de Haan, "Uit Oude Notarispapieren, II: Andries Cleyer," *Tijdschrift voor Indische taal-, land- en volkenkunde*, 46 (1903): 423–468.

appeared in the compilations that are the *Specimen* and *Clavis*, it is doubtful whether the question of "authorship" is meaningful in the simple way it is usually meant. The story of how two works on Chinese medicine came to appear in Germany, in Latin, in the 1680s, reflects instead on the aims and efforts of multiple people and groups. It is because many Europeans were wondering about the medicine of the Chinese that the books were published, and it is because so many people can be seen to have been involved in the production of them that an examination of their compilation can reveal connections between many places and interests.

Early Jesuit Interest in Chinese Medicine

The Jesuit father Michel Boym certainly did begin to translate some Chinese texts of medical interest into Latin. A study of his life and work published by Robert Chabrié in 1933, and an important commentary on it 2 years later by the great French Sinologist Paul Pelliot, remain the foundation for our understanding of him. His family originally came from Hungary but in the sixteenth century moved to places in the Polish-Lithuanian Commonwealth, perhaps in the train of the Hungarian-Polish princes Bátory: Michel's uncle, Georges Boym, had been a secretary to king Stéphane Bátory, and his father, Paul Boym, was physician to the sister of king Sigismond III and for a time "prévôt de Wilno" (Provost of the new Jesuit university established by the king in Vilnius). The family was certainly wealthy and devout, building an impressive chapel in the cathedral in 1609 in Lwow (then a self-governing city in the Commonwealth), where Michel-Pierre Boym was baptised 3 years later. He took up his noviciate in the Society of Jesus in Cracow on 16 August 1631, and after a visit to Rome travelled to Lisbon, from where he took a ship to the Far East, arriving in Portuguese Macao in 1642.[11]

The founder of the Society of Jesus, Francis Xavier, set out to convert Asia to the Catholic faith in 1552, and in his wake a successful mission to Japan was begun. He intended to visit China as well, although he died before reaching the mainland. In 1583, another group of Jesuits, including the famous Matteo Ricci, began the mission on the Chinese mainland, at the Portuguese trading town of Macao. As is well known, at first they copied many of the manners and dress of Buddhist monks, but in 1595 they changed tack and decided that they would be more successful by imitating the scholarly literati, who also constituted the higher levels of the imperial civil service. They did this not only by styling their hair and clothing in the manner of the mandarins but by pursuing a policy of "accommodationism". This was, first, an attempt to adapt Christian teachings to the language and customs of China, and second, a philosophically syncretic approach to find the essential commonalities between the various

---

[11] R. Chabrié, *Boym*, pp. 70–72; P. Pelliot, "Michel Boym," p. 96.

classical texts that formed the basis of the Chinese *ru*, or true learning, and their own Christian religion. Their main efforts undoubtedly went into trying to convert ordinary people, and they had many successes in that endeavour. But at the same time, their engagement with Chinese learning also involved them in intellectual exchange and controversy with the literati and even the imperial court, while at the same time it led to charges from within the Catholic Church that they had abandoned the true faith.[12]

The philosophical orientation of accommodationism can be noted in the Jesuits' reliance on both Thomistic views of natural reason and wide-spread assumptions at the time of a prisca theologia.[13] The belief in an original natural philosophy transmitted from God to Adam and his descendants meant that from the Jesuits' point of view there must be a fundamental compatibility between the learned teachings known as *ru* and Christian moral philosophy, so that the Chinese lacked only God's revelations and the teaching of Christ and his apostles. They identified the ancient teacher of the true natural philosophy as a figure from the beginning of Chinese history, Kongzi, whose name they Latinised as "Confutius" or Confucius, arguing that all the other texts, authors, and commentators took their principles from his original insights (just as in Europe it was argued that the Egyptian Hermes Trismegistus, sometimes identified with Moses, had been the first lawgiver, philosopher, and priest, from which all learning had descended). They also brought their philological expertise to Chinese literary history, making persuasive arguments about how missing texts could be reconstructed and others re-edited to bring out the essential meaning of *ru*, bringing what was arguably a multitudinous textual tradition into a harmonious system. In other words, the Jesuits created a kind of "Confucian monotheism," which they then tried to enrich by adding Christian teaching to it. At the same time, many of the late-Ming literati were already experimenting with syntheses of

---

[12] The literature on the early Jesuit mission is very large, but for recent views see esp. L. M. Brockey, *Journey to the East: The Jesuit Mission to China, 1579–1724*. Cambridge, MA: Harvard University Press, 2007; J. Gernet, *China and the Christian Impact: A Conflict of Cultures*, transl. by J. Lloyd, Cambridge; Paris: Cambridge University Press; Editions de la Maison des Sciences de l'Homme, 1985; L. M. Jensen, *Manufacturing Confucianism: Chinese Traditions and Universal Civilization* (Durham: Duke University Press, 1997); D. E. Mungello, *The Great Encounter of China and the West, 1500–1800*. Lanhamm, NY: Rowman and Littlefield, 1999; Erik Zürcher, "Jesuit Accomodation and the Chinese Cultural Imperative," in D. E. Mungello ed., *The Chinese Rites Controversy: Its History and Meaning*. Nettetal: Steyler Verlag, 1994, pp. 31–64.

[13] M. Lackner, "Jesuit Figurism," in T. H. C. Lee, ed., *China and Europe: Images and Influences in Sixteenth to Eighteenth Centuries*. Hong Kong: The Chinese University Press, 1991, pp. 129–149; J. Israel, "Admiration of China and Classical Chinese Thought in the Radical Enlightenment (1685–1740)," *Taiwan Journal of East Asian Studies*, 4 (2007): 1–25; and on the prisca theologia, the classic works remain F. A. Yates, *Giordano Bruno and the Hermetic Tradition* (New York: Vintage Books, 1969) and D. P. Walker, *The Ancient Theology: Studies in Christian Platonism from the Fifteenth to the Eighteenth Century* (London: Duckworth, 1972).

Confucianism, Buddhism, and Taoism, so that adding what the Jesuits told them about Christianity – which downplayed the crucifixion and resurrection – and at the same time taking on new views of their own literary and philosophical traditions, persuaded several high-ranking figures to convert. They in turn helped the Jesuits translate European works into elegant Chinese and Chinese works into polished Latin. For all this, the Jesuit missionaries needed a very good ability in Mandarin Chinese and the use of Chinese characters in reading and writing, which they acquired through intensive immersion from more experienced brethren upon arriving in Macao.[14]

Like other recruits, Boym no doubt went through a period of intensive study of Chinese in Macao, only after having mastered its spoken and written forms being assigned to a mission.[15] From 1645 he was apparently assigned to work with the head of the Tonkin mission preaching to eastern Indochina, Andre-Xavier Koffler, and until 1649 he probably spent most of his time on Hainan. But he then joined Koffler at the Ming court in exile. In 1644, the Manchus from the north had conquered Peking and founded the Qing dynasty, but in the south Ming resistance rallied about the Yong-li emperor in the provinces of Guangxi and Guangdong. Ming forces fought off sieges by the Qing in 1647 and 1648, and even retook the city of Guangdong (Canton). Several of the chief ministers and generals to the Yong-li emperor were converts to Christianity, and from the summer of 1646 Koffler had ministered to them, even baptising the three chief women at the court: Yong-li's sister, baptised as "Helen," his mother, baptised as "Maria," and his first wife, baptised as "Anne." At a moment when it seemed like the Ming might win their fight back against the Qing, a project was formulated to send Boym as ambassador to Rome with letters (dated 1 and 4 November 1650) from the two chief Christians at the court, the "Empress Dowager" Helen and the chief chancellor, the eunuch P'ang T'ien-cheou, asking for material support for their cause from the Pope.[16] Boym was accompanied by two Chinese Christians, Andreas Cheng An-to-le (aged 19 at the

---

[14] See esp. L. M. Jensen, *Manufacturing Confucianism*; T. H. C. Lee, "Christianity and Chinese Intellectuals: From the Chinese Point of View," in T. H. C. Lee ed., *China and Europe: Images and Influences in Sixteenth to Eighteenth Centuries*. Hong Kong: The Chinese University Press, 1991, pp. 1–27; D. E. Mungello, *Curious Land: Jesuit Accommodation and the Origins of Sinology*. Stuttgart: Franz Steiner Verlag, 1985; N. Standaert, *Yang Tingyun, Confucian and Christian in Late Ming China: His Life and Thought*. Leiden: Brill, 1988; Q. Zhang, "About God, Demons, and Miracles: The Jesuit Discourse on the Supernatural in Late Ming China," *Early Science and Medicine*, 4 (1998): 1–36; E. Zürcher, "Confucian and Christian Religiosity in Late Ming China," *The Catholic Historical Review*, 83 (1997): 614–653.

[15] The details of Boym's life are taken from P. Pelliot, "Michel Boym," and R. Chabrié, *Boym*, preferring Pelliot to Chabrié where there are differences of detail.

[16] The letters are published in two different English translations: E. H. Parker, "Letters from a Chinese Empress and a Chinese Eunuch to the Pope in the Year 1650," *The Contemporary Review* 101 (1912): 79–83, and I. Ying-Ki, "The Last Emperor of the Ming Dynasty and Catholicity," *Bulletin of the Catholic University of Peking*, 1 (1926): 23–28.

beginning) and Joseph Lo, the latter of whom fell ill on the way and had to abandon the trip. By the time Boym was underway, however, the Qing had recaptured the city of Guangdong (on 25 November), and because it was now clear that Boym was on a diplomatic mission for the losing side, the Portuguese authorities obstructed his journey all along the way. Boym and Cheng managed to sail to Goa, but then travelled overland to Smyrna (in Turkey), then made their way to Venice, and they were intercepted by members of his own Society who tried to keep Boym from going on to Rome since the Jesuit interest likewise was now to work with the Qing. With the help of the French ambassador and the Venetian Senate – at this time, neither were particular friends of the papacy – Boym and his companion managed to arrive in Rome in the first months of 1653, where he was kept waiting for nearly 3 years while the Church's diplomats discussed what kind of reply to give him.

During the period of Boym's European return he communicated much of what he had learned about China. A French version of the short account of Boym's mission that he had delivered in Smyrna was printed in Paris in 1654 by Sebastian Cramoisy, dedicated by the publisher to the Queen. In an addendum, Cramoisy declared that Boym had brought to the Holy Father letters from the court of China along with many other things both circumstantial and important, with the aim of printing them if his superiors approved. Cramoisy also listed the following seven titles as readied or almost readied by Boym for the press: (1) a Latin account of the Jesuit China mission, (2) a Chinese catechism, (3) a book on the fruits and trees of China, (4) a mathematics text in Chinese, (5) a collection of the works of Confucius ("Moralis Philosophia Sinarum"), (6) a work on Chinese medicine, and (7) a set of maps of the Empire of China.[17]

The existence of almost all the items listed by Cramoisy can be traced. The first item was apparently Boym's account of his mission that he delivered in Smyrna, and which constituted the basis not only for Cramoisy's edition but for several other contemporary editions, one in Latin, three in German, another in French, and two in Italian (and an eighteenth-century Polish edition). The third item is the only one to be published independently (in Vienna) by Boym himself, the *Flora Sinensis*, dedicated to Leopold Ignatius, son of the Holy Roman Emperor Ferdinand III, and King of Hungary, Archduke of Austria, etc. The contents contain a set of woodcuts with facing-page descriptions of over twenty plants, two birds, five animals, and an illustration and account of the Nestorian monument of the eighth century from "Hsi-an Fu" that had been brought to the attention of the Jesuits in 1625. (The city was also spelled Sian-fu, being modern Xian or Xi'an, in the northwestern province of Shensi or Shaanxi, a place in our own era made famous for the terracotta

---

[17] M. Boym, *Briefve Relation de la Notable Conversion des Personnes Royales, & de l'estat de la Religion Chrestienne en la Chine ... & recitée par luy-mesme dans l'Eglise de Smyrne, le 29 Septembre de l'an 1652*. Paris: Sebastian Cramoisy, Imprimeur ordinaire du Roy & de la Reyne, & Gabriel Cramoisy, 1654. Cramoisy's list is mispaginated at the end as pp. 72–75, but is six pages long.

warriors and horses.)[18] The last item are copies and transcriptions of Chinese maps and geographical descriptions in which Boym may have had a hand, but which are not necessarily of his own creation. Matteo Ricci had used one of the same major sources (Lu Ying-yang's Kuang-yü-chi) for his 1602 edition of a Chinese map of the world.[19] That Boym's versions of Chinese geography were never published is probably due to the return to Europe at about the same time of another Jesuit, Martinus Martini; but Martini sailed to Amsterdam, where he gave his somewhat less accurate collection of information to the renowned publisher Joan Blaeu, who used it for the sixth part of his Atlas Major (the Novus Atlas Sinensis, 1655).

But even Boym's own *Flora Sinensis* contained information from people other than himself, since there is no possibility that Boym had seen the Nestorian monument himself. The mathematics text and the collection of works of Chinese moral philosophy ("Moralis Philosophia Sinarum") are also clearly projects on which the Jesuits in China had been engaged for some time. In turn, Boym and his companion Cheng An-to-le were also clearly very important collaborators of the more famous Roman Jesuit compiler of natural knowledge, Athanasius Kircher, whose famous *China Illustrata* of 1667 contained a great deal of information identified as being from Boym. For instance, with the assistance of Boym (and no doubt Cheng), Kircher included a longer description of the Nestorian monument in his own work than Boym had done in his own *Flora Sinensis*. Kircher included a letter from Boym about it, an engraving of the inscription dated 1664 and done in a good Chinese hand, and a word-for-word Romanisation and translation with numbering of the Chinese characters by Boym but, given his poor calligraphy, probably with the help of Cheng.[20] Attribution of authorship is even more complicated, when one acknowledges that items two, four, and five on the list of Boym's works were clearly collaborative efforts of the China mission itself – the Chinese catechism must have been developed by the Jesuits for their missionary activities by at least the early seventeenth century – while a Latin version was again printed in Kircher's *China Illustrata*. Although Boym's role in shaping the final versions of these texts was probably important, it makes little sense to consider Boym as the individual author of any of them.

It is therefore likely that the "Medicus Sinicus," too—which is unknown in that form—was a work that had been begun by Jesuits in China and was

---

[18] M. Boym, *Flora Sinensis*. Vienne: Matthæi Rictii, 1656. On the Nestorian monument, see Mungello, *Curious Land*, pp. 164–165; L. M. Brockey, *Journey to the East*, p. 80.

[19] B. Szczesniak, "The Atlas and Geographical Description of China: A Manuscript of Michael Boym (1612–1659)," *Journal of the American Oriental Society*, 73 (1953): 65–77; B. Szczesniak, "The Mappa Imperii Sinarum of Michael Boym," *Imago Mundi*, 19 (1965): 113–115.

[20] W. Simon, "The Attribution to Michael Boym of Two Early Achievements of Western Sinology," *Asia Major*, 7 (1959): 165–169. On Kircher, see P. Findlen ed., *Athanasius Kircher: The Last Man Who Knew Everything*. New York, London: Routledge, 2004.

brought to Europe by Boym. The interest of the China mission in medicine is not well known. According to the recent work of Liam Brockey, Jesuit missionaries themselves avoided the use of medicine as a tool of conversion, despite the fact that by the 1630s they were gaining fame as exorcists and healers. As one Jesuit complained, "Too many begged to be baptized so they might 'attain bodily health, but they do not heal their souls.'" And yet ministering to the sick was an important duty for ordinary Christian converts as well as for the missionaries themselves, so that members of the confraternities they established were to visit the sick, sometimes "as encouragement to persuade non-Christians to submit to baptism".[21] Nevertheless, it is clear that a few of the early Jesuits considered it important to translate European views of medicine, anatomy, and physiology into Chinese as a part of their efforts to persuade the literati of the superiority of their knowledge.

Among them was Johann Schreck, known as Terrentius, who had studied medicine in Paris, Montpellier, Bologna, and Rome, and in 1611 became a member of the Italian scientific academy, the Lincei, for whom he worked on an edition of the natural history of New Spain. In the same year, he became a Jesuit (much to his friend Galileo's regret) and soon joined the China mission and took the name Teng Yü-Han. He chose the works on science and medicine that were a part of the large library his party carried out with them. Among his other accomplishments was the compilation of a Plinius Indicus of over 500 plants previously unknown in Europe, and a work in Chinese on human anatomy inspired by his friend Caspar Bauhin, which was composed in Hangzhou while working with Li Chih Tsao. Following Terrentius's death the manuscript passed to the chief of the mission in Peking, Adam Schall, who in 1634 showed it to a learned high functionary in Peking, Pi Kung Ch'En, who was so impressed that he asked Schall to quickly turn his attentions to composing a manual on Chinese medicine; because other projects intervened, Pi himself adapted Terrentius's work, which was published in 1635 as T'ai-hsi jên-shên shuo kai (the only known copy being in the national library in Rome). Pi was only one of the many Chinese scholars who appear to have considered European medical thought and practice to have something to offer.[22] In short, by the 1620s at the latest, some of the Jesuits and members of the literati were exchanging views and texts about medicine.

As the son of a physician himself, Boym is likely to have been particularly interested in the project. It would be rash to ascribe the edition and translation of the text on Chinese medicine called "Medicus Sinicus" in the *Briefve Relation* to him alone. Yet he no doubt had some sort of Latin version of one or more Chinese medical texts in his possession. The full title indicates that it was mainly

---

[21] L. M. Brockey, *Journey to the East*, quotations from pp. 359, 399.
[22] P. Huard, "La Diffusion de l'Anatomie Européenne dans Quelques Secteurs de L'Asie," *Archives Internationales d'Histoire des Sciences*, 32 (1953): 269–270; Hsing-Chun Fan, "La Médecine Occidentale en Chine vers la Fin des Ming (1644)," *Bulletin de l'Université l'Aurore*, S.III, 5, (1944): 677.

a work on the pulse interpreted according to the policy of accommodationism: "Chinese Medicine or the Singular Art of Exploring the Pulse to Prognosticate Future Symptoms and Ailments of Patients as Transmitted from many Years before Christ and Preserved by the Chinese; Which Indeed is a General and Admirable Art for Diverse Europeans." It might have been a version of one of the later published texts. If it was of a text identified by previous Jesuits as worth the effort of translation, it could have been something common to the literati of Peking or to cities in the south such as Guangsu (Canton). If it was mainly Boym's work, the Chinese version must have circulated in southern China or northern Indochina, since these were the only places visited by Boym.

The question of Boym's medical manuscripts is much complicated by his later relationships. After 3 years in Rome explaining his mission and helping other scholars like Kircher to learn about China, Boym finally obtained letters of reply and set off on the return journey. He embarked at Lisbon at the end of March 1656, traveling from there to Goa with eight younger Jesuits who had been recruited for the China mission, including three Belgians, one of whom, Philippe Couplet, would also be mentioned on the title page of the *Clavis* in 1686. Then, on his own, Boym travelled onward to the Bay of Bengal, then on to Ayuthia in Siam (modern Thailand), where he again met the three Belgians, who had come by a different route. While there, he wrote three letters to the physicians of Europe that would later be included in the *Clavis*. He managed to reach Tonkin by mid-July 1658 aboard a Chinese ship piloted by a Dutchman, where he remained until February 1659, hopeful that the (merely rumored) large army of Yong-li was about to conquer the Qing. But when Boym finally managed to reach Guangxi, he found all the passes guarded by armed Manchus, and while awaiting permission to return to Tonkin he fell ill, dying on 22 August 1659.[23] (As for the Ming court, it soon found itself pushed into the territory of the Burmese, where in 1662 the Yong-li Emperor was captured and died.)

The Jesuits and the Dutch

One of the Jesuits with whom Boym travelled east was Couplet. Born on 31 May 1623 in Mechlin (Malines), in the Spanish Netherlands, Couplet was educated by the Jesuits and entered onto a noviciate in 1640. At Leuven (Louvain) in 1654, he and two other young Jesuits, François de Rougemont and Ignatius Hartoghvelt, heard Martini speak about the China mission, from which Martini had just returned. They received permission to join it, were ordained in November, and to find passage to Lisbon made their way to Amsterdam, where they stayed for three months while waiting for a ship, making friends with the printer Joan Blaeu among others, who was then producing an edition of Martini's *Novus Atlas Sinensis*. Arriving in Lisbon in

---

[23] P. Pelliot, "Boym," pp. 129–132.

mid-April 1655, they found themselves too late for the fleet bound for Goa, and so spent the next year there. When they finally set sail at the end of March, they travelled with six other Jesuits: four Portuguese, one English, and Boym. At the beginning of November 1656 they arrived in Goa. Given that the port was then blockaded by the Dutch, they split up, their English companion finding a ship with a compatriot to take him to the Philippines, Boym and the three Portuguese working their way around to the Bay of Bengal on a series of coastal boats, and the three Netherlanders setting off overland in mid-1657, arriving probably in October in the large city of Ayuthia (in Siam, modern Thailand). There, perhaps by coincidence, they met Boym again. One of the group, Ignatius Hartoghvelt, who was originally from Amsterdam, fell seriously ill, and was carefully looked after by the head of the trading station run by the Dutch East India Company (VOC). Despite the care of Jan de Ryck and his wife, however, Hartoghvelt died there in May or June 1658.[24] Years later, Couplet told a Dutch admiral that, with De Ryck's assistance, they had sent a packet of Hartoghvelt's papers with a covering letter to the VOC's Governor-General Maetsuyker in Batavia to be passed on to his family and friends in Amsterdam, and that in the packet was also an account of their trip to China and "a little book on the Chinese method of taking the patient's pulse" (*een boekjen van de Sinoische manier van den pols der siecken te tasten*), perhaps meant for publication.[25] Because of the 1655 embassy of Maetsuyker to the court in Beijing to try to establish direct trade with China, the failure of which the Dutch blamed on the Jesuit astronomer Adam Schall, the project of publishing a work on the pulse had been abandoned.[26] But Boym had clearly passed on his interest in Chinese medicine, and perhaps some of his manuscripts, to Couplet. Couplet's subsequent "passion" to translate a Chinese medical work into Latin was noted in a contemporary Portuguese account referring to 1659 or 1661.[27]

---

[24] N. Golvers, "Philippe Couplet, S.J. (1623–1693) and the Authorship of *Specimen Medicinae Sinensis* and Some Other Western Writings on Chinese Medicine," *Medizin Historisches Journal*, 35 (2000): 175–182; Noël Golvers, "Ignatius Hartoghvelt, S.J. (1628 Amsterdam – 1658 Ayutthaya), un missionaire qui ne parvint jamais en Chine," *Courrier Verbiest*, 11 (1999): 4–6.

[25] Account from the Dagh-Register of Bort, quoted in E. Kraft, "Christian Mentzel et al.," pp. 185–186.

[26] "Miseram illius librum e regno Siami in novam Bataviam A. 1658 ut in Europam mitteretur. Sed quod offensi fuissent ob repulsum et successum insperatum legationis, quam in aulam Pekinensem adornaverant, attribuentque id P. Adamo Schall." Quoted from E. Kraft, ibid., p. 187. For an account of the 1655 embassy and the Dutch version of its failure due to Schall, see J. Nieuhof, *An Embassy from the East India Company of the United Provinces to the Grand Tartar Cham Emperor of China*, transl. by J. Ogilby, facs. of 1669 edition, Menston and Harrogate: Scholar Press and Palmyra Press, 1972, pp. 25–146.

[27] N. Golvers, *François de Rougement, S.J., Missionary in Ch'ang-Shu (Chiang-Nan): A Study of the Account Book (1674–1676) and the Elogium*. Leuven: Leuven University Press, Ferdinand Verbiest Foundation, 1999, pp. 530–531.

Couplet had further dealings with the Dutch a few years later, when he may have met his future collaborator, Cleyer. After a few months at the Jesuit school, he and other new arrivals travelled into China proper with Martini, and Couplet was posted to assist activities in Ganzhou in the province of Jiangxi before, in 1661, being transferred to the port city of Fuzhou in Fujian, on the mainland across from Taiwan. In October 1662, a large VOC fleet put into Fuzhou. In April 1661, a force led by a Chinese "admiral," Zheng Chenggong (known in European accounts as "Koxinga"), had laid siege to the main center of VOC activities on Taiwan, Fort Zeelandia, forcing its surrender in February 1662. Zheng was among the seafarers and merchants of the southeast coast of China who had remained opposed to the Qing dynasty. While the Manchus excelled in warfare on land, they could not control the seas, and so in 1660 they began forcibly removing the coastal population ten miles inland to deprive the "pirates" of their bases and supplies. Zheng therefore moved his forces to Taiwan, eventually expelling the Dutch.[28] The VOC in turn sent a fleet to see what could be done to inflict injury on Zheng's forces while at the same time using their common enemy as a lever for negotiating much-coveted trading privileges with the Qing. The fleet arrived in the city of Fuzhou in October 1662, and Admiral Balthasar Bort opened communications with Qing officials.[29] The main fleet returned to Batavia with the seasonal monsoon in February 1663, but a delegation and garrison under the merchant Constantijn Nobel stayed on until September to continue the negotiations with the Qing. Being from Mechlin, Couplet spoke a language almost the same as the more northern Dutch, and so via the VOC he sought current news from, and a channel for sending letters to, Europe (he also sought butter and cheese, and grape wine for celebrating the Mass.) In turn, he conveyed to the Dutch his impression of local officials and informed them of rumours both about the success of their negotiations and their Portuguese and English rivals. Perhaps he was also seeking a commercial arrangement with the VOC should they be successful in gaining trading concessions, since the Jesuit mission was chronically strapped for funds.

Cleyer is likely to have been serving with Nobel. Almost nothing is known about Cleyer before his first mention in the records in Batavia, in 1664. Eva Kraft was able to establish a few details from the church registers of "Alten-Stätter-Kirchen" of Kassel: his father was "Leutenant" Peter Kleier or Kleiier, who married the widow Agnes Amalia on 17 January 1633 (it was her third marriage), and Andreas came into their life on 27 June 1634. Kassel was the seat of the Prince and Elector of Hesse-Kassel, one of the bastions of Calvinism in Germany and an important ally of the Swedes and French against the Imperial forces during the on-going Thirty Years War. Andreas's father would have had many opportunities to put his military rank to the test in that conflict, which

---

[28] J. R. Shepherd, *Statecraft and Political Economy on the Taiwan Frontier, 1600–1800*. Stanford: Stanford University Press, 1993.

[29] J. E. Wills, Jr., *Pepper, Guns, and Parleys: The Dutch East India Company and China 1622–1681*. Cambridge, MA: Harvard University Press, 1974, pp. 25–104.

lasted until the boy was 14. Presumably the youngster had a good education in one or more of the local schools, and because the designation "V.M. Licent." ("utriusque medicinae licentiatus") was appended to Cleyer's name in the *Specimen*, it is often inferred that he studied medicine at a higher school or university. Perhaps he attended the local university at Marburg, for although there is no record of him in the archives there that would not be unusual. He is next recorded in the records of the Amsterdam Chamber of the VOC. Like many Germans and Scandinavians he took employment with the Dutch company, doing so on 22 November 1661, at the age of 27, signing an 8-year contract as an "adelborst," a kind of gentleman-soldier. He could have served as a military officer, and he later writes about having lived with the soldiers, but the rank also served as a catch-all title for VOC staff whose occupation was unusual enough not to have a separate designation.[30] He probably sailed on the "Amstelland," a 700-tonne fluit (a large ship, but not one of the great "retourships" meant to carry most of the cargo to the Indies and back), which left for the east on 9 January 1662. It put in at the Dutch station at the Cape of Good Hope from 27 April to 8 May, where it dropped off 20 soldiers and no doubt took on fresh water and supplies. If Cleyer continued on with the ship, he would have arrived in Batavia two months later, on 10 July.[31] He could then have gone on to China as part of the fleet headed by Bort.

The best evidence that Cleyer and Couplet met in Fuzhou is from a letter of 4 years later that mentions Cleyer as one of Couplet's "special friends." The letter was addressed to another delegation of the VOC, written by Couplet's companion Rougemont, asking for news of Europe and seeking their assistance in getting another letter to the Spanish Netherlands. It also passed on Couplet's greetings to "his special friends," Cleyer and Nobel.[32] Rougemont, Couplet, and almost all of the rest of the Jesuits were then under house arrest in Guangzhou (Canton). After Couplet's first encounter with the VOC, the chief of the China mission, Adam Schall, had assigned Couplet to Fuzhou so that Couplet could keep an eye on them (on the assumption that the Dutch would be allowed to trade there). But in January 1665, the government in Peking ordered the rounding up of all the Jesuits in China. One of the scholar-officials, Yang Guangxian (a rival astronomer), had brought charges against Schall and the rest of his order. All the Jesuits apart from Schall (who died a year later) and

---

[30] For what little is known of his early life, see F. de Haan, "Uit Oude Notarispapieren, II: Andries Cleyer," *Tijdschrift voor Indische taal-, land- en volkenkunde*, 46 (1903), and Eva S. Kraft, *Andreas Cleyer Tagebuch Des Kontors Zu Nagasaki Auf Der Insel Deschima*. Bonner Zeitschrift Für Japanologie, Band 6 (Bonn: 1985), p. 35.

[31] F. S. Gaastra, J. R. Bruijn, and I. Schöffer, eds., with assistance of and E. E. van Eyckvan Heslinga, *Dutch-Asiatic Shipping in the Seventeenth and Eighteenth Centuries, Vol. II: Outward-Bound Voyages From the Netherlands to Asia and the Cape (1595–1794)*. Rijks Gescheidkundige Publicatiën: Groote Ser., Vol. 166, The Hague: Martinus Nijhoff, 1979, voyage 0957.1.

[32] P. Demaerel, "Couplet and the Dutch," in J. Heyndrickx, ed., *Philippe Couplet, S.J. (1623–1693): The Man Who Brought China to Europe*, Nettetal: Steyler Verlag, 1990, p. 102.

three others, who remained in Peking, were ordered to Guangzhou, where they were placed under house arrest until September 1671. (They were then released following an astronomical contest ordered by the new Kangxi Emperor between Yang and Schall's successor, Ferdinand Verbiest, which Verbiest won.)[33] In the meantime, the VOC had continued to seek good formal relations with Peking, and in 1666 had sent another delegation to China, led again by Nobel, and it was to them that Rougemont directed the letter mentioning Cleyer.

Given Cleyer being named in this way, it is likely that he and Couplet had already met and become friends before 1666. It is just possible, but unlikely, that this happened during the time Couplet and his traveling companions stayed in Amsterdam – from late December 1654 to the beginning of March 1655 – while awaiting a ship to Lisbon. It was then that Couplet met the printer Blaeu, and Cleyer was also well-known to Blaeu. But Couplet did not mention Cleyer among his Amsterdam friends; Cleyer would then have been 20 years old and probably still resided in Germany; the first time we know of him in Amsterdam is 6 years later, in 1661. Additionally, the pairing of the names of Cleyer and Nobel in Rougemont's letter of 1666 suggests that Cleyer had been part of the delegation under Nobel's direction that stayed in Fuzhou for much of 1663. I have looked at the microfilm copy of the Dutch accounts from their official in China during 1662–1663 (in the form of Dagh-Registers), and do not find Cleyer mentioned, but that would only indicate that he was not involved in negotiations or special events: that he did nothing important enough to be mentioned in dispatches, as the later phrase has it.[34] But I think it likely that he met Couplet while serving in China either as a military officer or as a medical advisor to Nobel.

Two years later, Couplet was again in touch with Cleyer. From September 1668 to March 1669, while the fathers were still under house arrest, Nobel was once more in China negotiating for trading privileges, this time in Guangzhou, and again Couplet made contact with him. Via letters smuggled back and forth, Couplet helped Nobel deal with both the Chinese and the Portuguese authorities for obtaining the release of a group of ship-wrecked Dutch sailors imprisoned in Fuzhou, and not long before leaving the city the Dutch delegation found means to visit the Jesuits in their confinement. When nearing departure, Nobel received a number of items from them for conveyance to Batavia and Amsterdam, among which was a Chinese Herbarium for Cleyer together with a letter that, among other things, requested medicines.[35] It is very likely that, through Nobel, Couplet also managed to send to Europe a book on the pulse edited by one of his companions. In 1671, *Les Secrets de la Médecine des Chinois*

---

[33] For a recent account of this period, L. M. Brockey, *Journey to the East*, pp. 125–136.
[34] Nationaal Archief, Den Haag, VOC archives, 1.04.02.1243 and 1.04.02.1244.
[35] J. Bartens, "Hollandse Kooplieden Op Bezoek Bij Concilievaders," *Archief voor de Geschiedenis van de Katholieke Kern in Nederland*, 12 (1970): 75–120; E. Kraft, "Christian Mentzel et al.," p. 179.

Consistent en la Parfaite Connaissance du Pouls was published at Grenoble (and translated into Italian at Milan 1676).[36] Following a publisher's preface dedicated to the physicians of the College of Medicine of Grenoble (which makes a case based on Galen and Hippocrates for taking an interest of the pulse), there is a general "Avis du Lecteur" signed from Canton, 21 October 1668. It said that the work had been conveyed by a Frenchman desirous of peace and the good of his compatriots, who had been banished to Canton by sentence of the court of Peking; now that they have time between their devotions, they are setting down knowledge prized in France, which they have learned to be very useful for the preservation of health, including this, which is the discovery the secrets of the pulse. The author had translated the Latin into French. Perhaps, then, this text is the little book on the pulse Boym mentioned as having been in his own possession when he was in Rome, but clearly it had been worked on further by the Jesuit fathers collectively, and then put into his native tongue by one of those among them who was French.

By that time, Cleyer was enmeshed in a grand project to collect medical information from all parts of Asia. According to the biographical details provided by Frits de Haan over a century ago, Cleyer had quickly grown into Batavia society due to inheriting much of the wealth of Johannes Ammanus and his wife Anna. Johannes had been a lawyer and an important official in The Company who, through a good marriage, his work as an official, and his trade in wood and bamboo, had gained a very comfortable social position (a canal in Batavia took its name from his residence, the Ammanusgracht). At some time Anna was widowed, and on 8 July 1664, the now blind Anna named her nephew and lodger, Andreas Cleyer, as her heir; after her death on 27 August, Cleyer inherited her house and land on the Tijgergracht, the street with the grandest houses in Batavia. By leveraging his expertise and his social position, "Doctor" Cleyer (as he was called) was appointed by the local government on 15 December 1665 to distill chemical medicines for 1 year. The contract allowed him to charge fifty percent more than in the Netherlands for the medicines, and also contained permission to purchase what he needed for distillation from the VOC's stores at half price (this apparently being typical of the contracts written up after the visitation of Pieter van Hoorn, who was sent out to Batavia in 1663 by the governors of the VOC, the Lords Seventeen, to provide for the independent economic development of the city). In addition to contracting to supply chemically-prepared medicines, Cleyer worked for Pieter Berthem, the apothecary to the Castel of Batavia. Berthem not only ran the Medical Shop established in the port in 1663 (which provided all the medical consumables to both local people and visiting ships) but also had oversight over all local medical assistance and the medical personnel in Batavia and aboard ship. By 1666 Cleyer was

---

[36] *Les Secrets de la Médecine des Chinois, Consistent en la Parfaite Connaissance du Pouls, Envoyez de la Chine par un François, Homme de grand merite*. Grenoble: Philippes Charuys, Marchand Librarire, en la Place de Mal-Conseil, 1671. The most extensive comment on *Les Secrets* is in M. D. Grmek, "Les Reflets De La Sphygmologie Chinoise," pp. 59–64.

starting to appear on the lists of senior VOC functionaries, was given charge of the local Latin school, and upon the death of Berthem in May 1667 succeeded him in the post of Apothecary to the Castle and Head of Surgery. By the time his daughter was baptised on the following 6 October, he had been accepted among the social elite: His Most Nobile Joan Maetsuyker, Governor of the Dutch East Indies, was one of the witnesses.[37]

As the chief medical officer of the VOC government in the Indies, Cleyer was deeply involved in a new project to collect information about the local medicines in use in Asia, so that his shop could use fresh and efficacious medicines gathered in the region rather than rely on stale imports of medicines from Europe. It was a project initiated partly by the efforts of the Directors of the VOC in The Netherlands trying to cut costs and discover new products to import, and partly by the investigations of a physician turned VOC merchant in Persia, Malabar, and Ceylon, Robert Padbrugge, who in 1668 wrote a report on the medical riches of the Indies that prompted letters to the Directors and back. As early as the beginning of the 1630s, a physician of Batavia, Jacobus Bontius, had written about the beneficial uses of foods and drugs he had learned of from the local people, and his work had been reprinted as part of a grand and beautiful edition about the medicines and natural history of the East and West Indies in 1658, just 3 years before Padbrugge enrolled in the medical faculty at Leiden. From at least 1668, Cleyer had been involved in exchanges of seeds and plants with the VOC station at the Cape of Good Hope, and by early 1669 he was supporting the efforts of the government in Batavia to look into the possible use of local medicinals available at their main posts ("factories"). According to a later statement of his own, Cleyer tried to find out about unknown medicinals and medical practices from informants within the large Chinese community of Batavia, as well.[38]

In response to Couplet's manuscript and letter sent via Nobel, then, Cleyer replied on 28 June 1669 with thanks. He promised to forward Rougemont's letter to Europe, and asked Couplet for additional printed Chinese books on medicine, especially anything with illustrations. In addition, anything Couplet might have that had been translated from Chinese into Latin for the sake of the medical art, he would spare no expense to acquire. In particular, in order to help Cleyer understand Chinese, he wanted to acquire a translated example of a work on the Chinese method of touching the pulse, which he considered to be fundamental to their practice and a proof that they knew of the circulation of the blood. He also sent Couplet a packet of medicines, newspapers, and pamphlets ("blue books"), as well as information of his own on current events.[39] The relationship between Couplet and Cleyer developed into a regular "Holland-

---

[37] F. de Haan, "Uit Oude Notarispapieren," and E. Kraft, *Andreas Cleyer Tagebuch*, p. 35.
[38] See H. J. Cook, *Matters of Exchange*, pp. 305–309.
[39] P. Begheyn, "A Letter from Andries Cleyer, Head Surgeon of the United East India Company at Batavia, to Father Philips Couplet, S.J., Missionary in China, 1669," *Lias*, 20 (1993): 245–249; P. Demaerel, "Couplet and the Dutch," p. 108.

connection" that allowed the Jesuits to send packages of their letters to Cleyer, who posted them on to Amsterdam, from where they would be sent to their local destinations in Europe, often via Antwerp.[40]

Over the next few years, Cleyer must have worked on the collection of translations and commentaries that he collected, probably helped by some of the many Chinese who resided in Batavia.[41] What would later become the *Specimen* was ready for the press by 1676. About 300 years later, Eva Kraft, of the Staatsbibliothek of Berlin, discovered some additional manuscript material related to it. These include a dedication to the Governors General in Batavia that begins with a reference to his sailing to the East Indies "Tria lustra effluxerunt" (three 5-year periods ago), and a celebratory poem to be published with the book, written in German and dated 24 May 1676.[42] The work was probably sent to The Netherlands but could not find a publisher until it was forwarded to Germany in 1680 or 1681. It later appeared from a printer in Frankfurt because of the efforts of members of the German Academia Naturae Curiosorum, a "scientific" society founded in 1652 and composed almost entirely of physicians living in the Holy Roman Empire.[43] One of their publications, Jacob Breyne's *Centuria plantarum*, had mentioned that Cleyer made an oil of camphor from the root of a wild plant that grew in Ceylon (Sri Lanka) that was good in cases of arthritis, and because the Great Elector Friedrich Wilhelm suffered from the condition, his physician and librarian, Christian Mentzel, noticed, and in 1677 wrote to Cleyer. By 1680, Mentzel had managed Cleyer's membership into the Academia (under the name of "Dioscorides"), and by the end of 1681 another member of the Academia, Sebastian Scheffer, had seen Cleyer's *Specimen* through the press.[44]

But Cleyer had collected additional texts, and enlisted the help of a well-educated Dutch physician, Willem ten Rhijne, who in turn sought advice about their interpretation from physicians in Japan.[45] Ten Rhijne was sent by the VOC to Japan in the expectation that he would become a physician to the Shogun. On his way to Japan, in early 1674 – some months before Cleyer probably sent the manuscript for the *Specimen* to Europe – Ten Rhijne stopped at Batavia for a few weeks, and met Cleyer: we know that Cleyer invited him to lecture in the anatomy theatre he had recently established (but which did not

---

[40] P. Demaerel, "Couplet and the Dutch," pp. 111–112.

[41] On the mixed community of Batavia, see J. G. Taylor, *The Social World of Batavia: European and Eurasian in Dutch Asia*. Madison: University of Wisconsin Press, 1983, and L. Blussé, *Strange Company: Chinese Settlers, Mestizo Women and the Dutch in Voc Batavia*, Verhandelingen KITLV 122, Dordrecht/Riverton: Foris, 1986.

[42] E. Kraft, "Christian Mentzel, et al., " pp. 192, 168.

[43] F. M. Barnett, *Medical Authority and Princely Patronage: The Academia Naturae Curiosorum, 1652-1693*. m PhD dissertation, University of North Carolina at Chapel Hill, 1995.

[44] Ibid., pp. 163-167.

[45] For this section on Ten Rhijne, I rely on information published in H. J. Cook, *Matters of Exchange*.

last much longer).⁴⁶ Ten Rhijne then went to Japan, where he spent the next 2 years. During his time there, Ten Rhijne spent several periods from a few days to a few weeks working closely with Japanese medical practitioners, sharing his knowledge of anatomy, botany, and chemistry with them and receiving in turn some instruction in their own texts and methods of treatment. He had with him some Chinese medical texts, including authentic diagrams of the channels and points "which," he later wrote "had long been neglected and ignored through want of an interpreter." Ten Rhijne had taken possession of them "after I was assigned to Japan and sought out these representations for myself."⁴⁷ While it is not entirely clear whether these were manuscripts that had been sent back to The Netherlands as part of the trade in Asian *curiosa*, that he brought to Asia with him, whether they were manuscripts he found in Japan, or whether Cleyer lent them to him on his way to Japan, it seems likely to have been the latter.

"In order that this treasure, which had been entrusted to but [was] unappreciated by its former owner, might not lie idle in my possession," Ten Rhijne later wrote, "I made every effort to meet a Japanese physician ... with a knowledge of Chinese .... My efforts succeeded as I wished." The Japanese physician trained in Chinese medicine, Iwanaga Sōko, did his best to translate the texts into Japanese, and the interpreter, Shōdayū Motogi, then tried to render the result into Dutch. In Ten Rhijne's view, Shōdayū spoke only "faltering Dutch in half-words and fragmentary expressions," and although he also possessed "more experience in medical matters than all the other interpreters ... he was also more cunning."⁴⁸ Most of what he gleaned was a brief account of the history and practice of medicine in East Asia, although given his method of working, and his peculiar Latinisation of names that the Japanese used for Chinese characters, it is not clear what the original text was.⁴⁹ Not surprisingly, translating from Chinese through Japanese into Dutch and then into Latin made for far more mystery than going from Chinese directly into Latin.

After his hopes of becoming a physician to the Shogun were dashed, Ten Rhijne returned to Batavia in 1676 without further instructions, and while the government of the VOC was deciding how to make use of him, and he was writing to friends in Amsterdam trying to get the directors of the VOC to give him his back pay, he made good use of his time by writing a number of treatises about his medical experiences. These included the first long tract in a European language on acupuncture and moxibustion, which in turn included the remarks

---

⁴⁶ J. M. H. van Dorssen, "Willem Ten Rhijne," *Geneeskunde Tijdschrift voor Nederlands Indië*, 51 (1911): 150–152.

⁴⁷ R. W. Carrubba and J. Z. Bowers, "The Western World's First Detailed Treatise on Acupuncture: Willem Ten Rhijne's *De Acupunctura*," *Journal of the History of Medicine*, 29 (1974), p. 376.

⁴⁸ Ibid., pp. 376–378.

⁴⁹ See, for example, the comment about how this is "no small puzzle" in G-D. Lu and J. Needham, *Celestial Lancets: A History and Rationale of Acupuncture and Moxa*. With new intro. by V. Lo, London: Routledge Curzon, 2002, pp. 271–276.

on Chinese medicine he had gleaned from his Japanese colleagues and four charts of the acupuncture points. At the same time, he also assisted Hendrik van Reede with the editing of his huge and impressive work on the botany of the Malabar coast, the *Hortus Malabaricus* (1682). He also agreed to help Cleyer sort and edit some of the Chinese texts he had been collecting. But by the time that Cleyer had decided on the publication of the materials for the *Clavis*, they had fallen out. In a letter of 25 March 1681, Ten Rhijne complained to a friend that Cleyer had been making life difficult for him for some time on account of "the well-known Chinese work on the pulse," even trying to denigrate him in the eyes of his friends in The Netherlands. He could nevertheless prove from Cleyer's own letters that it was Cleyer who had broken their mutual agreement (*stipulatie*), and that Ten Rhijne had written the commentary on the text without the help of anyone else (*en ik niet anders als myn eygen commentaria sonder iemants hulpe of toedoen*), a copy of which he had sent to The Netherlands. Given that the times were delicate, however, this was all he could say for the moment. In a PS, he noted that he understood the same ship that would carry his letter was also carrying a manuscript from Cleyer, which contained a more complete version of the Chinese work on the pulse along with other items – clearly the material that would become the *Clavis*.[50] At about the same time (after Cleyer had entered into correspondence with Scheffer in Frankfurt), Cleyer wrote that Ten Rhijne's slaves in Amsterdam "mancipiis Amstelodami" were the ones responsible for suppressing the earlier *Specimen*.[51] In later letters, Ten Rhijne still spoke of Cleyer as having done "no more than purchase a Translation by force of money, which yet without an Interpreter can do Little Good in Europe. I have Taken some pains about it but there are Mysteries not yet to be revealed which yet I hope the Learned world Shall some time partake of."[52] And he continued to say that he did much of the work to get the herbal and treatise on pulses into shape, although without his commentary he thought that the book would be of little value.[53] After hearing that the now-printed *Specimen* contained a commentary, he added that the "Tractatus de pulsuum tacta sinensium in quem commentariolum" had not been fully completed by him, but that it had "appeared in another edition under another name. I expected that my views and commitment of time (to the project) would be openly demonstrated as was agreed in advance, but political and domestic affairs have currently worked against this. The Jesuit Couplet, who has recently returned to Europe, who is deep in this mystery, promises to explain all in due course."[54]

---

[50] Sloane MS. 2729 fol. 73, Ten Rhijne to Sibellius, 25 March 1681.

[51] M. B. Valentini, *India Literata*, 2nd ed., Frankfurt am Main: Prostat apud Haeredes Zunnerianos, 1716, pp. 432–433, which prints an undated letter of Cleyer to Scheffer; also see Szczesniak, "The Writings of Michael Boym," pp. 516–517.

[52] 21 March 1682/3: letter of Ten Rhijne to Joannes Groenevelt, excerpted in English in Letter books of the Royal Society, LBC.8, fols. 447–448.

[53] Sloane 2729, fols. 130–131, Ten Rhyne to Sibellius, 25 February 1683.

[54] Letter of 25 August 1683: Royal Society, LBC.9, fol. 374.

Indeed, Couplet and one of their learned converts, Michael Shen Fuzong, had been chosen by the Chinese mission to represent the Jesuits in Europe during the growing controversy over the strategy and tactics of their accommodationist policy.[55] On the way back, at the end of December 1681 his ship stopped at Bantam, and Couplet received permission to come to nearby Batavia, where he stayed over a year (until March 1683). Cleyer left for Japan on 27 June 1682, but the two had four months to visit. Evidence indicates that Couplet checked and corrected Cleyer's herbarium (collection of dried plant specimens), saw a Chinese-Latin medical text sent to Cleyer from the naturalist, Georgius Everhardus Rumphius, of Ambon, and saw a copy of the manuscripts that composed the *Specimen*. Later, in Europe, Couplet wrote to Mentzel that all the parts composing the second publication, the *Clavis*, had been sent to the Dutch from himself, based on the work of Boym, but because the VOC had blamed the Jesuits for keeping them out of China, the project for publishing it had been completely opposed ("totum opus obiecerunt"), but that Cleyer had collected together the fragments of it. He therefore considered Boym to be the author of the *Clavis*. That he did not say the same about the *Specimen* suggests confirmation of Ten Rhijne's claim to have had a hand in composing some parts of it.[56]

It is almost certain, then, that the four letters published in Cleyer's *Specimen* from an anonymous "erudit" sent to Cleyer from Canton and dated 12 February and 20 October 1669, and 5 and 15 November 1670 (which concern pulse diagnosis and the circulation, and are followed by nine diagrams and tables), are from Couplet.[57]

Conclusion

The best evidence therefore points overwhelmingly to a network of Europeans in Asia being quite interested in Chinese methods of medical practice, particularly their manner of feeling the pulse. They tried to learn more about these methods, and to understand and translate some of the classic works on the subject. Since there was no single author or editor of any of the translations, it makes little sense to enter into discussions about whether Cleyer (or anyone else) plagiarized them. Clearly, however, Europeans would have remained almost entirely ignorant of Chinese medicine had not the Jesuit missionaries

---

[55] Th. N. Foss, "The European Sojourn of Philippe Couplet and Michael Shen Fuzong, 1683–1692," in J. Heyndrickx ed., *Philippe Couplet, S.J. (1623–1693): The Man Who Brought China to Europe*. Nettetal: Steyler Verlag, 1990, pp. 121–140; D. E. Mungello, "An Introduction to the Chinese Rites Controversy," in D. E. Mungello ed., *The Chinese Rites Controversy: Its History and Meaning*. Nettetal: Steyler Verlag, 1994, pp. 3–14.

[56] For the evidence and Kraft's interpretation of them, which was written without knowledge of Ten Rhijne's letters but is otherwise in close agreement, E. Kraft, "Christian Mentzel et al.," pp. 181–188. Also see P. Demaerel, "Couplet and the Dutch," p. 117.

[57] P. Demaerel, "Couplet and the Dutch," p. 117–118.

studied the Chinese language and some of their ancient textual traditions, with which the medical works seemed entirely in keeping.

By the middle of the seventeenth century, Boym was following in the footsteps of his predecessors and collecting and possibly translating some of the medical texts into Latin. But he was not happy enough with the results to publish the little work on the pulse that he had with him during his sojourn in Europe from 1653 to 1656. By the time he had arrived in Ayuthia, in 1658, he had finished at least enough of the work for a little booklet on the subject to be forwarded to the main European settlement in the region, Dutch Batavia. But given the recent setback in the VOC's hopes for trade relations, which they blamed on the Jesuits in Beijing, whatever Boym had sent on was dispersed or destroyed. But Boym's younger colleagues, especially Couplet, kept up an interest in the subject. During their period of internal exile in Canton from 1666 to 1671, the Jesuit fathers worked on a number of translations, including their later famous edition of the works of Confucius,[58] and made some effort with medical works as well. By this time, Couplet's connections with the Dutch were very good, and he sold a number of manuscripts to Cleyer, who in Batavia was investigating various aspects of Asian medicine. It would seem to be then, in 1668, that the manuscript of what was published as *Les Secrets de la Médecine Chinoise* in 1671 started its journey to Europe from Canton. By the mid-1670s, Cleyer had assembled a number of texts and illustrations pertaining to Chinese medicine, and helped by his colleague and subsequent antagonist, Ten Rhijne – who had tried to decode at least one of them in partnership with Japanese physicians – sent a group of them back to Europe, which would eventually be published in 1681–1682 as the *Specimen*. Without the help of a learned physician like Ten Rhijne, or an expert in Chinese, Cleyer also managed to collect together all of the papers he could lay his hands on from Boym's project of the 1650s, which was published as the *Clavis* in 1684 (acknowledging Boym as the author and Couplet as the "editor"). Ten Rhijne himself published a work in 1683 that included his interpretation of ancient Chinese medicine as best he could understand from his period in Japan.[59] More effort needs to go into understanding the nature of the process of translation and editing, but a provisional view is that the Jesuits looked for the ancient wisdom of the *prisci theoligii* in Chinese texts, including medical ones, while the employees of the VOC cared mainly for efficacious therapies.

Whatever the motivations, however, the result of the efforts of both groups meant that from the mid-1680s, European readers could begin to imagine that the Chinese possessed special medical skills, particularly in their methods of touching the pulse, which they could use not only for diagnosis but for treatment. This was an ancient art that extended thousands of years into antiquity,

---

[58] D. E. Mungello, "The Seventeenth-Century Jesuit Translation Project of the Confucian Four Books," in Ch. E. Ronan, *East Meets West: The Jesuits in China, 1582–1773*. Bonnie B. C. Oh ed., Chicago: Loyola University Press, 1988, pp. 252–272.

[59] Willem ten Rhijne, *Dissertatio De Arthritide*. London: R. Chiswell, 1683.

perhaps even before the Flood, and even seemed to have anticipated William Harvey's discovery of the circulation of the blood. A look at the content of the translations must await another time, but it seems that Ten Rhijne, Couplet, and Boym all equated *Yang* with the medieval Latin medical term *calidum innatum*, or innate heat; wrote of *Yin* as the medieval *humidum radicale*, or radical moisture; and identified *qi* with "vital spirit." Many people who found the new mechanical philosophy to be inadequate would therefore have reason to appeal to both the classical past *and* to other ancient learned traditions such as the Chinese as more holistic alternatives to the new mechanical philosophy. All this fit well with the developing view of China as an ideal state ruled by philosopher-kings, a huge, wealthy, and well-run Empire worthy of emulation even from the Sun King of France, lacking only Christianity to make it the most glorious place on earth.

Such equivalences drew criticism. Within a year of the publication of the *Clavis*, for instance, the famous *savant* Pierre Bayle published a review of it in which he said it gave a good description of the things at which Chinese medicine was supposed to do very well, although the truth of their principles were not the clearest in the world. Unfortunately, he went on, the new mechanical principles recently discovered by modern physicians threw great doubt on the faculties, natural heat, radical moisture, and other great foundational principles of Chinese medicine just as much as they challenged those of the Aristotelians.[60] From here, the debate about Chinese medicine got drawn into the vociferous debate about whether the learning of the ancients or moderns was superior.

We might today say that Bayle was not entirely correct about his point: most of the Latin terms signified not ancient medical ideas but medieval ones – and perhaps they worked to some extent as bridging terms because they were derived from Islamic medicine, which in turn drew some influence from Chinese medicine. But of course here we venture into high-flown speculation about that troublesome thing called "influence." Any attempts to create equivalences between different medical cultures run into trouble. Shigehisa Kuriyama put the problem this way: when early modern Europeans examined Chinese medical methods, they thought they saw practitioners "feel the pulse." But Europeans who felt the pulse and Chinese who felt the *mo* were experiencing their bodies and their clients' bodies in very different ways. "By the evidence of the eyes, *qiemo*, palpating the *mo*, was unmistakably pulse diagnosis. Chinese writings [however,] testified that the eyes were wrong. The hermeneutics of the *Mojue* were unlike any dialect of the pulse language known in Europe." He goes on to ask: "How can gestures look the same, yet differ entirely in the experience?" People lived in very different worlds, and therefore gave explanations for what they were doing that were not commensurable. Or, as Bayle and his like who defended the modern knowledge of Europeans would have it, Chinese

---

[60] M. D. Grmek, *Reflects de la Sphygmologie*, 74; the quotation is given in English in Lu and Needham, *Celestial Lancets*, 286.

explanations about their methods – something many people referred to as their "philosophy," as we would now speak of their "theory" – were not simply "mistaken," but were "ridiculous," "phantastical," "chaotic," or "absurd."[61] But as Kuriyama himself knows, despite the conceptual gulfs some Europeans took a different view and made considerable efforts to understand East Asian medical practices and concepts. Perhaps it was a naive quest, but it was carried out often under great duress because they believed that this form of communication exchange was very important. Some thought it was a pragmatic search for efficacious treatments, others a quest for the recovery of an ancient and holistic medicine that would have beneficial effects for all peoples. The efforts and interpretations of both parties have left a powerful legacy even today.

---

[61] S. Kuriyama, *The Expressiveness of the Body and the Divergence of Greek and Chinese Medicine*. New York: Zone Books, 1999, pp. 21–22.

Adoption and Adaption: A Study of Medical Ideas and Techniques in Colonial India

Deepak Kumar

Diffusionist models of knowledge transmission have had a long sway. These were probably useful in the context of centre-periphery explanations. Recent scholarship has moved ahead. Transmission is no longer seen as emanating from a powerful centre. Now more attention is paid to the changes and adaptations that occur in the process. Adoption and adaptation are not always imposed or induced from above or outside; these can come from within and on their own.

As postcolonial theorists argue, these could be both multi-sited and hybrid. To quote a contemporary critique,

> A postcolonial perspective suggests fresh ways to study the changing political economies of capitalism and science, the mutual reorganisation of the global and the local, the increasing trans-national traffic of people, practices, technologies, and contemporary contests over 'intellectual property'.[1]

It is important to emphasise the value of 'dialogues' and 'conversations'. But these alone do not explain knowledge 'production' or 'dissemination'. It is a much more complex process. It could be socio-cultural as well as politico-economic. 'Dialogues' or 'negotiations' do not negate the equation of power. Analyses do not always lead to synthesis; similarly negotiation does not necessarily mean integration. Domination and marginalisation lurk in the background and mostly appear as the two sides of the same coin.

To illustrate the above arguments this paper would focus on the examples from India's medical tradition and the challenges it faced when the modern medical system entered the country as part of the colonial technologies of control. Indian society has always been 'knowledge-oriented', and never a xenophobic one; it did sometimes see significant shifts at least in medical

---

D. Kumar (✉)
ZHCES, Jawaharlal Nehru University, New Delhi, India
e-mail: deepak_jnu@yahoo.co.in

[1] W. Anderson, "Postcolonial Technoscience," *Social Studies of Science*, 32, 5–6 (2002): 643–658.

practices. Conceptually knowledge revolved round canonical texts and advanced, albeit slowly, through commentaries and discussions. Ayurveda as 'science of (living to a ripe) age' originated in a magico-religious milieu but gradually evolved alongside with Buddhist rationalism.[2] Its protagonist might have been inaccurate in their knowledge of human physiology but they were extremely good at plant morphology, its medical functions and therapeutics.[3]

Medicine in the Pre-colonial India

The canonical texts like the *Samhitas* of *Charaka* and *Sushruta* were there to guide but knowledge advanced through commentaries and even replications.[4] In the Southern India the *Siddhas* had developed the knowledge of pulse and the methods of diagnosis by eight kinds of clinical examination.[5] By the end of the thirteenth century, thanks to the works of Chakrapanidata, Vangasena and Sarangdhara, a new change had come in the shape of the blending of Ayurveda and Rasashastra (medicine and alchemy). These developments have been mostly internal but the possibilities of external influences from other cultures can not be ruled out.[6] But unfortunately these texts and later commentaries have no anatomical or surgical illustration.[7] Dissection became a taboo and knowledge became secretive, confined to families or caste networks. *Todarananda* (a sixteenth-century encyclopaedic work), for example, says that *rasavidya* (alchemy) is to be kept 'as secret as mother's genitals'.[8] Even mineral classification reflected caste divisions. For example, diamonds are described as of four types, Brahmana, Kshatriya, Vaisya, and Sudra; they are also male, female or *napunsaka*'.[9] The *Anandakanda* (The Root of Bliss) encompasses almost the entire Hindu alchemical tradition in six thousands verses. It is the sole attempt to fuse *rasa siddhas* (alchemists) with the *nath siddhas* (spiritual tantries/yogis).[10]

---

[2] D. Wujastyk, *The Roots of Ayurveda*, New Delhi: Penguin, 1998, pp. 1–38.

[3] G. P. Majumdar, "Health and Hygiene," *Indian Culture*, II, 1–4 (1935–1936): 633–654.

[4] G. J. Meulenbeld, "The Many Faces of Ayurveda," *Journal of European Ayurvedic Society*, 4 (1995): 1–10.

[5] C. K. Sampath, "Evolution and Development of Siddha Medicine," in S. V. Subramanian and V. R. Madhavan, eds., *Siddha Medicine*, Madras: I.I.T.S, 1983, pp. 1–20.

[6] Fascination with alchemy, for example, most probably arose out of early contacts with China where Taoist speculative alchemical tradition had been developing since the second century CE. D. G. White, *The Alchemical Body*, New Delhi: Munshiram Manoharlal, 2004, p. 53.

[7] D. Wujastyk, "Indian Medical Thought on the Eve of Colonialism," *IIAS Newsletter*, 31 July 2003, p. 21.

[8] G. J. Meulenbeld, *A History of Indian Medical Literature*, vol. II A, Groningen: Forsten, 2000, p. 278.

[9] Ibid., p. 286.

[10] White, op. cit., p. 167, Note 6.

Apart from attempts for synthesis, there may be some examples of total repudiation or departure. Dominik Wujastyk has recently brought to our notice one such rare example, a text titled *Rogarogvada* (Debate on Illness and Health) composed by Viresvara in 1669. This text explains systematically the principal theories of Ayurveda and then refutes them one by one. For example, the authorities define disease as identical to an imbalance in the humours and yet they accept that the humours may naturally exist in different quantities without causing illness. Phlegm predominates at the start of the day or after a meal but one is not always ill after a meal! And so the central doctrine that humoral imbalance is identical with disease must be wrong.[11] Thereafter Viresvara gives a curious explanation, diseases come and go for no apparent reason. Just like the rising and setting of stars, disease is the pain of the mind, body or sense organs and it arises for no reason. This explanation may appear quixotic but it is nevertheless offered in 'a spirit of intellectual rigour and debate which speaks of an original if impulsive mind'.[12]

The scenario becomes even more interesting when Islamic medical scholars introduced the Galenic tradition. There gradually appeared a hybrid of Muslim–Hindu system known as the *Tibb*. They differed in theory, but in practice both traditions seem to have interacted and borrowed from each other. A fine example of this interaction is *Ma'din al-shifa-I-Sikandarshahi* 1512 CE, which was authored by Miyan Bhuwah.[13] He leaned heavily on the Sanskrit sources and even thought that the Greek system was not suitable for the Indian constitution and climate. The concept of *arka* from Islamic medicine entered Ayurveda. Several Sanskrit medical texts were translated into Arabic and Persian, but instances of Islamic works being translated into Sanskrit are rare. The eighteenth century is significant because of the appearance of two Sanskrit texts *Hikmatprakasa* and *Hikmatparadipa* which refer to the Islamic medicine and use numerous Arabic and Persian medical terms.[14] The concept of individual case studies and hospitals (*bimaristans*) also came from the *Unāni* practitioners.[15] In 1595 Quli Shah had built a huge *Dar-us-Shifa* (House of Cures) in Hyderabad.[16] During the reign of Muhammad Shah (1719–1748) a large hospital was constructed in Delhi, and its annual expenditure was more than Rupees three hundred thousands. Numerous medical texts, mostly commentaries, were written during this century, for example, Akbar Arzani's *Tibb-I-Akbari* (1700), Jafar Yar Khan's *Talim-I-Ilaj* (1719–1725), Madhava's *Ayurveda Prakasha* (1734), and *Bhaisajya Ratnavali* of Govind Das. A Christian

---

[11] D. Wujastyk, "An Argument with Medicine," *Friends of the Wellcome Library Newsletter*, no. 32, 2004, pp. 6–7.

[12] Ibid.

[13] The manuscript was first printed by Nawal Kishore Press, Lucknow, in 1877.

[14] So is Krishnarama's *Siddhabhesajamanimala* of the nineteenth century. Meulenbeld, "The Many Faces of Ayurveda," pp. 1–9, Note 4.

[15] S. H. Askari, "Medicines and Hospitals in Muslim India," *Journal of Bihar Research Society*, 43 (1957): 7–21.

[16] D. V. Subba Reddy, "Dar-us-Shifa built by Sultan Muhammad Quli: The first Unani teaching hospital in Deccan," *Indian Journal of History of Medicine*, II (1957): 102–105.

medical practitioner of Mughal origin, Dominic Gregory, wrote *Tuhafatul-Masiha* (1749), which, alongwith the descriptions of diseases, anatomy, and surgery, contains important notes in Persian and Portuguese on alchemy and the properties of various plants, along with drawings of instruments, and interestingly, a horoscope.[17] An outstanding physician of this century, Mirza Alavi Khan, wrote seven texts of which *Jami-ul-Jawami* is a masterpiece embodying all the branches of medicine then known in India.[18] Another great physician during the period of Shah Alam II (1759–1806) was Hakim Sharif Khan who wrote ten important texts and enriched *unani* medicines and indigenous Ayurvedic herbs.[19] Some works were unique and ahead of their time. For example, Nurul Haq's *Ainul-Hayat* (1691) is a rare Persian text on plague, and Pandit Mahadeva's *Rajsimhasudhasindhu* (1787) refers to cowpox and inoculation.[20]

Caravans of men and streams of thought constantly moved and flowed between India, Iran and Central Asia resulting in intimate cultural relations. The Mughal court patronised notable hakims. Shams-al-din-Gilani was *Hakim al-Mulk* during Akbar's reign (1556–1605), Abul Qasim under Jehangir (1605–1627), Mir Muhammad Hashim during Aurangzeb's period (1658–1707) and Mir Muhammad Jafar under Muhammad Shah (1719–1748). Hakim Gilani had authored the famous commentary (4 vols.) in Arabic on Ibn Sina' *Qanun* called *Sharh-I-Gilani*.[21] As for the Mughal translations and compositions, a recent critic says:

> there is a pathos in their efforts and waste in their labours. The system of medicine on which they spent hours and hours translating and transcribing was already dying and outdated. Discoveries were being made everyday which rendered their views obsolete. Had they looked outside their circle, had they listened to the strangers who were coming to their country in such large numbers, they could have learnt that their physiology had been proved by practical experiment to be full of error, that their pathology was being overturned by such new inventions as the microscope and the test tube, and that most of their anatomy was pure imagination...[22]

A number of European physicians visited Mughal India. Francois Bernier, Niocolao Manucci, Garcia d'Orta, and John Ovington wrote extensively on Indian medical practices. The Western medical episteme was not radically different from that of Indian physicians; both were humoral, and both prescribed almost similar drugs but their practices differed greatly. Neither of them was able

---

[17] A. Rahman, ed., *Science and Technology in Medieval India: A Bibliography of Source Materials in Sanskrit, Arabic and Persian*, New Delhi: INSA, 1982, p. 57.

[18] R. L. Verma and N. H. Keswani, 'Unani Medicine in Medieval India: Its Teachers and Texts', in N. H. Keswani, ed., *The Science of Medicine in Ancient and Medieval India*, New Delhi, 1974, pp. 127–142.

[19] H. A. Hameed, *Exchanges between India and Central Asia in the Field of Medicine*, New Delhi, 1986, p. 41.

[20] Rahman, op. cit., pp. 129, 165, Note 17.

[21] Hameed, op. cit., Note 19.

[22] C. Elgood, *Safavid Medical Practice*, London: Luzac & Co., 1970, p. 88.

Adoption and Adaption

to develop a comprehensive theory of disease causation, but there seems to be a general agreement among both Indian and foreign practitioners that the Indian diseases were environmentally determined and should be treated by Indian methods. Europeans, however, continued to look at the Indian practices with curiosity and disdain.[23] They preferred blood letting whereas the *vaidyas* prescribed urine analysis and urine therapy. But in the use of drugs Europeans and Indians learned from each other, as the works of van Rheede, Sassetti, and d'Orta would testify.[24] The Europeans introduced new plants in India that were gradually incorporated into the Indian pharmacopoeia. They had brought venereal diseases such as syphilis which was noticed as early as the sixteenth century by Bhava Misra, a noted vaidya in Benaras, who called it *Firangi roga* (disease of the Europeans). Indian diseases received graphic description in Ovington's travelogue.[25] The best account of smallpox and the Indian method of variolation was given by J. Z. Holwell in 1767. To him this method although quasi-religious, still appeared 'rational enough and well-founded'.[26] The travellers depicted Indian medical practices more as a craft and one that was governed by caste rules and wrapped in superstition. Yet they could not help admiring the wonder called rhinoplasty (on which modern plastic surgery is founded), nor could they deny the efficacy of Indian drugs. The Indians on their part did not completely insulate themselves from the 'other' practices. This was reflected in a nineteenth-century text *Brhannighanturatnakara*. Other examples of the impact of the West on the Indian nosology are Govindadasa's *Bhaisajyaratnavali* (eighteenth century) and Binod Lal Sen's *Ayurvedavijnana* (nineteenth century). This reference to Ayurveda as *vijnana* (science) is significant. As the interaction grew in the eighteenth century, the *vaidyas* even took to bleeding in a large number of cases. Yet while the European medical men were gradually moving, thanks to the works of Vesalius and Harvey, from a humoral to a chemical or mechanical view of the body, Indians remained faithful to their texts. Asia kept doing what it had been doing for centuries; Europe changed basically.[27]

---

[23] A European traveller, Edward Ives (1755–1757) thus writes of the Indian belief that 'man was divided into two or three hundred thousand part; ten thousand of which were made up of veins, ten thousand of nerves; seventeen thousand of blood, and a certain number of bones, choler, lymph, etc. and all this was laid down without form or order, either of history, disease or treatment'. Quoted in H. K. Kaul, *Travellers India: An Anthology*, Delhi, 1979, p. 299.

[24] For details see John M. de Figueredo, "Ayurvedic Medicine in Goa according to European sources in the sixteenth and seventeenth centuries," *Bulletin of History of Medicine*, 58, 2 (1984): 225–235.

[25] A. Neelmeghan, "Medical Notes in John Ovington's Travelogue," *Indian Journal of History of Medicine*, VII (1962): 12–21.

[26] J. Z. Holwell, *An Account of the Manner of Inoculating for the Small Pox in the East Indies*, London, 1767, p. 24.

[27] M. N. Pearson, "The Thin End of the Wedge: Medical Relativities as a Paradigm of Early Modern Indian-European Relations," *Modern Asian Studies*, 29, 1 (1995): 141–170.

The Colonial Watershed

As for the use of the prefix colonial, post-colonial theorists and diaspora scholars have tried to challenge, even reject, the binary division between the coloniser and the colonised. A recent work talks of the 'tensions of empire' based on the universalising claims of European ideology and the limitation faced by the rulers.[28] This is smart subterfuge. The meaning of 'colonial' is neither elusive nor shifting. What makes colonisation real is that even in its rejection there is an implicit acceptance of the standards set by the coloniser. By arguing that the colonial power and discourse is shared by both the coloniser and the colonised, and by putting emphasis on 'ambivalence' and 'indeterminacy' in this relationship, are the post-colonial scholars trying to take the (political and economic) sting out of colonialism?[29]

Western medical discourse functioned in several ways: as an instrument of control which would swing between coercion and persuasion, as the exigencies demanded, and as a site for interaction and often resistance. This discourse was mediated not only by consideration of political economy but also by several other factors. Polity, biology, ecology, the circumstances of material life and new knowledge interacted and produced this discourse. The emergence of tropical medicine at the turn of the last century may be seen in this light. It may be argued that tropical medicine itself was a cultural construct, 'the scientific step child of colonial domination and control'.[30] In now burgeoning literature, terms like tropical medicine, imperial medicine and colonial medicine have often been used interchangeably. But they have specific connotations. Tropical medicine and imperial medicine emphasise the tropics and the empire as units of analysis while colonial medicine stresses the colony. Each may attract different sets of questions. In tropical medicine what ought to be the determining factor – climate, race, geography or all taken together? What was carried over from the old medicine of tropical civilisations into the new tropical medicine? What attempts were made outside Europe to reconcile the older discourse of body humours and environmental miasmas with the new language of microbes and germs? Interestingly enough, a medical historian described colonialism as 'literally a health hazard'.[31] But to dismiss the colonial doctors reductively as the handmaidens of colonialism or capitalism would also be to ignore a more complex, and more

---

[28] F. Cooper and A. L. Stoler, eds., *Tensions of Empire: Colonial Cultures in a Bourgeois World*, Berkeley: University of California Press, 1997.

[29] See, D. Kumar, *Science and the Raj*, Delhi: Oxford University Press, 1995, pp. 180–182; also idem, "Medical Encounters in British India," in A. Cunningham and B. Andrews, eds., *Western Medicine as Contested Knowledge*, Manchester: Manchester University Press, 1997, pp. 172–188.

[30] L. Manderson, *Sickness and the State: Health and Illness in Colonial Malaya 1870–1940*, Cambridge: Cambridge University Press, 1996, pp. 10–14.

[31] D. Denoon, *Public Health in Papua New Guinea*, Cambridge: Cambridge University Press, 1989, p. 52.

interesting, reality.[32] The doctors had to assume multiple roles. They had little choice. Still one can ask, what role did the 'peripherals' play? Could a synergetic relationship between the core and the periphery develop? These questions assume special significance when viewed against the four centuries of European's struggles in the 'torrid zones' and their transition from early explorers, travellers, and traders to conquerors and ultimate arbiters of the trampled tropics. Earlier the 'tropical discourse' was viewed through its pioneers; now issues and dichotomies have been given primacy. However, these still abound in metropolitan theorizations and do not include the study of indigenous (non-settler) societies through their own literature and practitioners.

From the Indian point of view the mid-nineteenth century was a period of looking for fresh opportunities and acquiring new knowledge. Syncretism, not revivalism, was the agenda. Even among the British officials there were some who wanted the government to attempt a fusion of 'both exotic principles and local practices, European theory and Indian experience', and thereby 'revive, invigorate, enlighten and liberalise the native medical profession in the mofussil'. Similar views were echoed by the emerging Indian intelligentsia in ample measure. To illustrate, we cite three relatively less known (though important) Indians from the three presidency areas: Raja Serfoji (1798–1832, Tanjore), B. G. Jambhekar (1802–1846, Bombay) and S. C. G. Chuckerbutty (1826–1874, Calcutta).

Raja Serfoji, the last Maratha ruler of Tanjore, having surrendered real power to the British Resident, spent his time in the pursuit of knowledge. Father Schwartz, a German missionary, was his friend, philosopher and guide. Fascinated by the different medical system, he had opened an institution for research in medical science and called it the Dhanvantri Mahal (abode of Lord Dhanvantri, the God of medicine). He assembled leading physicians from the Ayurvedic, Yunani, Siddha and Western systems. As a result of their interactions and investigations, the best among the tried and effective remedies were collected in a series of works named *Sarabendra Vaidya Muraiga*.[33] These were composed by the court poet in Tamil verse to facilitate easy memorising and popularisation. With the help of Father Schwartz and the British Resident, Serfoji procured hundreds of European medical books and even surgical instruments. He already had a large collection of Tamil and Sanskrit manuscripts. Some of them dealt with diseases of animals and even birds. Ahead of his time, Serfoji also organised a hand-painted herbarium of medicinal plants in natural colours.[34] In

---

[32] H. Bell, *Frontiers of Medicine in the Anglo-Egyptian Sudan, 1899–1940*, Oxford: Clarendon Press, 1999, p. 10.

[33] S. G. Rao, "Dhanvantari Mahal," *Journal of the Tanjore Saraswati Mahal Library*, 30, 1977, I–IV, Numerous books, instruments and medical case sheets survive as the Modi Raj Records at Saraswati Mahal Library, Thanjavur.

[34] R. Venkatraman, "The Impact of Modern Science on a Tamil Traditional System in the Eighteenth and Nineteenth Centuries," Paper sent to Seminar on Science and Empire, New Delhi, January 1985.

the eye wing of his Dhanvantri Mahal he maintained a set of opthalmic case sheets in an album, with authentic pictures of the eye and its defects for research purposes. This is perhaps a very early example of 'methodical clinical research' under 'native' patronage, and must have induced the traditional physicians to take cognisance of the new therapies and methods.

Serforji was not an intellectual. He was a man of resources with a genuine interest in medicine, perhaps a self-taught doctor, who is said to have learnt the art of cataract removal. In contrast, Bal Gangadhar Jambhekar was the first Indian to teach mathematics at the Elphinstone College in Bombay. He was also perhaps the first Indian to start a journal for popularising science (*Bombay Durpan* in 1831) and established the Native Education Society, which later did a commendable job of translating some European works into Marathi and Sanskrit works (like the nosology of Madhav and the anatomy of Susruta) into English. He wanted the native practitioners to improve by studying 'anatomy from the natural subject', even though touching a dead body was taboo at the time.[35] In 1837 his opinion was sought by the Bombay government on the desirability of a medical school in Bombay and the nature of medical education to be given to the natives. In a written reply Jambhekar asked for (1) the education of a limited number of natives in all branches of the science, and (2) the dissemination of the elements of medical knowledge among the vaidyas, hakims and the community of the interior in general through the means of local languages.[36] This dissemination was to be achieved through translations or writing synthetic books specifically for the purpose. Ordinary vaidyas and hakims, he felt, would respond better than the more 'learned' native practitioners, as the latter were quite convinced of their own superiority and were unlikely to compromise their status. Jambhekar wanted the government to go slowly, without ruffling feelings, and be 'as little offensive as possible'. He argued that the repugnance of the Brahmins at dissection, etc., could be overcome 'by a little perseverance'.[37] How right he was!

S. C. Chuckerbutty came from a Brahmin family. He graduated from Calcutta Medical College and was one of the first four Indian medical graduates sent to England for higher studies, in 1945. He was so charmed by Western values and people that he even embraced Christianity before leaving for England, and put his teacher's name before his surname (be became Soorjo Coomar Goodeve Chuckerbutty). Later he pronounced 'a day in London' of more value than 'a month in Calcutta'.[38] True to his training, he lambasted indigenous practitioners: 'Every Boydo (vaidya) was a born Koberaj (physician)...' To suppose that a Boydo could not be a physician unless he

---

[35] *The Bombay Durpan*, 9 January 1835, pp. 119–120.

[36] Home, Public, No. 18, K.W. Pt. A, 18 July 1838, preserved at National Archive of India (NAI).

[37] Ibid.

[38] S. C. G. Chuckerbutty, *Popular Lectures on Subjects of Indian Interest*, Calcutta, 1870, p. 56.

Adoption and Adaption

passed an examination, was to question the ruling of Manu (an ancient lawgiver).[39] He was not in favour of medical education through Sanskrit or Arabic. He called it 'oriental mania'. But Chuckerbutty's perceptions later changed. He came to support the vernacular medium fully and criticised Calcutta University for representing 'only European opinion and interests' and ignoring 'the national element'.[40]

We have thus seen the views of a local king, a cultural interlocutor and a 'modern' doctor. The first was action-oriented, the second persuasive, and the third served the colonial state without being servile.[41] The emerging educated class showed great aptitude for change and new knowledge. But not the traditional vaidyas! When Madhusudan Gupta dissected a dead body in 1835 a vaidya of high repute, Gangadhar Ray, is said to have left Calcutta in disgust.[42] The traditionalists were convinced that an alien government would not help them.[43] Earlier the government had abolished medical classes at the Calcutta Madarsa and the Sanskrit College. Several thousand signatures were collected in protest.[44] But nothing could stop, or even dilute, the Anglicists' victory.

In average public esteem, however, the indigenous practitioners continued to hold sway. In Calcutta Gangaprasad Sen and Neelamber Sen were extremely popular.[45] They introduced fixed consultation fees, priced medicine, the publication of sacred texts and publicity through advertisements. Gangaprasad started the first Ayurvedic journal in Bengali, *Ayurveda Sanjivani*, and even exported Ayurvedic medicines to Europe and America.[46] These were the indications that certain European practices could be internalised and turned to the advantage of practitioners of indigenous medicine. Even at the conceptual level the then reigning miasmatic theories and the humoral pathology (of the vaidyas and hakims) were not very incompatible. What the Westerners were averse to was the oriental 'process', not its substance. Almost all of them did recognise the

---

[39] "Lecture on the present state of the medical profession in India," dated 2 February 1864: Ibid., 138.

[40] "Lecture on the necessity of forming a Medical Association in Bengal," dated 27 May 1863: Ibid., 135.

[41] Chuckerbutty may have been uncomfortable with the government. Later his professorship of *materia medica* was lumped with medical storekeeping and his salary was temporarily stopped. General, Medical, No. 30, June 1867, West Bengal State Archive.

[42] B. Gupta, "Indigenous Medicine in Nineteenth and Twentieth Century Bengal," in Charles Leslie, ed., *Asian Medical Systems*, California, 1977, p. 371.

[43] G. Mukhopadhyay, *History of Indian Medicine II*, Calcutta, 1923, p. 18.

[44] Home, Public, No. 9, 13 March 1835, and Nos. 44–45, 8 April 1835 (NAI).

[45] Chuckerbutty records that one of his serious patients asked for Neelamber Sen. When the vaidya arrived people lined up to see him. The patient could not be saved but the day and hour of death foretold by the vaidya proved to be correct. Chuckerbutty, *Popular Lectures:* 139.

[46] Gupta, "Indigenous medicine," pp. 372–373, Note 42.

importance of and later emphasised the use of indigenous drugs.[47] But diagnostic procedures and, of course, surgery were to remain major areas of difference for a long time to come.

The Ayurvedic texts written during the early twentieth century show a large number of Sanskrit equivalents for terms borrowed from western medicine, e.g. *jivanu* (micro-organism), *samkranti* (transmission of contagious diseases), *svasyantra* (the respiratory tract), etc. In his *Siddhantanidana*, Gananath Sen (1877–1945) refers to new types of fevers such as *antrikajvara* (enteric or typhoid fever), *granthikajvar* (plague or bubonic fever), *slesmakjvara* (influenza), *kalajvara* (kalazar).[48] Similar impact can be seen in P. S. Varier's *Astangsarira* which discusses modern anatomy and physiology in 2045 Sanskrit verses.[49]

Medicine did emerge as a nationalist issue in India. Even at the height of colonial power, voices against the dominant medical discourse were heard. The indigenous practitioners vehemently denied that their system was unscientific or irrational, yet they did not see anything wrong in learning and benefiting from the new knowledge. Their emphasis was on reforming the system by adopting the 'scientific' method and not on changing the fundamentals of the system. A critical anti-colonial spirit permeated the indigenous response. To quote a verse from a Hakeem in 1910:

> Kuch-ilaj Aya na Kuch Charagiri Ayee:
> Tibb-e-Unan Ke Munh Doctory Ayee:
> Band Sheeshe Mein Vilayat se Pari Ayee:
> Lal-Peeli Hui, Gusse mein Bhari Ayee:
> Chaman-e-Tibb se Guldasta Uda Kar Layee:
> Nayee Tarkeeb se Bandish Saja Kar Layee.

(Knows no method of treatment, but Doctors dared to challenge *Unâni*. In a closed bottle a fairy has come full of anger from foreign lands. The bouquet stolen from the garden of Unani Tibb has been rearranged in a new fashion).[50]

There were several areas in which the Western and indigenous systems could collaborate but did not. The former put emphasis on the cause of the disease, the latter on *nidana* (treatment). Microbes and microscopes constituted the new medical spectacle.[51] But the *vaidyas* emphasised the power of resistance in the human body. The Westerners were forced to take cognisance

---

[47] W. B. O'Shaughnessy (Professor of Chemistry, Calcutta Medical College, 1835–1849) compiled a *Bengal Pharmacopoeia* to facilitate greater use of locally available drug materials and reduce expensive imports from Europe. Later Dr. Waring published the *Pharmacopoeia of India* in 1868.

[48] Meulenbeld, *A History of Indian Medical Literature*, pp. 402–404, Note 8.

[49] Ibid., p. 377.

[50] J. Dakani, quoted in Neshat Qaiser, "Colonial Politics of Medicine and Popular Unani Resistance," *Indian Horizons*, April–June 2000, pp. 29–42.

[51] W. Anderson, "Laboratory Medicine as Colonial Discourse," *Critical Inquiry*, 18 (1992): 506–529.

of indigenous drugs and the *vaidyas* took to anatomy, ready delivery of medicine, quick relief, and so forth (as in case with P. S. Varier, 1869–1958 and Hakim Ajmal Khan, 1868–1927). But the comparison ends there. As a recent critique argues, 'they were inclined to borrow but could not create a dialogue between the two epistemics'.[52] Borrowed knowledge seldom develops into organic knowledge.

---

[52] K. N. Panikkar, "Indigenous Medicine and Cultural Hegemony: A Study of the Revitalisation Movement in Keralam," *Studies in History*, VIII, 2 (1992): 283–307.

How Electricity Energizes the Body: Electrotherapeutics and its Analogy of Life in the Japanese Medical Context

Akiko Ito

Electricity is life – we are in the grip of the network of this invisible power. Appearing in many guises such as lights, the telegraph, motors, and more recently as electronics and computers, electrical technology has propelled modern life and created new cultures all over the world. Electricity is a lifeline which can never be relinquished. Before the era of electrification, however, the juxtaposition of electricity and life meant something different: electricity was literally an essence of the human body. Dating back to Luigi Galvani's experiments which appeared to confirm the existence of 'animal electricity' in the last decade of the eighteenth century, electricity was linked to the mystery of life. It identified with nervous force or nervous fluid circulating throughout the body and providing vital energy necessary to maintain individual life. The notion of electricity as an essence of life widely spread in both European medical research and popular interest. Physicians, electricians, and entrepreneurs, considering the medical field as a good business venture, accelerated the application of electricity in medicine.

The history of electrotherapeutics fall upon the point where the history of science, technology and medicine intersect. Early studies of electricity were almost indistinguishable from those of medicine.[1] Since electricity was deemed

---

A. Ito (✉)
Department of World Languages and International Studies, Morgan State University, Baltimore, MD, USA
e-mail: itoxx015@gmail.com

[1] Margaret Rowbottom and Charles Susskind provide the most extensive historical overview of the medical application of electricity from early electrical experiments in the seventeenth century to electromagnetic radiation invented in the twentieth century. Rowbottom and Susskind, *Electricity and Medicine: History of Their Interaction*, San Francisco: San Francisco Press, 1984. David Charles Schechter also wrote a meticulous chronological research of the history of the study of electricity and medical electricity. D. Ch. Schechter, 'Origins of Electrotherapy Part I,' *New York State Journal of Medicine*, 71, 9 (1971): 997–1008; Schechter, 'Origins of Electrotherapy Part II,' *New York State Journal of Medicine*, 71, 10 (1971): 1114–1124.

a linchpin of the nature and function of the nervous system in Cartesian physiology, electricity and life had always been interwoven in scientific and philosophical thoughts. Galvani, an anatomist and obstetrician who conducted many electrical experiments, and Michael Faraday, highly regarded for his discovery of electromagnetism and who also speculated how his discovery would advance medicine, exemplify that the early development of electrical technology was driven by the desire to reveal the mystery of life and to contribute to medicine and health. As I shall describe, the early records and books of the Tokugawa period (1603–1868) on electrotherapeutics, which were introduced into Japan from Holland in the middle of the eighteenth century, also reflect how electrical technology was interwoven with medicine under the influence of Dutch science and technology.

The history of electrotherapeutics also sheds light on another intersection where Japanese medical culture and tradition met Western medicine. At the point of its introduction into Japan, electrotherapeutics was regarded as one of the latest medical treatments in the West. It aroused great interest particularly among Japanese intellectuals who were eager to learn Western science and technology. Not surprisingly, Japanese people marveled at unconventional phenomena like sparks and shocks which electrotherapeutic appliances produced. What is curious enough here is to trace the notion of electricity and life within Japan. The intimate relationship between electricity and life, no matter how it changed – from 'animal spirits' and 'animal electricity' to 'nervous force' – had been nourished within European medical and philosophical discussions for several centuries. It provided physiological and even anatomical rationales for electrotherapeutics, explaining to practitioners and patients how the human body functioned by electricity and how electricity could cure diseases and symptoms. Now, if we admit that no medical treatment can be simply distilled techniques and methods, but that it must encompass the medical tradition and thoughts in it develops, how did electrotherapeutics travel from its native culture to Japan which did not share the notion of electricity and life? Did the Japanese naively believe that electricity was an essence of their life? If not, in order to position itself within the Japanese medical culture, did electrotherapeutics create a new cultural context within which the electricity was applied to the body?

Japan's Encounter with Electrotherapeutics

A merchant vessel of the Dutch East India Company sailed into the port of Nagasaki in 1753. It carried trade goods as well as some gifts for the Tokugawa shogunate. Among the gifts was an electrotherapeutic appliance. The invention of the Leyden jar in 1745, which enabled the accumulation of static electricity, had promoted applying electricity to medical treatment. Researchers throughout Europe soon started to deliver strong shocks from Leyden jars to patients with paralysis while engaging in spirited discussion over clinical cases of medical electricity.[2] Dutch scholars also joined the discussion and, responding to demand, instrument makers produced electrotherapeutic appliances, typically simple static electricity generators attached to a small Leyden jar.[3] An electrotherapeutic appliance was thus a worthy tribute for the shogunate, expressing Dutch friendship as a trading partner and displaying at the same time their supremacy over this far away Eastern country. Strictly refusing contact with foreign countries except Holland and China to protect itself against the destabilizing influence of Christianity, Japan had not known yet what electricity was. More electrotherapeutic appliances were transported from Holland in following years.

The earliest publication to mention electrotherapeutics was *Oranda banashi* (Stories of Holland). Written by Gotō Rishun in 1765, it was the first introductory text of Dutch science and technology. In the book, an electrotherapeutic appliance called *erekiteru* was recognized as a medical instrument which "can remove fire from diseased parts of the body, and thus mitigate all pain." Gotō mistakenly wrote that 'electricitejit,' the Dutch word for electricity, was the name of the inventor of the device. This episode tells us that even Gotō who was a prominent leader of *Rangaku* (Dutch learning) lacked a basic grasp of electricity at that time. Luck fell upon Hiraga Gennai, a former samurai, who was a natural philosopher and entrepreneur. He read *Oranda banashi* and developed a strong interest in *erekiteru*. Hiraga found a broken *erekiteru* on his visit to Nagasaki in 1770.[4] After spending 6 years repairing it and then making several replicas of *erekiteru*, he opened a salon in Edo (present-day Tokyo). At his salon, customers were entertained with electrical experiments such as moving a paper doll and making sparks by means of static electricity while enjoying meals and drinks. He also administered static electricity from his *erekiteru* to sick people for treatment. Like Gotō, Hiraga believed that *erekiteru* could remove the 'fire' causing fever and headache from the body. Although Hiraga's *erekiteru* was able to discharge only a small amount of electricity for a very short time and his performances were rudimentary, his customers, who had never experienced any sort of electrical phenomena until then, looked with wonder upon this Western device. As Hiraga's fame and popularity of his *erekiteru* spread, many followed in his footsteps. Craftsmen began copying the *erekiteru*,

---

[2] For the invention of the Leyden jar and its introduction into medical applications see Rowbottom and Susskind, *op. cit.*, Chapter 2.

[3] Lissa Roberts explores the Dutch interaction with electrical machines in the eighteenth century in terms of both Dutch culture and its identities and uses. Roberts, 'Science Becomes Electric: Dutch Interaction with the Electrical Machine during the Eighteenth Century,' *Isis*, 90 (1999): 699–707.

[4] For a biography of Hiraga Gennai, see Isamu Jofuku, *Hiraga Gennai*, Tokyo: Yoshikawa Kobunkan, 1971; Tessa Morris-Suzuki, *The Technological Transformation of Japan: from the Seventeenth Century to the Twenty-first Century*, Cambridge: Cambridge University Press, 1994, pp. 23–27.

and affluent people imported them from Holland. Electrical demonstrations by means of *erekiteru* also took place in major cities captivating many spectators.[5]

While *erekiteru* sparked the Japanese interest on the distant European continent, a new era dawned in the science of electricity. In 1791 Luigi Galvani, a professor of obstetrics at the Bologna Institute of Science and Art, published an account of his experiments in which he formed a circuit by connecting a dissected frog's leg nerves and muscles with two dissimilar metals. The occurance of muscle contraction in a closed circuit led Galvani to conclude that electricity was produced between the muscles and nerves resulting in spontaneous movements. This marked the discovery of the existence of animal electricity long postulated by many of his predecessors. His discovery caused a great stir among physiologists and physicians. Physiologists were delighted that finally a long sought-after vital principle was within their grasp. New methods of treatment seemed to be revealed to physicians. At a feverish pace, countless researchers reported new work in which they expanded Galvani's results within their own fields. Strong interest in Galvani's work did not abate even when Italian physicist Alessandro Volta falsified Galvani's recognition of animal electricity. He revealed that the electricity which Galvani perceived in the frog's muscles and nerves was actually produced by an interaction between the two different metals used to form the circuit.[6] Ironically enough, Volta's refutation provoked even more speculation on the mysterious relationship between electricity and life and thrust this topic into the realm of cultural phenomena far beyond scientific discussion. The eponymous term 'galvanism' was created, and various kinds of electrical devices named after Galvani – such as the galvanometer, galvanic battery and galvanic charger – were invented. Galvanism also became synonymous with electrotherapeutics. By the mid-1830s when electrophysiology, mostly initiated by Emil du Bois-Reymond, was well developed, it was conclusively determined that animal electricity did not exist. Yet even after that, the notion of electricity and life survived as a rhetorical and biological analogy. In late nineteenth-century Europe and the United States where great social interest in 'nervous ailments' developed alongside rapid modernization and urbanization, the public image of the human body was often likened to an electrical machine or battery and nervous force to electricity.[7] In the domain of electrotherapeutic practice, the human body could be revitalized by passing electricity through the nerves.

---

[5] These demonstrations which retailers usually held to snare customers were called *erekiteru kogyo* (a show of electricity). Kikuchi Toshihiko, '*Kaisetsu*,' in *Edo koten kagaku sosho* vol. 11, Tokyo: Towa Shuppan, 1988, p. 57.

[6] In doing so, Volta essentially invented the Volta battery which contained piled metal plates and electrolyte solution (1800). For a detailed description on Galvani's experiments and the subsequent debate presented by Volta, see Rowbottom and Susskind, *op. cit.*, Chapter 3.

[7] To describe the popular faith in electrical medicine in late nineteenth-century America, David E. Nye pointed out that many technical terms rooted in electricity permeated American people's language. He gave examples: 'energetic,' 'electrifying,' 'recharge his battery,' 'blow a fuse' and so on. See Nye, *Electrifying America: Social and Cultural Meanings of a New Technology*, Massachusetts: MIT Press, 1990, p. 155.

These developments in the study of electricity soon bubbled over European borders. *Zukai ransetsu sansaikikan* (Dutch Illustrations of Three Dimensions of the Universe Viewed through a Scope), written in 1808 by Hirose Shūhaku, provided updated information about *erekiteru*. While covering a wide range of topics about Dutch science and technology such as astronomy, meteorology and medicine, Hirose, as a doctor of Western medicine, paid particular interest to apoplexy and the paralysis attributed to it. With the explanation that paralysis was casued by the malfunction of nerves which governed all of the senses and motions of the body, the author valued *erekiteru* as a "great art" which could work on the nerves, moving the paralyzed parts. Since the 1770s when Western anatomy was introduced into Japan through the considerable efforts of Sugita Genpaku and his colleagues, new concepts of the nerves and the nervous system took wings among Japanese doctors of Western medicine.[8] Compared to the past descriptions in which *erekiteru* was simply considered to suppress excessive fire inside the body, *Sansaikikan* relocated *erekiteru* within the new anatomical knowledge of Western medicine.

Galvani's name and galvanism appeared in various scientific fields. One of the earliest examples is *Seisetsu igensūyō* (Fundamentals of Western Medicine) by Takano Chōei, a prominent doctor.[9] Written in 1832 this was the first comprehensive work on Western physiology. In the section on the brain and spinal cord, he described the characteristics of electricity in detail: electricity was unseen, untouchable, weightless and continually wavering between two poles; it worked "in a magical and unexpected way" when it was applied to the nerves; and it existed in both the nerves and the atmosphere. For Takano, who recognized the value of physiology, acquiring the fundamentals of electricity was important to understand the latest physiological debates in the European community. Furthermore, Takano for the first time categorized electricity into three kinds – *erekiteru* (static electricity), galvanism and magnetism – and explained that each form of electricity had a different curative efficacy.[10]

---

[8] Matsumura Noriaki, 'The Similarity between Two Medicines: Oriental Medical Thought of Meridian and Western Medical Knowledge of Nerve,' *Historia Scientiarum*, vol.13, 3 (2004): 224–231.

[9] This book was based on Dutch translations of works in German by Blumenbach and T. G. A. Roose and in French by J. de la Faye. See Ellen Gardner Nakamura, 'Physicians and Famine in Japan: Takano Chōei in the 1830s,' *Social History of Medicine*, 13, 3 (2000): 429–445 (p. 431).

[10] These three divisions of electricity were very common in the second half of the nineteenth century. Iwan Rhys Morus pointed out that these divisions were not regarded as being different in themselves but were treated separately because they were produced by either a static electricity generator with a Leyden jar, a Galvanic apparatus or a magnetic apparatus. I. R. Morus, *Frankenstein's Children: Electricity, Exhibition, and Experiment in Early-Ninenteenth Century London*, Princeton: Princeton University Press, 1998, p. 126. Hirose's categorization of electricity explicity followed the trend of scientific discussions in Europe during the period.

Fig. 1 A design plan for a galvanic electrotherapeutic appliance in *Naifukudokō* (1858). This appliance consists of a battery, an induction coil, a switch and two electrodes. The section also includes detailed design plans for constructing an induction coil and switch

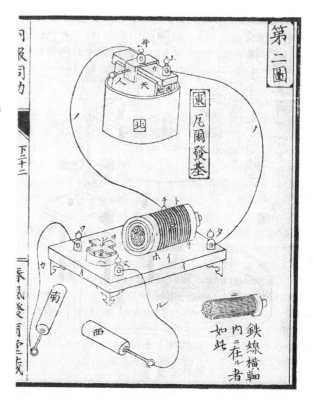

Naifukudōkō (Medical Instruments as Effective as Drugs) in 1858 is another medical book which discussed both the conventional *erekiteru*, which simply discharged static electricity, and galvanism, which produced electric current by an induction coil and a Volta battery (Fig. 1). Authors, who had learned Western medicine from Dutch physicians in Nagasaki, wrote anonymously in this work intending to provide practical information to their fellow physicians. Its contents included a long list of diseases which galvanic electricity could cure such as rheumatism, paralysis, asthma, deafness, mutism and so on; where to place the electrodes on patient's bodies; and the duration for which the patient should be administered electricity. The authors added: "both *erekiteru* and galvanism are very effective in treating regional diseases because they are favorably related to the nerve."

In addition to the books mentioned, there were more publications on medicine, chemistry and physics which discussed electrotherapeutics and Galvani's work.[11] Along with the proliferation of Western scientific and

---

[11] For example, Shimamura Teiho, *Seiri hatsumō* (Introductory of Physiology, 1866) and Kawamoto Komin, *Kikaikairan Kōgi* (Broad Definition of the View of the Earth's Atmosphere, 1851) described Galvani's biography, his experiments on frogs and his findings of animal electricity.

medical information in the middle of the nineteenth century, more advanced, practical and clinical knowledge about electrotherapeutics circulated within the Japanese intellectual community. *Erekiteru* and electrotherapeutics were no longer a novelty just stirring people's curiosity as they had been at their inception in the eighteenth century. Japanese discussions slowly but steadfastly reflected the European trend in the study of medical electricity.

Looking back to the notion of electricity and life which considered electricity as vital energy circulating through the nerves, however, we notice that it dropped off the nineteenth-century Japanese intellectual map. As we have already seen, several Japanese scholars like Hirose and Takano remarked about the efficacy of electrotherapeutics as related to the nerves and nervous system, but the point as to why nothing other than electricity could affect the human body totally slipped away. Indeed, Japanese intellectuals, unlike their European counterparts, had never closely debated the issue. Eventually, electrotherapeutics faded from Japanese medical discussions in the early Meiji period (1868–1912). When the Meiji administration promoted Western medicine in earnest, adopting German medicine as a model and establishing Westernized medical schools, electrotherapeutics failed to be legitimized having a position neither in medical education nor in emerging medical disciplines. The German medical professors who had been summoned by the Japanese government to engage in medical education actually employed electrotherapeutics in their own clinical practices, and in a draft of a medical school curriculum, electrotherapeutics was listed as a subject.[12] Some of the newly opened private clinics also contained electrotherapeutics facilities. Nonetheless, these examples were very limited, and electrotherapeutics could not prevail in the emerging medical system. Considering these points, Japanese medical students and researchers simply did not embrace the mystery of electricity and life which provided its own physiological and anatomical rationales for electrotherapeutics. Meanwhile, as new knowledge, theories and practices deluged the Japanese medical community, electrotherapeutics was lost in the flood. It was not until the 1910s when physiotherapy was introduced into Japan that electrotherapeutics was reconsidered and reevaluated as one of its techniques.

Although it failed to continue attracting the interest of Japanese medical professionals, electrotherapeutic appliances targeting ordinary people for home care gradually penetrated the healthcare market emerging at the turn of the century. Amidst the considerable upswing in the exchange of people and products between Japan and Western countries, several European and American companies which produced and sold electrotherapeutic appliances entered this new market. How did these foreign companies promote their products in the Japanese market? Did they think that the notion of electricity and life was as persuasive in explaining how electricity could affect the body in Japan as it was in their countries?

---

[12] *Chikenroku* (Records of Clinical Trials) published in 1882 recorded several cases of electrotherapeutics employed by Professor of Internal Medicine Theodor Hoffman, a German army surgeon, and Professor of Surgery Leopold Müller, a German naval surgeon at the Medical School of Tokyo University.

Electrotherapeutics in the Market

Around 1903 an American business enterprise, the Sanden Electric Company, opened its first Japanese branch in Yokohama under the Japanese name Sanden Denki Shōkai. Founded by Albert T. Sanden in 1885, the company had its headquarters on Broadway Avenue in New York City and manufactured an eponymously named product – Dr. Sanden's Electric Belt. They expanded their business very quickly, opening offices in major cities in the United States and eventually went on to Canada, England, France, Sweden, Germany, Poland, Finland and Japan.[13] Late nineteenth-century European and American consumers were overly familiar with electric or galvanic belts. Typical electric belts were made of a wide, thick piece of fabric with built-in metal plates which presumably produced electric current when the metal plates were moistened. Devices of this sort had existed in the European market since the 1840s but the popularity of electric belts took off when in the 1870s George Miller Beard, an American neurologist and proponent of electrotherapeutics, proclaimed his theory of neurasthenia or nervousness.[14] Rising middle class 'brain' workers who felt exhausted with the complex modern life became good patrons of electric belts which could charge a reduced nervous force and revitalize their bodies.

The Sanden Electric Company achieved great success in America and set their sights on an untapped market, Japan. To cultivate Japanese consumer's interest in their Sanden Electric Belt (the Japanese product name: *Sanden Denki Obi*), the company employed the same marketing methods as it did in the United States: it ran advertisements in periodicals, offered free consultation on health issues at its offices or by mail and distributed without cost a Japanese translation of Dr. Sanden's Book, *Tennen ni okeru kenkō* (Fig. 2).[15]

The title page is written in both English and Japanese and on the imprint page the translator's name, W.C. Watoson (Watson?) appears. Although the work appears to be derived from an English original entitled *Health in Nature: A Practical Treatise on Self-treatment with Galvanic Electricity for the Cure of Disease and the Restoration of Vital Strength*, the original has not been found. It is reasonable here to examine *Tennen ni okeru kenkō* by comparing it to the text of two similar Sanden publications in English, because the structure, content and illustrations of these

---

[13] Zachary Ross, 'Linked by Nervousness: Albert Pinkham Ryder and Dr. Albert T. Sanden,' *American Art*, 17, 2 (2003): 88–89. Ross addressed Sanden's brief biography through his friendship and patronage of painter Albert Pinkham Ryder.

[14] Beard saw the nervous system as a self-contained circuit through which electricity as nervous force passed. He held that the quantity of this nervous force was fixed and explained that the pressures of a rapidly advancing civilization drained it to the point of a 'nervous bankruptcy'. This ultimately caused physical and mental symptoms (neurasthenia). George M. Beard, *American Nervousness, Its Causes and Consequences: Supplement to Nervous Exhaustion* (*Neurasthenia*), New York: G.P. Putnam's Sons, 1881.

[15] Albert T. Sanden, trans., W.C. Watoson, *Tennen ni okeru kenkō* (Health in Nature: A Practical Treatise on Self-treatment with Galvanic Electricity for the Cure of Disease and the Restoration of Vital Strength), Yokohama: Sanden Denki Shōkai, 1905.

Fig. 2 Title page of *Tennen ni okeru kenkō* (Health in Nature: A Practical Treatise on Self-treatment with Galvanic Electricity for the Cure of Disease and the Restoration of Vital Strength), Sanden Denki Shōkai (1903)

HEALTH IN NATURE

A practical treatise on Self-Treatment with Galvanic Electricity for the cure of disease and the restoration of vital strength.

BY

DR. A. T. SANDEN.

PUBLISHED BY

THE SANDEN ELECTRIC Co.,
No. 51, MAIN STREET, YOKOHAMA.

［ドクトル、サンデン］博士講述

天然ニ於ケル健康

サンデン電氣帶ノ説明

横濱市山下町五拾壹番舘

サンデン電氣商會藏版

works show little difference.[16] By so doing, we will discuss how the Sanden Electric Company presented its product and promoted electrotherapeutic cures in the Japanese market.

In his *Three Classes of Man*, Sanden exhorted that electricity was a 'life principle.' The intimate relationship between electricity and life is not suppressed here:

> When we consider that our nervous system, which is the fountain of life to the kidneys, liver, stomach, brain, sexual organs and the various organic functions of the body, depends for its sustenance upon the vitalizing element of electricity, and that without this it is impossible to keep up a normal condition of health in the body, it is easy to understand that a waste of this *life principle* will be followed by weakness and disease, and it is also easy to understand why the natural restoration of this electric force in the nervous system, will saturate the various vital functions which have become weakened, with a new energy which will place every vital organ of the body in a state of natural *health*.[17]

---

[16] The Bakken Library and Museum, Minneapolis, MN holds two Sanden books: A. T. Sanden, *Three Classes of Men*, New York: Sanden Electric Co., ca. 1896; Sanden, *A.T. Sanden, Originator of the Celebrated Home Treatment for the Cure of All Chronic, Nervous and Wasting Diseases, without Drugs or Medicines*, New York: Prout & Ward, (undated).

[17] Sanden, *Three Classes of Men*, p. 10.

Sanden's 'life principle' theory appeared straightforward and consistent to his readers. Following this statement, he inventoried diseases and sicknesses which his belt could cure – from neurasthenia, spermatorrhea and general debility to rheumatism, lumbago and sciatica – and offered articulate explanations regarding why people acquire these diseases. Certainly, the excess of mental and physical activities wasted nervous force and inevitably caused nervous disorders which led to dysfunctions of each organ.

Then, if his theory of life principle was indispensable to his electric belt, how did the Sanden Denki Shōkai propose it to Japanese consumers? Here are some tactics: they borrowed certain words and concepts from Japanese traditional medical thought and imbedded them within the translation of Sanden's book. What is the most glaring is the replacement of 'nervous force' in English with a Japanese indigenous concept, *genki* (inborn, original *qi* energy). In the Japanese traditional view of the human body, *genki* was considered as a source of vitality circulating within the body and governing each function. Clogged or decreased flow of *genki* was believed to cause all health problems and ultimately the end of life. In *Tennen ni okeru kenkō*, they only occasionally cited *shinkei ryoku*, a literal translation of nervous force which might have sounded unclear to their Japanese readers. After adroitly employing this traditional idea, the Japanese translation modified the nerve center (*shinkei tyūsū*) as the place where *genki* resided (p. 3). From there, *genki* flowed to the organs by means of the nerves. By administering electric current the Sanden Denki Obi could bolster *genki* when it was deficient and tone the nerves when they were debilitated (p. 39). The indigenous concept of *genki* was persuasively inlaid within the network of the nervous system, and there *genki* and electric current were interchangeable and passing together.

Subsequent explanations of the nervous system used the telegraph as a metaphor: as the telegraph network extended throughout the country, the nervous system covered all parts of the body (p. 50). For the Japanese who witnessed the rise of the strong centralized government utilizing the telegraph as a state apparatus, the metaphor of the telegraph and the nervous system was powerful enough to imagine that their body and health were irresistibly governed by the functioning of the nervous system. By harnessing both electricity to traditional concepts and modernization to the human body, they constructed and presented to their customers a new image of the body in which electricity should be applied.

In addition to *Tennen ni okeru kenkō*, the company also published small booklets every 4 to 6 months in which testimonial letters from their users were collected, complete with their name and full home address. Let us assume that the testimonial letters were not faked as prospective customers were invited to contact the respondent directly to inquire about their personal experiences with the Sanden Electric Belt. According to the testimonial letters, there was much interest among users who purchased the electric belt in the hope that it would cure neurasthenia (*shinkei suizyaku*). In accordance with the rhetorical exchange between *genki* and electricity which *Tennen ni okeru kenkō* drew upon, one user of the electric belt eagerly shared his experience: "As the pleasant electric current from the Sanden Electric Belt stimulated my nerves, I gradually felt *genki* filled my body and eased the neurasthenic symptoms."[18] Detailed records about its corporate performance are not available but the Sanden Denki Shōkai operated their business at least until 1908.[19]

Approaching the late 1910s, engineering companies, large or small, broke into a new industry of electric household products which teemed with great potential. Amongst these products were electrotherapeutic appliances for home use. As is evident from the many newspaper and magazine advertisements, a wide variety of electrotherapeutic appliances for home care which were domestically produced sprung up. These domestic electrotherapeutic appliances shared a certain character which merits attention; they manifested strong affinities with traditional practices of medicine such as acupuncture, moxibustion and massage (*anma*).

Affinities between electrotherapeutics and traditional medical practices were first embodied in the electrification of practical techniques. Electro-moxibustion, electro-acupuncture machines and similar variations can be included within this category. Following the basic principles of traditional medicine, these products were designed to simplify processes and techniques of treatment as well as to achieve better results. For example, moxibustion was familiar and recognizable to everyone but electro-moxibustion was novel in that, instead of burning a small cone of moxa herb on the skin, it employed heat produced through electric current. An advertisement which appeared under the pretense of a magazine article for an electro-moxibustion appliance, the *Erekutera*, emphasized its superiority over conventional moxibustion practice; it could transmit the heat easily and quickly to deep points within the body where traditional moxibustion with moxa herbs could not reach. The users did not even have to know the location of *tsubo* (acupuncture points) as, "The electric current can automatically strike out and affect them."[20]

It is worthwhile to recall here that for the Japanese, electrotherapeutics had been a symbol of Western medicine from the time of the introduction of *erekiteru*. Even the Sanden Electric Belt, in which indigenous concepts were strategically used, still remained within the framework of Western medicine. One cannot simply consider electrotherapeutic appliances like electro-moxibustion or electro-acupuncture devices as derived from Western medicine,

---

[18] Sanden Denki Shōkai, *Sōkō raijō shū* (Collection of Testimonial Letters about the Effectiveness), the tenth edition, Yokohama: Sanden Denki Shōkai, 1907.

[19] In 1914, Sanden and his company were accused by the U.S. Postmaster General in New York of using the mail to defraud customers of over one million dollars. He was acquitted but his company closed in or around 1915. Ross, *op.cit.*, p. 89.

[20] 'Kateiyō denki onkyūki Erekutera no himitsu (The Secret of the Electromoxibustion Appliance Erekutera),' *Katei no denki* (Home Electrification), 6, 2(1929): 47–48.

however. In many cases, the companies who entered into the production of electrotherapeutic appliances for home use were neither makers of medical instruments or equipment nor did they have business experience in the healthcare industry. Although the primacy of Western medicine had been firmly established, traditional thoughts and concepts of medicine still endured in the public mind. Thus, for engineering companies and entrepreneurs who sought a good business opportunity in the electrical household product industry, traditional medical practices like moxibustion and acupuncture appeared to provide them many ideas for making products targeting the broader public. Their electrotherapeutic appliances readily appealed to consumers both by their technological advancement as well as by a sense of comfort and familiarity nurtured within their own culture.

In addition to the appliances that electrified traditional medical practices, diathermy and low-frequency therapeutic devices were also launched in the market. In general, these appliances issued heat, shocks or shakes and sometimes mild pain to the body by two attached electrodes. Electrified points varied depending on diseases and symptoms so that the users had to refer to instruction manuals identifying the location for the proper placement of electrodes (Fig. 3). While it seems that the theories and methods of the new appliances had little in common with traditional medical practices, frequently advertisements and manuals associated the electrified points with *tsubo* (acupuncture points) and *keimyaku* (meridians).

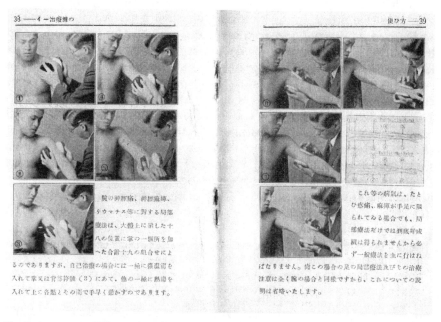

Fig. 3 A sequence of illustrations from an instruction manual on diathermy, *Ichiryōki* (1930)

Some advertisements used the terminology of electrophysiology like motor points, intimating to the user that they were more advanced and effective than their rivals. Acupuncture points functioned as an analogue for motor points to help the users comprehend the treatment and how it could solve their health issues; as long as they knew that motor points worked as acupuncture points did, users could forgo any deeper understanding. As electrotherapeutics became more complicated and technologically sophisticated, it needed to adopt such analogies to an ever greater extent. Herein, affinities between electrotherapeutics and traditional medicine were strengthened.

Between the Traditional and the Modern

The history of electrotherapeutics in Japan poses a puzzle for us: what divides the traditional and the modern? We casually assume that the two exist in different spheres, one irreconcilable from the other. When we look into history, however, the traditional and modern are not always so distant. In this essay I addressed the Japanese history of electrotherapeutics from the middle of the eighteenth century to the early twentieth century. Through this period, the shape of electrotherapeutics in Japan was transformed intrinsically: from *erekiteru* which fascinated Japanese scholars as a symbol of Western modern science and technology, to electrotherapeutic appliances for home use such as electro-moxibustion and diathermy, which displayed cogent affinities with traditional medicine. While the transformation of electrotherapeutics seemed radical, it occurred in a seamless sequence as if there was no clear boundary separating the traditional from the modern. The ascendance of electrical engineering propelled electrotherapeutics in a new direction. After electrotherapeutics lost the favor of medical professionals, it mostly followed market politics and consumer orientations in health and medicine. Electrotherapeutic appliances for home use took modern technological shapes, while at the same time inheriting most of the fundamentals of traditional medical practice. In the Japanese history of medicine, thus, electrotherapeutics can be labeled neither Western, modern nor traditional. It evades such categorization.

When electrotherapeutics traveled from Europe and, eventually, from the United States to Japan, a new image of the body upon which electricity was applied needed to be invented. Unlike in late nineteenth century Europe and the United States where electrical analogies for the nervous force and nervous system were indispensable to electrotherapeutic practices, the Japanese never embraced the notion of electricity and life. Instead, indigenous concepts of medicine like *genki* were adopted as analogies for electricity and the nervous system, offering users of electrotherapeutic appliances a rationale for how electricity could energize and revitalize their body. These analogies between vital energy – whether it was called nervous force or *genki* – and electricity linked across cultures, enabled electrotherapeutics to be incorporated within the Japanese medical context. The history of electrotherapeutics indicates to us that medical treatment and medical technology encompass the tradition and thoughts in which they develop. When they travel over cultural boundaries they need to locate a new medical context in which they can position themselves.

Today, over a century after the Sanden Electric Company started their business in Japan, we can still observe the strong presence of electrotherapeutic appliances in the Japanese medical market. Even more kinds of therapeutic devices are appealing to health-minded consumers who hope to restore the resilience of their bodies as did customers of the Sanden Electric Belt. By continuing to recreate the medical context in which electricity is applied to the body, electrotherapeutics can also travel across the time.

What is 'Islamic' in Islamic Medicine? An Overview

Hormoz Ebrahimnejad

What is 'Islamic medicine' and how did it develop? What is 'Islamic' in Islamic medicine and when did this term appear in the medical historiography, considering that it was never used during the medieval period in the Islamic countries or by any Muslim physicians or historians? With the exception of specific literatures on spiritual healing or the books of Medicine of the Prophet, the *Tebb al-Nabi*, or the *Tebb al-A'emma* (medicine of the Imams) which consist of the sayings and living habits (*sunnah*) of the Prophet and the Shiites Imams, there are no titles of medical treatises written by Islamic scholars which allude to Islam or religion. The major books of Râzi (tenth century) were titled the *Kitab al-Mansuri* (dedicated to Caliph al-Mansur), and the *al-Hâwi* (Comprehensive, or Continent). The book of Majusi (tenth–eleventh century) was called *Kâmel al-Senâ'a* (The Perfect Art) or *Tebb al-Maleki* (The Royal Medicine). Avicenna's master work was the Canon, and the first classical medical treatise written in Persian was the *Zakhira* (Treasure) of Khawrazmi (thirteenth century). And this was the same throughout the following centuries up to the present time.

The embryo of what is called Islamic medicine was forged by a number of scholars who were mostly Christian, or Jewish. By the time when the Islamic scholars like Razi, Avicenna or Zahrawi wrote their books, almost all medical literature these physicians used in their work had already been translated into Arabic by non-Muslims, such as Hunayn b. Ishaq (809–73), Sâbt b. Qurra (b. 835), Shapur b. Sahl (d. 869), and the members of the Bokhtishu' family who provided physicians to the Omayyad and Abbasid Caliphs for several generations.[1]

H. Ebrahimnejad (✉)
Faculty of Humanities, University of Southampton, Southampton, UK
e-mail: h.ebrahimnejad@soton.ac.uk

[1] L. Leclerc, *Histoire de la Médecine Arabe; Exposé Complet des Traductions du Grec, les Sciences en Orient, leur Transmission à l'Occident par les Traductions Latines*, 2 vols., Paris: E. Leroux, 1876; See also articles of M. Mayerhof, edited by P. Johnstone in *Studies in Medieval Arabic Medicine: Theory and Practice*. London: Variorum Reprints, 1984.

The term 'Islamic Medicine' was first coined by modern Western historians principally because it was developed under Islam and particularly because of the role of later Islamic scholars in the elaboration of a core medical literature translated into Arabic from Greek. The term 'Islamic medicine' can also be explained by the fact that the Western historians of Islamic countries tend to identify all aspects of these societies by their religion without taking into account socio-political, cultural, linguistic and anthropological differences between them. The term 'Islamic' medicine has also been used by historians of Islamic faith for similar reasons. For example, although Islam is not the dominant religion in the Indian subcontinent, the identification of *Unâni*, i.e. Greek, medicine with Islam is apparent in various historiographical works.[2] Since the last decades of the nineteenth century, the link between what the Indian hakims called *Unâni tebb* and Islam was emphasised seemingly in an attempt to confirm the identity of their medicine against what they called Western or colonial medicine. The Islamic or religious overtone of Galenic medicine is also illustrated in the term *tebb-e sonnati* (traditional medicine) in Iran. When Khomeini, the founder of the Islamic regime, took power, he advocated the revival of traditional medicine that, according to him, had been abandoned due to Western influence. In other words, the revival of 'traditional medicine' was associated with the revival of Islamic values and culture. It is worth mentioning that while by the end of the nineteenth century in the West, modern medicine was identified with microbiology,[3] in contemporary Iran the term modern was attributed to be any kind of medical knowledge introduced from Europe regardless of their conceptual or epistemological variations. The fact that the theoretical foundation of what Khomeini called 'traditional medicine' was humoral and based on Greek medicine, indicates the extent to which Greek medicine was assimilated by Islam. The purpose of this paper is to examine this assimilation through two processes: the creation[4] of Islamic power and the elaboration of Islamic theology and cosmology.

---

[2]See for example S. K. Hamarneh, *Background of Yunani (Unani), Arabic and Islamic Medicine and Pharmacy*, ed. Hakim M. Said, Karachi: MAS Printers, 1997.

[3]A.-M. Moulin, 'Le Dialogue Médical Franco-persan au XIXe Siècle,' in N. Pourjavadi et Z. Vesel, eds., *Sciences, Techniques et Instruments dans le Monde Iranien (Xe-XIXe siècle)*, Tehran: IFRI and Tehran University Press, 2004, pp. 305–329 (p. 305).

[4]By "creation" we mean the social and political development of Islam and not the chronological creation of Islamic power that occurred under the Prophet and the first caliph. What we mean by "Islamic science" based on "Islamic power" is certainly a later development and did not belong to the time of the Prophet. In other words, "Islamic sciences" developed, or were "created" after the territorial expansion of Islam.

Formation of Islamic State

The development of what is called 'Islamic sciences' was closely linked to the establishment of central state in Islam. The Umayyad established their state in Damascus in the mid-seventh century by adopting elements of Byzantine and Sasanid administrative systems. As the Umayyad and the Abbasid caliphs needed the expertise and know-how of the elites of the conquered countries, they developed court patronage and sponsored men of sciences of all creeds, Arab and non-Arab alike; most of them were naturally non-Muslim or had recently converted to Islam, such as Abd-Allah Ibn Moqaffa' (died ca. 756) and Barmak (ca. 685–725),[5] both of them Zoroastrian converts to Islam.[6]

Although court patronage was not a new phenomenon, it developed significantly due to the need of the nascent Islamic empire. It was this patronage, aimed at strengthening the caliphate that eventually ensured that non-Islamic sciences were systematically translated and integrated into Arabic literature and, in the long run created what is called 'Islamic sciences'. The development of Islamic sciences, according to D. Gutas, was mainly the consequence of the 'imperial ideology' that the Abbassids inherited from the Sasanians. The elaboration of the 'imperial ideology' necessitated the translation of scientific texts, such as medicine, astronomy and philosophy, from around the world and their collection in a royal library that was called 'house of wisdom'. Gutas maintained that this appropriation made through translation created a 'culture of translation' that in turn was transmitted to the Abbasid caliphate when the caliphs borrowed the Sasanid 'imperial ideology' along with its administrative system. The culture of translation was thus at the origin of the translation of Greek sciences into Arabic.

The 'appropriation of knowledge' was not, however, exclusive to Zoroastrianism as it can be seen in other religions and cultures as well. According to Islamic religion, all human knowledge belongs to God and therefore the sciences are the manifestation of divine wisdom. They exist on the divine Table, *lowh-e mahfuz* (lit. conserved table), before coming to the minds of the men of science, whether Muslim or not. Astarâbâdi, a cleric-doctor in nineteenth-century Iran admonished the modern-educated doctors in the following terms: 'You do not appreciate the merit of your medicine [i.e. traditional

---

[5] The father of Khaled b. Barmak (d. ca. 783) and ancestor of the Barmakid dynasty. See Ali-Akbar Dehkhoda, *Loghatnâmeh*, second edition, 1377/2000, Tehran University Press, vol. I, p. 189.

[6] About the role of these men in the administrative reforms under the Abbasids, see M. Q. Zaman, *Religion and Politics under the Early 'Abbasis: The Emergence of the Proto-Sunni Elite*, Leiden, New York, Koln: Brill, 1997, cf. especially Ch. 3. On Ibn Moqaffa' and a translation of his book to Caliph al-Ma'mun for the state and religious reform, see C. Pellat, *Ibn al Muqaffa: 'Conseilleur' du Calife*, Paris: Maisonneuve et Larose, 1976.

medicine]. Observe the history of the *Haramân* dome.[7] This [medical] knowledge is the heritage of the Prophet Idris and all other prophets used it until the last Prophet Mohammad, who perfected it. The *Tebb al-Nabi* (medicine of the Prophet), the *Tebb al-Rezâ* (medicine of Imam Rezâ) and the *Tebb al-A'emma* (medicine of the Imams), are all available and in fact the source of physicians such as Avicenna were the right Traditions of the Prophets...'.[8] In another case, Ibn Rizwân al-Misrî (from Cairo) (died ca. 1067), the eminent physician under the Fatimid in Egypt, claimed that it was Egyptian medicine that was carried to the Greeks. In other words, the Greek authors, such as Hippocrates, Plato, Aristotle, and Theophrastos, who were praised by the Arabs were in fact the bearers of Egyptian [identified with Arab] knowledge.[9] As a human phenomenon, science and knowledge necessarily become a common heritage. The negative connotation of the term appropriation should not conceal the more important fact that the appropriation of knowledge indicates the approval of that knowledge. Acquisition of sciences has always been made through various discursive strategies.

As far as the Sasanians are concerned, the construction of 'imperial ideology,' was a means of consolidating the central state and in this sense it motivated the acquisition of other sciences and techniques. The mere translation of scientific texts and syllabus for practical or educational as well as ideological reasons existed outside Sasanid Iran. From the fifth century on, the Nestorians and Jacobite scholars, who were not under Sasanian authority, translated Greek writings into Syriac; but this did not create a culture or movement of translation. Under the Sasanians, on the other hand, the house of wisdom was part of the state apparatus and it was within such a framework that the translation of scientific texts acquired significance and had a long-lasting effect on the transmission of knowledge.

As mentioned above, it is important to examine various strategies employed for the acquisition of knowledge. If Islam was going to expand beyond the Arabian Desert, it needed to enter in peaceful relationship or in cultural dialogue with the populations of other countries. Insofar as principally all non-Islamic creeds, whether monotheist or not, were refuted by Islam, the message of tolerance of the Prophet towards 'the peoples of the Book,' namely, Christians, Zoroastrians and Jews, was a strategy aimed at reducing the enemies

---

[7]Heramân or Haramân, (two domes) refer to two of the pyramids in Egypt that are said to have been built by the Prophet Idris (from Egypt), or according to some, to Hermes, the Greek, in order to protect sciences from deluge, tornado or storm. See Dehkhodâ, *Loghatnâ-meh*, second edition, 1377/2000, Tehran University Press, vol. I, pp. 1571–1573 and vol. XV, p. 23540.

[8]Astarâbâdi, *Safineh-ye Nuh*, Quoted by H. Ebrahimnejad, 'Religion and Medicine in Qajar Iran,' in, R. Gleave, ed., *Religion and Society in Qajar Iran*. London: Routledge, 2004, pp. 401–428 (p. 419).

[9]S. K. Hamarneh, *Background of Yunani (Unani), Arabic and Islamic Medicine and Pharmacy*, Hakim M. Said, ed., Karachi: MAS Printers, 1997, pp. 135, 137.

of Islam. Another *Hadith* of the Prophet enjoined the believers to 'seek ... learning though it be in China' or to 'seek knowledge from the cradle to grave'.[10] Such traditions obviously justified or favoured the acquisition of knowledge be it non-Islamic.

Nevertheless, 'the peoples of the Book' should pay *jaziya* or poll-tax if they refused to convert to Islam. And even when they converted to Islam, they were treated as an inferior class and called *mawâli* (plural of *mawlâ*), meaning clients, protected people, freed slaves. The social segregation between Arab and non-Arab that was created in the aftermath of the early conquests gradually petered out as the conflict of interest arose among the Arabs themselves. As long as the *mawâli* (or the converts) were small in number, they did not dare to stand against the Arab domination. With their number growing, however, they raised their voice against the social and racial discrimination practiced by the Umayyad. The Arab dissidents relied on discontent *mawâli* to fight against the Umayyad's sovereignty. The resulting political alliance between the Arab dissenters and the *mawâli* provided a favourable ground for social and cultural integration and furthered the assimilation of knowledge and sciences belonging to the *mawâli*.

Formation of Islamic *shari'at*

The territorial expansion of Islam in a relatively short period between seventh and ninth centuries was not synonymous with the establishment of the Islamic orthodoxy. The Islamic schools of law were founded in the early years of Islam after the death of the Prophet in the regions close to the birthplace of Islam, but they were far from constituting an orthodoxy not only because there was no central authority able to ensure their dominance but also because the theoretical demarcation of these new born schools had yet to be achieved.[11]

The primitive Islamic creed could hardly find an audience in the newly conquered territories without a well elaborated corpus of law and religious authority. Although the main schools of Islamic law (Mâleki, Hanafi, Shâfi'i and Hanbali) were principally based on the Koran and the *hadith* (traditions of the Prophet), the Islamic scholars made also a selective use of non-Islamic sciences. Greek philosophy was of particular interest to them because it could

---

[10] M. I. H. Farooqi, *Medicinal Plants in the Traditions of Prophet Muhammad*, Lucknow: Sidrah Publishers, 1998.
[11] On the absence of orthodoxy in Islam, see G. R. Hawting, *The First Dynasty of Islam: The Umayyad Caliphate A.D. 661–750*, London, Croom Helm, 1986, p. 6.

help them to defend their faith against many Christians living in the newly conquered lands that were Hellenised.[12] It was the need for providing Islamic *shari'at* (religious law and jurisprudence) with intellectual and philosophical consistency that led to the assimilation of non-Islamic sciences.

Aristotelian and Galenic ideas were incorporated into Islam as part of the overall formation of the Islamic philosophy, illustrated, among others, in the teaching of Avicenna. The assimilation of humoral theories took place because it belonged to a worldview that for example could explain the distinction between the intellectual faculties of the soul, which transcend the body and the sensitive faculties, which are lodged in the body and disappear with death. Such a distinction borrowed from Aristotelian philosophy provided explanatory tools for prophecy and the separability of the soul after death, a principal element of eschatology.[13] The relationship, in philosophical as well as medical terms, between spirit, soul and body in Galenic medicine, on the one hand, and the relation between soul, knowledge and God, in existential or theological viewpoints in Islam, on the other, produced over time an intrinsic interrelation between religion and medicine in Islam as seen at different levels in both medicine of Avicenna and medicine of the Prophet.[14] It is for this reason that in the medieval period, whether in Islam, Christianity or Judaism, but particularly in Islam, medicine and philosophy went together and usually learned physicians were also philosophers, hence the term *hakim*, that signifies both physician and philosopher.[15]

Moreover, the first generation of Islamic scholars were Christian, Zoroastrian or Jewish converts and the science of their forefather or cultural legacy informed their input in their new religion. Therefore the link between Islam and Greek sciences was both intellectually and sociologically established and conferred to these sciences a sort of legitimacy rendering them unavoidable for later scholars who, accordingly, divided sciences into two categories: *'Olum-e Avâyel* (pre-Islamic sciences, such as philosophy, mathematics, medicine, music and so forth; and *'Olum-e avâkher* (or Islamic sciences, such as *hadith, shari'at,*

---

[12]H. F. Cohen, *The Scientific Revolution: A Historiographical enquiry*, Chicago, London: University of Chicago Press, 1994. See also G. E. von Grunebaume, *Islam: Essays in the Nature and Growth of a Cultural Tradition*, London: Routledge, 1969; J. J. Saunders, 'The Problem of Islamic Decadence,' *Journal of World History*, 7 (1963): 701–720.

[13]D. Gutas, 'Intuition and Thinking: The Evolving Structure of Avicenna's Epistemology,' in R. Wisnowsky, ed., *Aspects of Avicenna*, Princeton: Markus Wiener Publishers, 2001. For a study of relation between surgery and Shafi'i School of Law in Islam, see E. Savage-Smith, 'Attitudes toward dissection in medieval Islam,' *Journal of the History of Medicine and Allied Sciences*, 50, 1 (1995): 67–110.

[14]See preface of S. H. Nasr to *Ibn Qayyim al-Jawziyya, Medicine of the Prophet*. P. Johnstone, ed., Cambridge: Islamic Texts Society, 1998.

[15]S. H. Nasr, *Science and Civilization in Islam*, Cambridge, Massachusetts: Harvard University Press, 1968, p. 184.

commentary on Koran, etc.).[16] This further legitimised Greek sciences that helped to blunt or even to remove contradiction and antagonism between the latter and Islam. The involvement of non-Muslims in the formation of the Islamic sciences created a sociological background underpinning cultural or conceptual reconciliation. These two corpuses of knowledge were therefore taught and discussed in parallel in the Islamic schools. Medicine was one of the pre-Islamic sciences together with mathematics, astrology, music, and others. This was also useful for a non-Islamic science such as Greek medicine based on humoral physiology could then be integrated institutionally into Islamic scholarship and therefore be assimilated into, or added to, magic and religious healing. Conversely, this institutional devise helped the medicine of the Prophet and the Imams to integrate elements of humoral physiology. The result of this mutual integration was the creation of a vast literature that incorporated religion and magic healing as well as Galenic medicine.[17]

Once the Islamic community was permeated by Islamic faith, everything was to be justified by the *hadith* and by the Koran. In such a situation everything including medical knowledge, whether it belonged to the Islamic community or not, was to be blessed or approved by the words of Koran or the Prophet and other saints. In this light, prophetic medical instructions or precepts seem to be rather the result of the observation or experiences mostly made by others and collected posthumously in the book called 'medicine of the Prophet'. Just as in other prophetic 'traditions', the medical instructions were narrated down to the contemporary *ommah* (Islamic community) through several narrators who were considered the great authorities in Islamic knowledge.

For example, a *hadith* (lit. An event or experience that occurred in the life of the Prophet or an advice of the Prophet to his companions) about the utility of one of the medicinal plants, such as safran, was narrated by Zaid b. Arqam, who narrated from Tirmizi, who narrated from Ibn Maja, who narrated from Masnad Ahmad, who narrated from the Prophet.[18] It seems unlikely that the Prophet systematically experienced or collected this information. The fact that some of these traditions talk about medical qualities of herbs or foods that were unknown to the contemporary community of the Prophet sustains this idea. In a tradition, for instance, the Prophet says: 'rice has healing powers'. This was narrated by 'Ayesha and Al-Soyuti. However, rice, a principal cereal of the wet regions of the tropics was not the staple food of the Arabs during the time of the Prophet.[19]

---

[16]Shams al-din Mohammad b. Mahmud Âmoli, *Nafâyes al-fonûn fi 'arâyes al-'oyûn* (The precious branches of learning in the quintessential sources of knowledge), Hâj Mîrzâ Abul-Hasan Sha'rânî, ed., 3 vols., Tehran: Library Eslâmiyeh, 1958.

[17]See for example: Mirzâ Musâ Sâveji, *Dastur al-Atebbâ*, and many other tracts on cholera or plague.

[18]Farooqi, *Medicinal Plants*, p. 78.

[19]Farooqi, op. cit., p. 150.

Islam and Medicine in the Modern Period

The intimate link between the formation of the Islamic religion and the development of sciences in Islamic countries not only led to the Islamisation of the sciences and philosophy, but also to the monopoly of the religious establishment on education. Consequently, when modern sciences were introduced into Islamic countries the religious scholars were practically among the first who became familiar with modern medicine. As education was controlled by the religious establishment, most learned physicians in Islamic countries were religious scholars, the *Mullahs*. Therefore it should not come as surprise that the *Unâni* medicine in the Indian subcontinent has preserved its faith in religion despite the fact that it has integrated many concepts and techniques of modern medicine, and undergone institutional and professional transformation similar to what happened in the West.[20] The inherent link between Galenico-Islamic medicine with religion can be illustrated in the words of the physicians of *Unâni* medicine who believe that traditional medicine is not only a science but also an art and many of its principles cannot be explained, while modern Western medicine is meant to be a pure science.[21] Nevertheless, as medical institutions and professions developed, religious scholars could not assume, neither theoretically nor institutionally the erstwhile continuity between Islam and 'modern' medicine. The intimate relation between religion and medicine in Islam as well as the growing professionalisation in the society at large made it no longer possible that religious scholars assume medical profession or education.[22]

As far as the Islamic societies in general are concerned, however, the introduction of both medieval Greek medicine and modern Western medicine into Islamic countries was conditioned by socio-cultural and political factors. State patronage played a fundamental role in their promotion and development. However, with regard to religion, the reactions of Galenico-Islamic and modern medicines were different. While in the medieval period, Greek medicine, as an integral part of the Greek sciences, became involved in the formation of Islamic cosmology,[23] and therefore crossed through religion, in the eighteenth and

---

[20]For an account on changes undergone by *Unâni* medicine see C. Liebeskind, 'Unani Medicine of the Subcontinent,' in Jan van Alphen et al., eds., *Oriental Medicine*, London: Serindia Publications, pp. 39–65.

[21]See foreword by Hakeem Abdul Hameed, to the book of Altaf Ahmad Azmi, *Basic Concepts of Unani Medicine: A Critical Study*. New Delhi: Hamdard Nagar, 1995.

[22]On this question see H. Ebrahimnejad, 'Religion and Medicine in Iran: From relationship to Dissociation,' *History of Science*, 40 (2002): 57–69.

[23]About the formation of Islamic cosmology, see S. H. Nasr, *An Introduction to Islamic Cosmological Doctrines*, Albany: State University of New York Press, 1993. About the integration of humoral medicine in the Islamic culture, an anthropological study see B. Good and M.-Jo DelVecchio Good, 'The Comparative Study of Greco-Islamic Medicine: The Integration of Medical Knowledge into Local Symbolic Contexts', in C. Leslie and A. Young, eds., *Paths to Asian Medical Knowledge*. Berkeley, Los Angeles, Oxford: University of California Press, 1992, pp. 257–271.

nineteenth centuries, modern medicine, founded on an entirely new worldview born of the Scientific Revolution, circumvented religion to make its way into Muslim societies. The different relationship between religion and medicine in ancient and modern times is an outcome of different worldviews. The dissection of the human body and anatomy were both principally recognised by medieval Galenism as a means of understanding the body and its illnesses. However, this principle became a dead letter in Islamic medicine because for Islam, the human body was sacred, and the human was the noblest creature of God. The situation was not much better in Christian countries. Surgery was markedly absent in Latin and Anglo-Saxon medical writings at the end of the first millennium.[24] Although bloodletting and bone setting were current in the Byzantine Empire, the use of the knife in surgery was not a major part of the medical practice there.[25] This was at the origin of the gap between theory and practice in medieval Galenic medicine, and the fact that the latter did not make progress in anatomy and pathology.

Modern medicine, on the other hand, posited the human body as an object of knowledge; the body was not sacred any more and therefore could be dissected. The bookish nature of early/classic medical knowledge affected both medicine and surgery and many physicians read them in the books and never even attempted to practice. It was even said that Hippocrates himself was responsible for such a custom or classification as he had commissioned his students to the practice of surgery and devoted his own work only to humoral medicine.[26] Most learned medical books, such as *Majma 'al-javâme'* of 'Muhammad Husayn ibn Muhammad Hâdi al-'Aqili al-'Alavi, a practitioner in India (eighteenth century), that contained large chapters on surgery, pharmacology and hospital treatment, were essentially based on theoretical knowledge and not practical anatomy and dissection.

Despite these contradictions between religion and modern medicine, however, they were absorbed by socio-political factors that underpinned the assimilation of modern medicine. The study of this question for all countries where Galenico-Islamic medicine was dominant at the turn of the nineteenth century falls outside the scope of this paper. The cases of Iran and India can illustrate this process.

In Iran, modern medicine was introduced into the country through state patronage. The Qâjâr elite in their endeavours to consolidate the central state needed to modernise the army and with it they introduced modern sciences and techniques without taking into account that they could come into conflict with

---

[24] A. Meaney, 'The Practice of Medicine in England about the Year 1000,' *Social History of Medicine*, 13, 2 (2000): 221–237; K.-D. Fischer, 'Dr Monk's Medical Digest,' *Social History of Medicine*, 13, 2 (2000): 239–251.
[25] E. Savage-Smith, 'The Practice of Surgery in Islamic Lands: Myth and Reality,' *Social History of Medicine*, 13, 2 (2000): 307–321 (p. 307).
[26] Anonymous Persian manuscript on the establishment of hospitals, ca. 1865 (Tehran, Majel Library, MS 505).

Islamic tenets or with the religious establishment. Nevertheless, for the latter, the threat of modern sciences was not immediately felt. Firstly because it was not widespread but limited to the activity of some court physicians and a few regiments. Secondly, the education in the modern sciences was also limited in the mid-nineteenth century to one state school, the Dâr al-Fonun (school of polytechnics). Moreover at the Dâr al-Fonun, both modern and traditional sciences were taught. Thirdly, the Qâjâr state was not under colonial domination and the education in the modern sciences imparted by Europeans did not provoke a strong anti-Western reaction. Accordingly, despite the fact that socio-political movements in Iran were more influenced by the religious establishment than in India, Galenico-Islamic medicine did not become a political instrument against Western influence. We can appreciate this point better if we bear in mind that basic medical education was imparted in the *madrasa* (Islamic college of theology) and that, as we have explained above, medicine was epistemologically linked to religion. The socio-political movements that developed towards the end of the nineteenth century were influenced by the religious establishment that aimed to make use of it to strengthen its power threatened by the centralisation of the Qâjâr state. Therefore, the opposition to modern Western medicine was not an ideological issue but the consequence of socio-political circumstances.

Furthermore, the Qâjâr state continued sponsoring traditional *tabib*s (physicians) even after it employed Western physicians at the court. This policy was due to the relatively strong presence of traditional medicine at the court. Even though there was a marked preference among the Qâjâr elite for Western medicine, this preference was not translated into any discriminatory policy towards local traditional physicians. As modernization of medicine was part of the state building process, the local elites, including the court physicians, participated in this modernization.[27] They were actively involved in the establishment of modern medical institutions, especially military hospitals and sanitary councils. This institutional involvement of traditional court physicians favoured an intellectual environment that thwarted the formation of ideological or political opposition to modern medicine.

Moreover, within such institutional and intellectual contexts, traditional physicians did not find it necessary to seek religious justification for the practice of modern medicine. While in the medieval period, Galenic medicine was taken on board within the framework of the formation of Islamic scholarship in general, in the modern period, modern Western medicine was accepted and assimilated as such without engaging with faith or religious factors.

In colonial India, on the other hand, the situation was different. The *Unâni* (or Greek) medicine was introduced into India since the twelfth century and it flourished especially under the Mughal emperors from the mid-sixteenth

---

[27] On this question see H. Ebrahimnejad, *Medicine, Public Health and the Qajar State: Patterns of Medical Modernization in Nineteenth-Century Iran*. Leiden, Boston: Brill, 2004.

century onward who also patronised some migrant Iranian physicians. The Mongol invasion and the destruction of the caliphate helped the revival of Persian which again become an official language at many Islamic courts, including in India. In general the patronage of physicians and the construction of hospitals in India occurred more extensively than in Iran. During the period of Mughal rule, some European physicians came to India during the seventeenth and eighteenth centuries and were also sponsored by the court. The Indian *hakim*s became acquainted with some aspects of modern (clinical) medicine. So far there was no fundamental opposition with Western modern ideas. The conflict began with the establishment of the British India government that favoured modern medicine but denied to support *Unâni tebb*. Traditional *Unâni* medicine continued to be taught at the Native Medical Institute in Calcutta until 1835 and at Lahore University up to 1907, and some *hakim*s were also employed in the rural areas but these measures were not aimed at further developing *Unâni* medicine.[28]

The end of state patronage by the colonial government was a bitter experience for the *hakim*s who were traditionally sponsored by Indian rulers for several centuries. Consequently, in the eyes of local *hakim*s Western modern medicine was associated with colonial power and the opposition to modern Western medicine in India took far more a nationalist overtone than in Iran.

Nevertheless, the anti-colonial stance of *Unâni* medicine and its revival especially during the struggle for Independence did not prevent the assimilation of modern medicine in India. In fact the *Unâni* medicine integrated elements of modern medicine in order to survive. However, this survival was only and necessarily at institutional level. It could not be epistemological because of the theoretical gaps and contrasts between biomedicine on the one hand and the *Unâni* medicine on the other. Consequently, in such a move, *Unâni* medicine ignored its ideological links to religion, just as did the traditional Galenico-Islamic medicine in Iran in the process of its transition to modern medicine in the nineteenth and twentieth centuries.

Conclusion

Modern historiography, perhaps following post-modernism in history, is experiencing a considerable amount of discussions about the revival of traditional or alternative medicine and especially the role of religion in medical practice and in healing. The question of antagonism between religion and modern medicine is also considered an outdated issue, for anthropological or even conceptual reasons. No doubt sacred healing, whether genuine or

---

[28] J.C. Hume, 'Rival Traditions: Western Medicine and Yunani Tibb in the Punjab, 1849–89,' *Bulletin of the History of Medicine*, 51 (1977): 214–231.

opportunistic, used even in modern European societies,[29] should be studied as a social and anthropological phenomenon. It is, however, undeniable that there is no epistemological link between religion and modern medicine. Setting aside theoretical issues, we can say that with modern sciences becoming pervasive in everyday life, unlike in the medieval period, it is no longer necessary or feasible to retain a relationship between religion and medicine or to justify the practice of modern medicine by reference to religion or tradition. Such a relationship was rather inherent in Galenico-Islamic medicine in that they were formed through parallel intellectual and ideological as well as socio-political processes. Paradoxically, this relationship was created in order to make possible the coexistence of Islam and Greek sciences. Otherwise there was nothing 'Islamic' in what is called 'Islamic medicine'.

---

[29]In Russia, for instance, faith healing has become a flourishing practice despite the statistics showing that it had an extremely poor record, and in some cases it has led to disaster.

Index

A
Abbas Vesim (d.1760), 121
Abdulhamid I, Sultan (r.1774-1789), 52
Abdulhamid II, Sultan (1842–1918), 128–129, 134–135
Abul Qasim (17th c.), 236
Accommodationism, 213–214, 219
Acupuncture, 210, 227–228, 255–257
Agricultural machinery, in China, 18
Agricultural revolution, in Britain, 18
Agriculture
  in Britain, 15–16
  in China, 18
Ainul-Hayat, 236
Ajā'ib al-makhlūqāt, 79
Akbar Arzani (17th–18th c.), 235
Akbar (r.1556–1605), 236
Al-Birjandi (d. after 1529), 121
Al-Biruni, Abu Rayhan (973–1048), 106
Alchemists, also *siddha*s, 19, 121, 234
Alchemy, India, 234, 236
Al-Dimashqī, Abū Bakr (d.1691), 81
Alemand, Louis Augustin (*c*.1653–1728), 211
Ali Pasha (d.1565), 126
Ali Rıza (d.1937), 188
Al-Kashi, Jamshid (1393–1449), 107
Al-Majusi, Ali Ibn Abbas (10th c.), 259
Al-Qazvīnī, Zakariyā'(d. 1283), 79
Al-Razi, Muhammed Ibn Zakariya (865–925), 259
Al-Sajawandī, Sirāj al-Dīn (d. 1210), 79
al-Ṭūsī, Naṣīr al-Dīn (1201–1274), 79, 114
Al-Zahrawi, Abu al-Qasim (936–1013), 259
Anandakanda, 234
Animal electricity, 245–246, 248, 250
Anma massage, 255
Antikythera mechanism, 106
Apparatus-keeper, 180
Appropriation, 1, 8, 160, 169, 193–206, 261–262
Archery, 29
Aristotelianism, 197, 202–203
Aristotelian (tradition, philosophy), 159, 205, 264
Aristotle (384 BCE–322 BC), 262
Arka, concept of, 235
Armorers, Ottoman, also *cebeci*s, 29
Armour, 19–20, 48, 57–58, 65, 68–70, 73
Artillery corps, quick-fire, Ottoman, 29, 42
Artillerymen, Ottoman, also *topçu*s, 29, 35, 37
Artillery, Ottoman, 2, 37, 41–42, 44–45, 48, 55
Astangsarira, 242
Astrolabes, 107, 113, 115, 120, 122, 125
Atlas
  Maior, 5, 81, 84, 86, 92, 94, 97, 100
  Minor, 5, 84, 86, 93, 97, 100
Aurangzeb (r.1658–1707), 236
Axenfeld, Auguste (1825–1876), 178
Ayurveda, 234–235, 237, 241
  Prakasha, 235
  Sanjivani, 241
Ayurvedavijnana, 237
Ayurvedic herbs, 236
Ayurvedic texts (20th c.), 242

B
Bacon, Francis (1561–1626), 200
Balasides, Dimitraki, 179
Balkans, 2, 28–29, 35–36, 38–39, 44, 52, 194–195, 197, 205
Barmak (*ca*. 685–725), 261
Battle of Kartal (Kagul), 42, 170
Beard, George Miller (1839–1883), 252

Index

Bellani, Angelo (1776–1852), 189
Benincasa, Grazioso (15th c.), 91
Bernard, Ch. Ambroise (1808–1844), 183
Bernier, François (1625–1688), 236
Beşir Fuad (1852–1887), 178
Bhaisajya Ratnavali, 235
Bhava Misra (16th c.), 237
Binod Lal Sen (19th c.), 237
Bion, Nicolas (1652–1733), 113
Blaeu, Joan (1596–1673), 84
Blaeu, Willem, J., (1571–1638), 84
Blood letting, 237
Bombardiers, Ottoman, also *humbaracı*s, 29, 32, 36, 44, 48, 51, 55
Bombay Durpan, 240
Bombshell factory, Ottoman, *humbarahane*, 51
Bonneval, Claude-Alexandre Comte de (1675–1747), also Humbaracı Ahmed Pasha, 32
The Book of Documents, 157
Boundaries, disciplinary, 175
Boundaries, regional and national, 167
Bows, composite, 65
Bows, crossbow, 65, 67, 69
Boym, Michael (1612–1659), 217
Brain gain, 28, 34
Brhannighanturatnakara, 237
Brodie, Benjamin (1817–1880), 187
Bruno, Jordano (1548–1600), 200
Buddhist church, 14
Busschoff, Herman (c.1620–1674), 209
Byzantine, 27, 78, 261, 267

C

Cabinet of curiosities, 111
Calendar, Gregorian, 4, 122, 126
Calendar, Islamic, 103–104, 121, 126
Calendrical methods, Chinese, 151–161
Calendrical methods, European, 6, 160
Calendrical techniques, *bu dou*, 158
Calendrical techniques, *fenye*, 158
Calendrical techniques, *santong*, 158
Campanella, Thommaso (1568–1639), 200
Cannons, Ottoman, *Balyemez*, 42, 44–45
Cannons, Ottoman, *Kolomborne*, 44
Cannons, Ottoman, *Şahi*, 44, 46–47, 49
Cardano, Girolamo (1501–1576), 200
Carré, Ferdinand (1824–1894), 188
Centuria plantarum, 226
Chakrapanidata (13th c.), 234
Charaka, 234

Chariot burial (Lchashen), 59
China*Illustrata*, 217
Chuckerbutty, S. C. G (1826–1874), 239
Cihān-nümā, 81, 83, 93, 95–97
Civilizing mission, 170, 172
Clavis, 211–213, 219, 228–231
Cleyer, Andreas (1634–1698), 221
Clockmaker, 4, 125–136, 138, 140
Clocks, mechanical, 1, 4, 125–128, 135, 137–149
Clüver, Philipp (1580–1622), 93
Collado, Louis (16th c.), 31
Commentarii bellici, 32–33
Concept of energy, 190
Confucian astronomers, 152–153, 160–161
Confucius (551 BCE–479 BCE), 214, 216, 230
Congreve, William (1772–1828), 48
Constantinople, 27, 30, 35, 182, 184, 195
Cosmology
   Chinese, 155
   Ptolemaic, 154–155
Count of Onsenbray (1678–1754), also Pajot, Louis Léon, 112
Couplet, Philippe (1623–1693), 222
Cultural appropriation, 169
Cultural redefinition, 165
Cursus mathematicus, 33

D

Dâr al-Fonun, 268
Darülfünun, 177
Darüşşafaka, 178
De La Hire, Philippe (1640–1718), 109
Delisle, Guillaume (1675–1726), 95
De re militari, 30
Descartes, René (1596–1650), 200
Dhanvantri Mahal Hospital, 239–240
Dharani texts, 21
Diagnosis methods, India, 234
Dias, Manuel (1549–1639), 154
Diathermy, 256–257
Diderot, Denis (1713–1784), 199
Dingli yuheng, 152–155, 158, 160
Discurso de la artillería, 31
Dissection, India, 234, 240, 267
Dominic Gregory (18thc.), 236
Du, angular measurements, 156
Du Bois-Reymond, Emil (1818–1896), 248
Dulcert, Angelino (14th c.), 78
Dutch East India Company, also VOC, 209–210, 220–221, 246

Index

E

Earth, size of, 156
Eclecticism, 2, 5, 54, 95–97, 99, 193–206
éclectism, 199–200
Eclipsarium, 104–105, 108–110, 120
Eclipse calculator, 103–106, 108–111, 113–122
Eclipse, lunar, 4, 105–108, 116, 118, 153
Eclipseometrum, 108, 119, 122
Eclipse prediction, 103–122
Eclipse, solar, 105–106, 109–110, 116, 118, 121–122
Ecumenical Patriarchate, 195
Education
  in colonial India, 167, 171, 175
  history of, 168, 171, 175
  science, in India, 172
  technical, in India, 173–175
Electrotherapeutic appliances, 246–247, 250–251, 255–258
Electrotherapeutics, 8, 245–258
Empiricism, 170
Enclosure Acts, 16
Encyclopédie, 199, 205
Energy conservation, the principle of, 190
Enlightenment, 3, 7, 29, 169, 195–198, 201–202, 214
Enlightenment vision, 7, 169
Equatorium, 107–108, 120
Erekiteru, 247–251, 255, 257
Erlitou (*ca.* 1900–1550 BCE), 60
Europe, 1–2, 4–5, 7, 9, 15, 17–20, 23–39, 47, 49, 52, 55, 57–58, 77–101, 103, 107–108, 121–122, 125, 127–128, 138–139, 152, 166–168, 170, 179, 182, 191, 195–197, 204
Expansion of European science, 7, 151, 165
Experimental philosophy, 197–198, 201, 203–204
Experiment(s), 3, 19, 28, 42, 111–112, 165, 180, 185, 193, 197–198, 200–201, 203–205, 214, 236, 245–250

F

Fazlı Necip (1863–1932), 178
Fenn-i Harb, 32–34
Fenn-i Harbiyye, 34
Fenn-i Lağım, 34
Fenn-i Muhasara, 34
Figuier, Louis (1819–1894), 178
Firangi roga, 237
Firearms, Ottoman, 2, 28–29, 36–38

Flammarion, Camille (1842–1925), 178
Flora Sinensis, 216–217
Force vive, 190
Foundry at Hasköy, 42
Foundry, Ottoman also *Tophane*, 35, 42
Foundry at Samakocuk, 52
French Academy of Medicine, 189
Fünun-i Harbiye ve Ebniye-i Harbiye Sanayii, 33
*Futuhname*s, 89

G

Galante, Henri, 189
Galen (129–201), 224
Galenic tradition, in India, 235
Galvani, Luigi (1737–1798), 248
Galvanism, 248–250
Gananath Sen (1877–1945), 242
Gangadhar Ray (19th c.), 241
Gangaprasad Sen (19th c.), 241
Ganot, Adolphe (1804–1887), 177
Garcia d'Orta (16thc.), 236
Gariel, Charles Marie (1841–1924), 187
Gastaldi, Giacomo (d. 1566), 86
Gavarret, Jules (1809–1890), 187
Gazette Médicale d'Orient, 184
Gazi Hasan Pasha, Cezayirli (1713–1790), 49
Genki, original *qi*, 254
Geography, Chinese, 217
Gırcikyan, Antranik (1819–1894), 178
Girindrasekhar Bose (1887–1953), 169
Gotō Rishun, 247
Govinda Das (18th c.), 237
Graham, Thomas (1805–1869), 187
Greek
  communities, 3
  orthodox populations, 194
  -speaking scholars, 193–206
Gribeauval, Jean Baptiste de (1715–1789), 41
Gufa, the ancient methods, 157
Gun carriage drivers, also *top arabacıları*, 29
Gun, carriages, Ottoman, 2, 29, 36, 41, 43–44, 47–51, 53–55
Gun, heavy, Ottoman, 45
Gun, Howitzer (Obus), 45
Gunpowder
  invention of, 5, 19
  social effects in China, 18, 20
  social effects in Europe, 20
  technology, in China, 2
  technology, in Europe, 2, 29, 37
  technology, Ottoman, 28–30

Gun, quick-fire, Ottoman, 42, 44–51, 53–55
Guo Shoujing (1231–1316), 157
Gutenbergs, Johannes (1398–1468), 224

H

Hacı İbrahim (Reşid) Efendi (d.1807), 43
Hafeẓ-e Abrū (d. 1429), 79
Hajjī Khalīfa (d. 1657), 81
Hakeem, also *hakim*, 242, 266
Hakim Ajmal Khan (1868–1927), 243
Hakim Sharif Khan, 236
*Hakim*s, in Mughal court, 236
Han dynasty (206 BCE–220 CE), 17, 67–68
Han-gul script, 22
Hapsburg Empire, 195
Hayrullah (1818–1866), 187
Heat, 188–190, 231, 255–256
Heaven, circumference of, 156
Herbarium, 223, 229, 239
Hermes Trismegistus, 214
Hernández, Francisco (1514–1587), 210
Hideyoshi Toyotomi (1536/1537–1598), 138
Hikmatparadipa, 235
Hikmatprakasa, 235
Hippocrates (460–370 BCE), 224, 262, 267
Hiraga Gennai (1729–1779), 247
Hirose Shuhaku, 249
Hisashige Tanaka (1799–1881), 146
Historiography of sciences, 165, 167, 176, 193
Hobbes, Thomas (1588–1679), 200
Holwell, John Zephaniah (1711–1789), 237
Holy War, 27
Homann, Baptist (1664–1724), 95
Hondius, Henricus (1597–1651), 84
Horology room, also *muvakkithane*, 126–127
Horse and chariot, 6, 63–65, 72
Horse, domestic, 57–59
Horse-riding, 6, 57, 67, 71
Hortus Malabaricus, 228
Hospitals, India, also *bimaristan* and *dar-us-shifa*, 235
Hours, unequal, 125, 127
Humoral, 8, 235–237, 241, 260, 264–267
Humoral, imbalance, 235
Humoral pathology, India, 241
Humours, Indian medicine, 241
Hunayn b. Ishaq (809–873), 259
Hüseyin Rıfkı Tamani (d.1817), 43

I

Ibn Moqaffa, Abd-Allah (d. *ca.* 756), 261
Ibn Sina (980–1037), 236
İbrahim b. Ahmad, 31
İbrahim Müteferrika (1674–1745/1747), 111
İbrāhīm Müteferriqa (d. 1744), 95
Ieyasu Tokugawa (1543–1616), 138
İlm-i Hikmet-i Tabiiye, 178, 180, 182, 186–191
Industrial Revolution, 14–18, 26
Inoculation, India, 236
Introductionem in universam geographicam tam veteram quam novam, 93
Iron industry, 26
Islamic philosophy, 264
Islamic sciences, 2, 78, 136, 184, 260–261, 263–265
Islamic *shari'at*, 263–264
İsmail Ali (1866–1913), 180
Istanbul observatory, 114
Italian peninsula, 195

J

Jafar Yar Khan (18[th]c.), 235
Jambhekar, Bal Gangadhar (1802–1846), 239
Jamin, Jules Célestin (1818–1886), 187
Jami-ul-Jawami, 236
Janissaries, 29, 44, 113
Jehangir (r.1605–1627), 236
Jesuits, 4, 7, 138–139, 148, 210–211, 213–229
Jörg of Nürnberg (15th c.), 35
Jörg of Nürnberg*Memoire della Guerra*, 32

K

Kangxi, the emperor (1661–1722), 223
Keimyaku, 256
Kircher, Athanasius (1602–1680), 217, 219
Knowledge
  diffusionist model of, 233
  imperial sociology of knowledge, 168
  indigenous, 171
  local, 9, 88, 167, 171
  naturalisation of, 171–172
  pre-modern forms and practices, 176
  regional adaptations, 166
  regional variations, 166
  transmission, 122, 262
  transmission of scientific knowledge, 167–168

Index

Koehler, George Frederick (1758–1800), 51
Kogury kingdom (37 BCE–668 CE), 70
Koran, 263, 265

L

Laboratory (ies), 112, 190–191, 197, 242
Language, in medical education, 183–184, 240
L'Artiglieria, 31
Lauh al-ıttısalat, 108
Lechunga, Cristóval (17th c.), 31
Leclerc, Jean (1657–1736), 200
Leibnitz, Gottfried Wilhelm (1646–1716), 200
Les Secrets de la Médecine Chinoise, 210, 223–224, 230
Les Travaux de Mars, 33
Levāmi' al-nūr, 93
Leyden jar, 246–247, 249
Li, the principle of, 159
Local apparent time (LAT), 125, 131–132
Localization of science, 175
Localization, strategies of, 172, 176
Logic, 166, 190, 201
Low-frequency therapeutic(s), 256

M

Madhava (18thc.), 235
Madhusudan Gupta (19th c.), 241
Ma'din al-shifa-i Sikandarshahi, 235
Makura dokei, 143, 146
Malaguti, Faustino (1802–1878), 178
Malebranche, Nicolas (1638–1715), 200
Mallet, Allain Manesson (1630–1706), 33
Manchus, 13, 215, 219, 221
Mannen Dokei, 147
Manual de Artilleria, 31
Manucci, Nicolao (17th c.), 236
Mapmakers, 86, 92–93, 95, 99
Mappa mundi, see World maps
Maps
 of al-Iṣṭakhrī (11th c.), 79
 of Anatolia, 87, 90, 92
 Arabic, 79
 of *Asia*, 87–88, 96
 of the Caspian Sea, 95–96
 Fatimid, 78
 in Ḥafeẓ-e Abrū's (d.1429) *Kitāb-e Tār īkh*, 79
 hemispheric, 81, 83–84

in Ibn Faḍ lallāh al-'Umarī's (d.1349) *Masālik al-abṣār*, 79–80
 of Ibn Ḥawqal (d. *ca.* 977), 79
 of Iran, 95
 in Niẓām al-Dīn Nīsābūrī's (d.1329) commentary on *Tadhkira*, 79
 non- Ptolemaic, 86
 Ottoman Turkish, 78–79, 81
 Persian, 78, 81, 89, 95, 108, 120
 Ptolemaic, 92
 in Sirāj al-Dīn al-Sajawandī's (d. 1210) text on folk astronomy, 79
Martini, Martinus (1614–1661), 217, 219, 221
Matter, 17, 22–23, 25–26, 43, 46, 196, 202–203, 206, 248
Mattioli, Pietro Andrea (1501–1587), 87
Medical ideas, India, 209–210, 233–243
Medical practices, India, 7, 209–210, 236–237, 267, 269
Medical techniques, India, 1
Medical texts, in Sanskrit, 235
Medicine
 Chinese, 7, 209–232
 colonial India, 7, 234–237, 268
 education, in Turkey, 183–184
 Galenic, 260, 264–265, 267–268
 Greek, 260, 265–266, 268
 Islamic, 8, 184, 231, 235
 modern, 7–8, 183, 233, 260, 266–270
 Pre-Colonial India, 7, 234–237
 of the Prophet, 259, 262, 264–265
 traditional, 8, 255, 257, 260, 266, 268
 tropical, 238
 western, 8, 236, 238, 242, 246, 249–251, 255–256, 266, 268–269
Medicus Sinicus, 212, 217–218
Mehmed Çelebi, Yirmisekiz (d.1732), 104
Mehmed Said (d.1761), also Said Pasha, 104, 111
Mehmet Ikhlāṣī, 93, 95, 99, 101, 178
Mehmet IV, Sultan (1642–1693), 99
Meiji period, 251
Mei Wending (1633–1721), 153
Membré, Michele (1505–1595), 88
Mercator, Gerard (1512–1594), 84
Meridian, 97, 115–116, 118, 131, 249, 256
Metaphysics, 201, 205
Meyer Company, 128
Meyer, Emile (1883–1954), 135
Meyer, Johann (1843–1920), 128
Meyer, Wolfgang (1909–1981), 135–136
Miasmatic theories, 241

Military Academy, Ottoman, 180, 188
Military acculturation, 37–39, 42
Military reforms, late 18th c, Ottoman, 42
Military treatises, 30–34, 38
Mines
   of Ergani, 50–51
   of Gümüşhane, 51
   of Kastamonu, 50–51
   of Sidrekapsi, 50–52
Ming dynasty (1368–1644), 23
Mir Muhammad Hashim (17th–18th c.), 236
Mir Muhammad Jafar (18th c.), 236
Mirza Alavi Khan (18th c.), 236
Miyan Bhuwah (16th c.), 235
Mongols, 13, 19, 23, 27, 59, 62, 269
Montecuccoli, Raimondo (1609–1680), 32
Mortar, Ottoman, also *havan*, 32, 45
Movable type printing, in China, 22
Moxa, 210, 227, 255
Moxibustion, 209, 227, 255–257
Mughals, 37–38, 236, 268–269
Muhammad Shah (r.1719–1748), 235–236
Muhasara-yı Kıla, 34
Mujintō, 147
Münster, Sebastian (1488–1552), 87
Mustafa Reşid Beğ (d.1819), 47–50
Mustafa Sıdkı (d. 1769–1770), 113

N
Naifukudōkō, 250
Napoleon Bonaparte (1769–1821), 41, 48–49
Nationalists, Indian, 169
Natural history museum of the medical school, Ottoman, 179
Nature, 2, 16, 28, 66, 77, 106, 112, 160, 176, 194, 197, 200–205
Needham, Joseph (1900–1995), 151
Needham question, 5, 14
Neelamber Sen (19th c.), 241
Negretti, Angelo Ludovico (1818–1879), 189
Neo-Confucian theory, 159
Nervous force, 245–246, 248, 252, 254, 257
Neurasthenia, 252, 254
Newtonian natural philosophy, 202
Newtonian physics, 202
Nobunaga Oda (1534–1582), 138
Nosology, India, 237
Novus Atlas Sinensis, 217, 219
Nurul Haq (17thc.), 236
Nuzhen, 13

O
Observatory(ies), 104–105, 107–109, 112, 114, 120, 123, 136, 197
Optics, 110, 112, 154, 172
Oranda Banashi, 247
Ortelius, Abraham (1527–1598), 92
Ovington, John (17th c.), 236
Ox cart, 62–63, 66, 68, 73

P
Pandit Mahadeva (18th c.), 236
Paris Observatory, 104, 108–109, 111–112, 120
Patriarchate (Ecumenical), 195
Paul, Constantin (1833–1896), 189
Paulhan, Fréderic (1856–1931), 178
Period of division (220–581), 22
Peter I (1682–1721), 95
Phanariots, 195
Philip II (1527–1598), 210
Physics, 3, 65, 111–112, 156, 177–180, 182–188, 190–191, 202, 205, 250
Physics instruments, 179, 185
Pirckheimer, Willibald (1470–1530), 87
Planets, size of, 154
Plática Manual de Artillería, 31
Plato (428/427–348/347 BCE), 262
Portolan charts
   Catalan Atlas, 78
   of Dulcert, 78
Portolan charts, in the *Geography* of Ptolemy, 86–88, 91–92
Post-colonial scholarship, 168
Post-colonial theory, 165
Post-colonial theory of science, 165
Prediction of eclipses, 105–106, 108
Pre-Islamic sciences, 264–265
Printing
   in China, 21–24
   in Europe, 23
   social effect, in China, 21
   social effect, in the West, 20
Prytanée National Militaire de la Flèche, 178
Psychoanalysis, 6, 168–170
   career of, 170
   cultural appropriation of, 169
   the decline of, in India, 170
   Freudian, in India, 168–169
Psychoanalysts, British, 169
Ptolemy (*ca.* 125), 78
Pulse, 210–211, 219–220, 223–225, 228–231, 234
   India, –220, 234

Index

Q
Qâjâr, 267–268
Qanun, 236
Qi, the vital force, 155
Qi, the vital spirit, 231
Qin dynasty (221–206 BCE), 63
Quadrants, 32, 120, 122, 125, 127
Quli Shah (16[th]c.), 235

R
Raja Serfoji (1798–1832), 239
Rajsimhasudhasindhu, 236
Ramusio, Giovanni Battista (1485–1557), 92
Rangaku, 247
Rasashastra, 234
Raşid Mehmed (18th c.), 109
Rationality, 6, 160, 170
Reformation, 24
Rhinoplasty, India, 237
Ribot, Théodore (1839–1916), 178
Ricci, Matteo (1552–1610), 217
Rice cultivation, social effects in China, 17–18, 20
Richardson, Benjamin Ward (1828–1896), 190
Rogarogvada, 235
Römer, Ole (1644–1710), 108
Rudigerus, Andreas (1673–1731), 200
Russo-Ottoman War of, 42, 176–177

S
Sâbt b. Qurra (b. 835), 259
Sağlam, Tevfik (1882–1963), 180
Saint-Remy, Charles Alexandre Louis Rouxel de (1746–1800), 48
Salih Zeki (1864–1921), 178
Samhitas, 234
Sanden, Albert T., 252
Sanden Denki Shōkai, 252–255
Sanden's electric belt, 252, 254–255, 258
Sarabendra Vaidya Muraiga, 239
Sarangdhara (13th c.), 234
Sardi, Pietro (17th c.), 31
Sasanian, 261–262
Sassetti, Filippo (16th c.), 237
Schall von Bell, Adam (1591–1666), 157
School of Medicine
    civil (Istanbul), 179–180, 183
    imperial (Istanbul), 179–180, 183
Schott, Gaspar (1608–1666), 33
Schreck, Johann (1576–1630), also Terrentius, 218

Scientific center (s), 193–194, 198
Scientific knowledge, 3, 7, 103, 167–168, 174, 176, 191, 203, 205–206
Scientific method (s), 1, 107, 193–206, 242
Scientific periphery(ies), 193–194, 198
Scientific revolution, 24, 166, 202, 204, 210, 267
Seasonal time system, 137–138, 140–144, 148–149
Seisetsu Igensūyō, 249
Selim III, Sultan (1761–1808), 32, 34, 42, 47, 54–55
Shah Alam II (1759–1806), 236
Shaku dokei, 143, 148–149
Shams-al-din-Gilani (16th–17th c.), 236
Shang dynasty (*ca.* 1250–1050 BCE), 64
Shapur b. Sahl (d. 869), 259
Sharh-i Gilani, 236
Shinkei suizyaku, 254
Shu, the numbers, 159–160
Siddhantanidana, 242
Six, James (1731–1793), 189
Societies
    Academia Naturae Curiosorum, 211, 226
    Leopoldina, 211
    Ottoman Medical Society (Istanbul), 183
    Société de Médecine (Paris), 189
    Société Impériale de Médecine de Constantinople, 184
    Society of Jesus, 139, 210, 213
Song dynasty (960–1279), 26
Specimen, 22, 78, 81, 94, 100, 112, 170, 210–213, 220, 222, 226, 228–230
Spices, trade of, 7
Stadlin, Franz (1658–1740), 139
Standardization
    of the caliber, 41
    of the measuring rod, 47
    of the ordnance, 29, 39, 41, 43, 46
Standard time (ST), 125, 131, 145
Stato militare dell'Imperio ottomano, 31
Steppe cultures
    Andronovo, 62
    Circum-Pontic, 64
    Lchashen, 59
Stirrup, 5, 57–73
Sugita Genpaku (1733–1817), 249
Sui dynasty (581–618), 13, 68, 70, 157
Sundials, 111, 125, 127, 131, 136
Sushruta, 234
Syncretism, 97, 199, 239
Syrbius, Johann Jacob (1674–1738), 200

T

Tabaq al-manateq, 107
Taccola (Ser Mariano di Giacomo Vanni) (1381–*ca*.1458), 30
Takano Chōei (1802–1850), 249
Talim-i-Ilaj, 235
Tang dynasty (618–907), 13
Taqi al-Din (d.1585), 105
Technological dialogue, 2, 27–39
Tempu, 141, 143, 145
Tennen ni okeru Kenkō, 252–255
Ten Rhijne, Willem (1647–1700), 226–231
Terminology, geographical, 86, 97, 100
Terrentius, 218
Tertib-i Ceng, 33
Textbook(s), 3, 6, 34, 79, 122, 165–176, 178, 186–187, 191, 196–197
Textbooks, on physics in Turkey, 178
Theatrum Orbis Terrarum, 87, 93
Theophrastos (*ca*. 371 BCE–*ca*. 287 BCE), 262
Theti, Carlo (1529–1589), 33
Thomasius, Christian (1655–1728), 200
Three Classes of Man, 253
Tıbb, 235
Tibb-i Akbari, 235
Time
  alla franca, 125
  alla turca also *ezânî*, Turkish, 126
  bell, 137–138, 144–146, 149
  equation, 132, 134
  International standard, 125
  keeper, also *muvakkit*, 105
  mean solar, 125, 134
  Turkish system, also *alla turca*, 125–136
Todarananda, 234
Tokisoba, 144
Tokugawa period, 137, 148, 246
Tott, Baron François de (1733–1793), 37
Traité de la construction et des principaux usages des instruments de mathématique, 115
Traité élémentaire de physique expérimentale et appliqué, 177
Tsubo, 255–256
Tuhafatul-Masiha, 236
Tyco Brahe (1546–1601), 154

U

Unani medicines, 7, 236, 266
Unani tibb, 242, 269
Universalization, 167
Universal Time (UT), 4, 126

University, Ottoman, also *Darülfünun*, 177
Usul-i Harbiyye, 34

V

*Vaidya*s, 237, 241–243
Valturio, Roberto (1413–1484), 30
Vangasena (13th *c*.), 234
Van Reede, Hendrik (1636–1691), 228
Van Verden, Carl (18th*c*.), 96
Varier, P.S. (19th–20th *c*.), 243
Variolation, India, 237
Vauban, Sébastien le Prestre de (1633–1707), 34
Vijnana, 237
Viresvara (17th*c*.), 235
VOC, 209–211, 220–227, 229–230
Volta, Alessandro (1745–1827), 248
Volvelles, 108

W

Wadokei, 138, 141–142
Warikoma, 142–144, 147–148
Watches, *Hamidiye*, 128–129
Watches, Longines, 128
Western anatomy, 249
Wheat cultivation, 17
World map
  Chinese, 216
  of al-Nawajī's (d.1455) poems, 79
  Catalan, 78
  hemispheric, 81, 83
  in Meḥmed b.'Alī Sipāhīzādeh's (d. 997/1589) *Awḍ aḥ al-masālik*, 80–81
  in Pseudo-Ibn al-Wardī's cosmography (1419), 79
  in Zakariyā' Qazvīnī's (d. 682/1283) *'Ajā'ib al-makhlūqāt*, 79

X

Xavier, Francis (1506–1552), 213
Xavier, Francisco (1506–1552), 138
Xianbei culture, 68, 70, 72
Xuanye cosmology, 155, 161

Y

Yang Guangxian (1597–1669), 157
Yanshi (*ca*. 1700–1600 BCE), 63
Yijing, 158
Ypsilanti, Konstantin (1760–1816), 34
Yuan dynasty (1271–1368), 13, 153, 156

Z

Zabarella, Jacopo (1532–1589), 203
Zambra, Joseph Warren (1822–1887), 189
Zhang Yongjing (17th c.), 152, 160
Zhoubi suanjing, 155–156
Zhou dynasty (*ca.* 1050–221 BCE), 62–63
Zoroastrians, 8, 261–262, 264
Zukai Ransetsu Sansaikikan, 249